IUTAM Symposium on Mechanical Behavior and Micro-Mechanics of Nanostructured Materials

SOLID MECHANICS AND ITS APPLICATIONS
Volume 144

Series Editor: G.M.L. GLADWELL
Department of Civil Engineering
University of Waterloo
Waterloo, Ontario, Canada N2L 3GI

Aims and Scope of the Series

The fundamental questions arising in mechanics are: *Why?*, *How?*, and *How much?* The aim of this series is to provide lucid accounts written by authoritative researchers giving vision and insight in answering these questions on the subject of mechanics as it relates to solids.

The scope of the series covers the entire spectrum of solid mechanics. Thus it includes the foundation of mechanics; variational formulations; computational mechanics; statics, kinematics and dynamics of rigid and elastic bodies: vibrations of solids and structures; dynamical systems and chaos; the theories of elasticity, plasticity and viscoelasticity; composite materials; rods, beams, shells and membranes; structural control and stability; soils, rocks and geomechanics; fracture; tribology; experimental mechanics; biomechanics and machine design.

The median level of presentation is the first year graduate student. Some texts are monographs defining the current state of the field; others are accessible to final year undergraduates; but essentially the emphasis is on readability and clarity.

For a list of related mechanics titles, see final pages.

IUTAM Symposium on Mechanical Behavior and Micro-Mechanics of Nanostructured Materials

Proceedings of the IUTAM Symposium held in Beijing, China, June 27–30, 2005

Edited by

Y.L. BAI
Institute of Mechanics, Chinese Academy of Sciences, Beijing, China

Q.S. ZHENG
Tsinghua University, Beijing, China

and

Y.G. WEI
Institute of Mechanics, Chinese Academy of Sciences, Beijing, China

Springer

A C.I.P. Catalogue record for this book is available from the Library of Congress.

ISBN-10 1-4020-5623-0 (HB)
ISBN-13 978-1-4020-5623-9 (HB)
ISBN-10 1-4020-5624-9 (e-book)
ISBN-13 978-1-4020-5624-6 (e-book)

Published by Springer,
P.O. Box 17, 3300 AA Dordrecht, The Netherlands.

www.springer.com

Printed on acid-free paper

Table of Contents

Part 3: Nanomechanics of Biomaterials

Part 4: Mechanical Behaviors of Carbon Nano-Tube, Nano-Wire, Nano-Layers

Part 5: Micro-Mechanics Models and Simulations for the Nanostructured Materials

Preface

Recently, we have seen that nano-structured materials demonstrate remarkable advantages over conventional ones, particularly in mechanical behaviors. This will have a great impact on new technology and engineering in the new millennium. In accordance with this need, IUTAM decided to hold a symposium to discuss the prospect of the wonderful mechanical behaviors of nanostructured materials and their relevant micro-mechanics, to encourage the mechanics community to explore the frontier of science and technology. The aim of the symposium was to provide such a forum to all experts in both mechanics and material sciences worldwide, to have an opportunity to exchange their research achievements and views in understanding these mechanical behaviors, exploring their applications, as well as enhancing their property designs of nanostrucutred materials.

The IUTAM Symposium on *Mechanical Behavior and Micro-mechanics of Nanostructured Materials* was held successfully in Beijing, China, June 27–30, 2005. Thirty participants with different research backgrounds including mechanics, materials sciences, physics, etc. from Australia, France, Germany, Japan, Singapore, UK and USA and China presented their contributions on various aspects closely related to the subjects, displaying wide research direction and progress.

In this proceedings volume, these contributions are collected in six parts:

(1) *Mechanical behaviors of nanocrystal materials*
(2) *Super-strength and ductility of nano-thin films*
(3) *Nanomechanics of biomaterials*
(4) *Mechanical behaviors of carbon nano-tube, nano-wire, nano-layers*
(5) *Micro-mechanics models and simulations for the nanostructured materials*
(6) *Mechanical behaviors of other nano-materials*

Actually, the theme of the lectures, discussions as well as the round table discussions in the symposium, was a proper evaluation of the perspectives and trends in the area of mechanical behaviors of nano-structured materials. In particular, in the two-hour round table discussion held in the final afternoon of the symposium, participants passionately discussed several hot topics, probable future developments, interdisciplinary research and young people's education, etc. In order to meet the challenges, some important questions were highlighted, such as:

How to promote interdisciplinary research and cooperation, for instance the stimulation of studying nano-structured materials from physics, chemistry to mechanics; How to highlight the frontier to guide its evolution/revolution; to create new micromechanics; How to develop innovative education programs for training young people in the emerging area? Etc.

To respond to these questions, some valuable suggestions and enlightening visions were proposed during the symposium.

Firstly, nano-structured materials possess nice geometrical patterns and demonstrate splendid behaviors, but they also form huge challenges to the mechanics community both theoretically and experimentally. For example, for some nano-structured materials, elasticity theory is valid to some extent, but for soft ones, few ideas are available. Facing these advances and challenges, the mechanics community should focus on building a bridge between the atomic reality in materials and the continuum in mechanics. Particularly, this should open up new opportunities in mechanics. During this course, perspectives for new ideas and methods are badly needed. New mechanics beyond conventional constitutive laws, integration of multiple length and time scales, etc. may emerge. So we cannot think of mechanics in traditional sense. In this aspect, the symposium covers a diversity of topics and methodologies, which will be very helpful in many related areas.

Nano-mechanical measurements to validate formulations, descriptions or concepts for nanosystems are very important. How to measure these phenomena precisely in the dimensionality at micro and nano-scales needs specific and novel techniques. For instance, the measurements of various inter-atomic forces, etc. become critically significant in mechanics at nanoscales. It is also necessary to overcome barriers between nano-scale mechanics and nano-manipulations. There are huge and urgent needs for precise measurements, to meet the needs new exciting tools will become available. Will we need new physics or will classical concepts be enough?

The combination of mechanics with other fields is important. Firstly, familiarity with other fields to properly define necessary theories/models is critical. In mechanics and materials communities, we are in close collaboration through modeling and simulations together with experimentation. New computational and theoretical micro-mechanics have been developed, and have made calculations of large systems possible in order to understand material behavior at microscales. Nano-scale experimental mechanical techniques made the understanding even more convincing. To exchange ideas and knowledge and provide valuable networking opportunities for researchers in the overlapping fields should be further encouraged. Secondly, a reasonable amount of materials are needed for mechanical tests and measurements. Regarding the exchange of material samples, sometimes the quantity of nano-structured materials is limited. This may lead to the lack of systematic studies on the production yield and quality control. Therefore, close cooperation to support enough and high quality experiments is needed. Finally, nano-manufacture with precision, control and large quantity needs further attention. In these aspects, what is the role of mechanics in nanomanufacturing? What are the emerging industries that will need large quantities of nano-structured materials? Anyway, the convergence of mechanics, materials science, as well industries is the other key in this area.

Since nanotechnology has emerged very rapidly but mechanical engineers are not well trained in the fields that bridge the nanosciences with engineering, the training in these fields is of great importance in providing society with the useful outcomes of these technologies. The mechanics community should be aware of what the significant components of nanotechnology are. What kind of nano-structured materials: nanoscale materials, nanoparticles, nanophase materials, etc. are needed in practice? What are their applications? From a young people's point of view, we are dealing with an exciting issue: nano is redefining science, in which mechanics is an active field. One of exciting things for mechanics people is to look at new materials and their new behaviors (including electrical, optical and biological). Here challenges and opportunities co-exist. For graduate students, it is an exciting time with great opportunities. To grasp the opportunities, one should be independent, creative and innovative. Mechanics people are not only clever but also quick in catching new opportunities. However, we still do not have good classes in these crossover fields/areas, especially when it comes to thinking about graduate level classes. So it is needed to integrate fundamental courses with these new challenges. Optimistic about the future, one needs to learn as much as possible, willing to adopt to new fields, ready and flexibility for change, but not to follow fashions.

The symposium was organized by the Chinese Society of Theoretical and Applied Mechanics (CSTAM), the Institute of Mechanics, the Chinese Academy of Sciences (CAS) and the department of engineering mechanics at Tsinghua University. The generous financial support of IUTAM is very much appreciated. We are also grateful for the support from CSTAM, NSFC (National Natural Science Foundation of China), the Institute of Mechanics Chinese Academy of Sciences and Tsinghua University. Special thanks should be given to Mr. Yang Yazheng and Ms. Tang Ya-Nan of the CSTAM office. In particular, we would like to take this opportunity to thank the authors for their cooperative efforts in quickly revising and returning their manuscripts. Finally, we would like to thank the staff of Springer for publishing the proceedings on time.

Yilong Bai
Quanshui Zheng
Yueguang Wei

IUTAM Symposium 2005 Beijing

Sponsors

The International Union of Theoretical and Applied Mechanics (IUTAM)
The National Science Founcation of China (NSFC)
Chinese Society of Theoretical and Applied Mechanics (CSTAM)
Institute of Mechanics, Chinese Academy of Sciences (CAS)
Department of Engineering Mechanics, Tsinghua University

International Scientific Committee of the Symposium

Y.L. Bai	(Chairman, China)
H.J. Gao	(Germany)
H.M. Jensen	(Denmark)
S. Suresh	(USA)
R.Z. Valiev	(Russia)
Z.L. Wang	(USA)
Q.S. Zheng	(China)
L.B. Freund	(IUTAM Representative)

Members of the Organizing Committee

Y.L. Bai (Institute of Mechanics, CAS)
Q.S. Zheng (Tsinghua University)
Y.G. Wei (Institute of Mechanics, CAS)
Y.Z. Yang (CSTAM)
Y.N. Tang (CSTAM)

Invited Presentations

L. Anand, *Department of Mechanical Engineering, Massachusetts Institute of Technology, Cambridge, MA 02139, U.S.A.*

G.H. Chen, *Department of Chemistry, The University of Hong Kong, Hong Kong, China*

Y.T. Cheng, *Materials and Processes Laboratory, General Motors Research and Development Center, U.S.A.*

S. Forest, *Ecole des Mines de Paris / CNRS, Centre des Matériaux / UMR 7633, BP 87, F-91003 Evry Cedex, France*

H.J. Gao, *Max Planck Institute for Metals Research, Heisenbergstr. 3, D-70569 Stuttgart, Germany*

W.L. Guo, *Institute of Nano Science, Nanjing University of Aeronautics and Astronautics, Nanjing 210016, China*

J.M. Hill, *ARC Australian Professorial Fellow, School of Mathematics and Applied Statistics, University of Wollongong, Wollongong, NSW 2522, Australia*

Y.S. Hong, *LNM, Institute of Mechanics, Chinese Academy of Sciences, Beijing 100080, China*

K.J. Hsia, *Department of Theoretical and Applied Mechanics, University of Illinois at Urbana-Champaign, Urbana, IL 61801, U.S.A.*

Y. Huang, *University of Illinois at Urbana-Champaign, Urbana, IL 61801, U.S.A.*

Z.P. Huang, *Department of Mechanics and Engineering Science, Peking University, China*

C.S. Lu, *Centre for Advanced Materials Technology (CAMT), School of Aerospace, Mechanical and Mechatronic Engineering, The University of Sydney, NSW 2006, Australia*

L. Lu, *Shenyang National Laboratory for Materials Science, Institute of Metal Research Chinese Academy of Sciences, Shenyang 110016, China*

T.J. Lu, *Xian Jiaotong University, P.R. China, Department of Engineering, University of Cambridge, UK*

E. Ma, *Department of Materials Science and Engineering, Johns Hopkins University, Baltimore, MD 21218, U.S.A.*

X.S. Mao, *Department of Mechanical Engineering, University of Pittsburgh, Pittsburgh, PA 15261, U.S.A.*

A. Nakatani, *Department of Adaptive Machine Systems, Osaka University, 2-1 Yamadaoka, Suita, Osaka 565-0871, Japan*

Z.C. Ou-Yang, *Institute of Theoretical Physics, The Chinese Academy of Sciences, P.O..Box 2735, Beijing 100080, China, Center for Advanced study, Tsinghua University, Beijing 100084, China*

J.M. Qu, *School of Mechanical Engineering, Georgia Institute of Technology, Atlanta, GA 30332-0245, U.S.A.*

A.K. Soh, *Department of Mechanical Engineering, The University of Hong Kong, Pokfulam Road, Hong Kong, China*

H.Y. Wang, *LNM, Institute of Mechanics, Chinese Academy of Sciences, Beijing 100080, China*

T.C. Wang, *LNM, Institute of Mechanics, Chinese Academy of Sciences, Beijing 100080, China*

Z.L. Wang, *School of Materials Science and Engineering, Georgia Institute of Technology, Atlanta, GA 30332-0245, U.S.A.*

Y.G. Wei, *LNM, Institute of Mechanics, Chinese Academy of Sciences, Beijing 100080, P.R. China*

G. Xu, *Department of Mechanical Engineering, University of California at Riverside, Riverside, CA 92521, U.S.A.*

W. Yang, *Department of Engineering Mechanics, Tsinghua University, Beijing 100084, China*

T.Y. Zhang, *Department of Mechanical Engineering, Hong Kong University of Science and Technology Clear Water Bay, Kowloon, Hong Kong, China*

Y.W. Zhang, *Department of Materials Science and Engineering, National University of Singapore, 119260 Singapore*

Q.S. Zheng, *Department of Engineering Mechanics, Tsinghua University, Beijing 100084, China*

M. Zhou, *The George W. Woodruff School of Mechanical Engineering, Georgia Institute of Technology, Atlanta, GA 30332-0405, U.S.A.*

Part 1

Mechanical Behaviors of Nanocrystal Materials

Mesoscopic Modeling of the Deformation and Fracture of Nanocrystalline Metals

Lallit Anand* and Yujie Wei

*Department of Mechanical Engineering, Massachusetts Institute of Technology, Cambridge, MA 02139, U.S.A.; *E-mail: anand@mit.edu*

Abstract. In order to model the effects of grain boundaries in nanocrystalline materials we have coupled a crystal-plasticity model for the grain interiors with a new elastic-plastic grain-boundary interface model which accounts for both reversible elastic, as well irreversible inelastic sliding-separation deformations at the grain boundaries prior to failure. We have used this new computational capability to study the deformation and fracture response of nanocrystalline nickel. The results from the simulations capture the macroscopic experimentally-observed tensile stress-strain curves, and the dominant microstructural inelastic deformation and fracture mechanisms in this material. The macroscopically-observed nonlinearity in the stress-strain response is mainly due to the inelastic response of the grain boundaries. Plastic deformation in the interior of the grains prior to the formation of grain-boundary cracks was rarely observed. The stress concentrations at the tips of the distributed grain-boundary cracks, and at grain-boundary triple junctions, cause a limited amount of plastic deformation in the high-strength grain interiors. The competition of grain-boundary deformation with that in the grain interiors determines the observed macroscopic stress-strain response, and the overall ductility. In nanocrystalline nickel, the high yield strength of the grain interiors and relatively weaker grain-boundary interfaces account for the low ductility of this material in tension.

Key words: crystal plasticity, interface failure, finite elements.

1 Introduction

In contrast to conventional grain-sized materials, nanocrystalline (nc) materials typically exhibit ultrahigh strength/hardness, but with an attendant much-reduced ductility (e.g., Gleiter, 1989; Suryanarayana, 1995; Torre *et al.*, 2002). For example, while the ultimate tensile strength levels in (nominally fully-dense) electro-deposited nc-nickel, approach $\approx$ 1500 MPa, the ductility that can be obtained in this material is generally low and usually does not exceed $\approx 3\%$ (Torre *et al.*, 2002). Kumar *et al.* (2003) have recently reported on an extensive set of experiments that they conducted to observe the deformation mechanisms in electro-deposited nc-nickel both after and during deformation using transmission electron microscopy (TEM). Briefly, they concluded (a) the density of dislocations visible in their deformed specimens

Yilong Bai et al. (eds), IUTAM Symposium on Mechanical Behavior and Micro-Mechanics of Nanostructured Materials, 3–10.

could not account for the macroscopically-imposed plastic strains; (b) observations of tensile specimens which contained perforations in the center of their gage sections and which were incrementally strained in discrete steps *in-situ* in a TEM, provided evidence of nucleation and growth of grain-boundary cracks and triple junction voids ahead of a growing crack. Their experiments show that the grain-boundary slip and separation, together with dislocation-based plasticity in the interior of grains adjacent to the propagating crack, are the dominant mechanisms of inelastic deformation in the "process-zone" associated with the tip of a crack in nc-nickel.

Atomistic computer simulations of nc-metals have also been widely reported in the recent literature (e.g., Schiotz *et al.*, 1998; Swygenhoven *et al.*, 2002; Yamakov *et al.*, 2001; Farkas *et al.*, 2002). These atomistic simulations collectively show that as the grain size decreases, intra-granular plasticity driven by dislocation mechanisms becomes more difficult at nanometer-scale grain sizes, and that the macroscopically-imposed deformation is accommodated by grain-boundary sliding and separation.[1]

Thus, from the physical experiments and atomistic simulations reported in the literature, it is clear that grain-boundary-related slip and separation phenomena begin to play an important role in the overall inelastic response of a polycrystalline material when the grain-size decreases to diameters under ≈ 100 nm, and dislocation activity within the grain interiors becomes more difficult. One possible computationally-tractable modelling approach to account for the combined effects of grain-boundary-related deformation as well as plasticity within the grains, is to couple a single-crystal plasticity constitutive model for the grain interior with an appropriate cohesive interface constitutive model to account for grain-boundary sliding and separation phenomena.

The purpose of this brief paper is to introduce a new numerical simulation capability to study the deformation and fracture response of a poly-crystalline aggregate with an accounting of the grain-interior using standard crystal plasticity, and an accounting for grain-boundary response using a new elastic-plastic traction-separation constitutive model (Wei and Anand, 2004). Such a computational capability serves as a *meso-scale simulation tool* to study the deformation and fracture response of nanocrystalline materials.

2 Modeling Methodology

In recent years, cohesive interface models have been widely used to numerically simulate fracture initiation and growth by the finite element method (e.g., Xu and Needleman, 1994); however, most such models are limited to traction-separation relations which are *nonlinear elastic* and therefore *reversible* in nature. Since accounting for dissipation due to grain-boundary sliding and separation is of paramount importance, we have developed an *elastic-plastic interface constitutive model* to simulate such grain boundary phenomena in nc-materials. The reader is referred to Su

[1] We note that since atomistic calculations involve only a limited number of grains and nearly instantaneous (in a matter of picoseconds) loading, they only provide a qualitative understanding of the real mechanical response of materials at laboratory length and time-scales.

et al. (2004) and Wei and Anand (2004) for the details of the interface model. As to the deformation in grain interior, we shall employ the now classical single-crystal plasticity theory.

Both the interface model and the single crystal plasticity model have been implemented in the finite element program ABAQUS/Explicit (ABAQUS, 2002) by writing a USER INTERFACE subroutine and USER MATERIAL subroutine, respectively. We have used this numerical simulation capability to model the response of nc-nickel.

3 Application to nc-Nickel

Experimentally-measured stress-strain curves in simple tension from specimens of electro-deposited nc-nickel with grain sizes ranging from 15nm to 40nm from different groups show a highly nonlinear stress-strain response up to an ultimate tensile strength between $\approx$ 1.3 to 1.7 GPa, and final fracture at strain levels ranging from $\approx$ 2.5 to 5% (e.g., Wang *et al.*, 1997; Yin and Whang, 2001; Xiao *et al.*, 2001; Torre *et al.*, 2002). We shall use this representative experimental information to *estimate* the material parameters for the grain interiors as well as the grain boundaries by judiciously adjusting the values of the material parameters in our constitutive model to approximately match these stress-strain curves.

For reasons of computational efficiency, in our simulations we have used a *quasi-three-dimensional* polycrystalline aggregate consisting of a collection of columnar grains.[2] A finite element model of a polycrystalline aggregate containing 50 columnar grains is shown in Figures 1a and 1b. *Note that in order to capture the possible heterogeneous nature of deformation in each grain, the grain interiors are modelled by a number of finite elements.* Figure 1c shows an experimentally-measured (111) pole figure of as-received electrodeposited nc-nickel (Xiao *et al.*, 2001); the pole figure projection direction is normal to the plane of the sheet specimen. The corresponding (111) pole figure from the discrete grain orientations assigned to the 50 grains used in the polycrystal simulation is shown in Figure 1d.

For the grain interiors, the anisotropic elasticity tensor C is specified in terms of three stiffness parameters, $C11$, $C12$ and $C44$. The values of the elastic parameters for nickel are taken as (Simmons and Wang, 1971):

$$C11 = 247 \text{ Gpa}, \quad C12 = 147 \text{ Gpa}, \quad C44 = 125 \text{ Gpa}.$$

For f.c.c. crystals, crystallographic slip is assumed to occur on the standard twelve $\{111\} < 110 >$ slip systems.

Molecular dynamics simulations show that when grain-boundary deformation cannot be accommodated due to geometric restrictions, local stress concentrations develop to cause the emission of a few partial dislocations from grain boundaries, and these high stresses drive the partial dislocations across the grain interiors to be

[2] Electro-deposited nc-nickel does possess an approximately columnar grain structure; see Kumar *et al.* (2003, figure 1c).

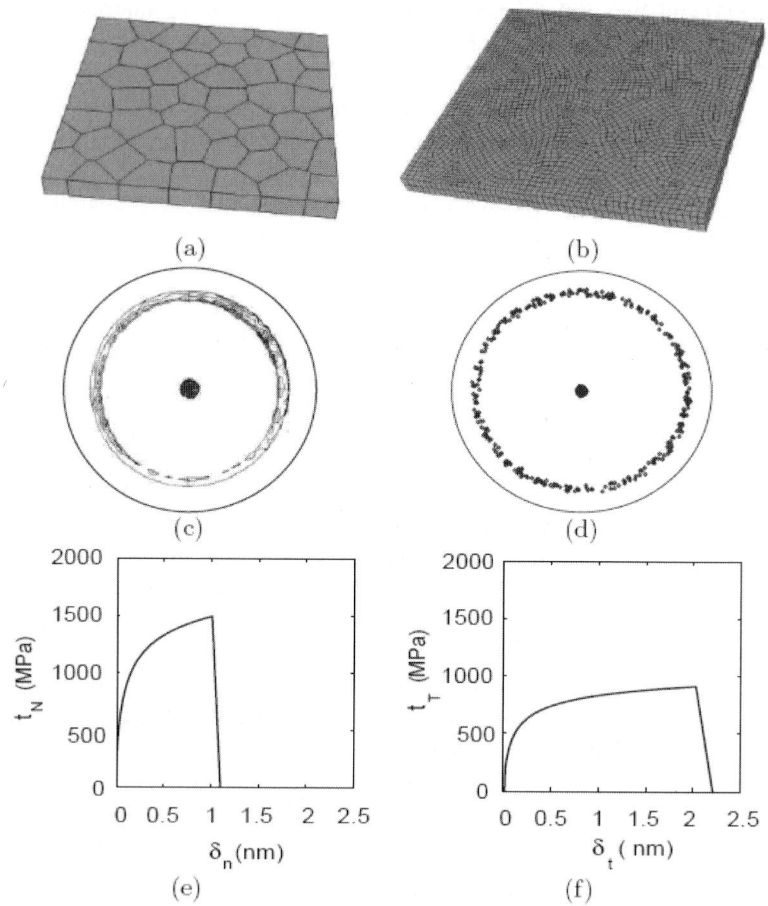

Fig. 1. (a) Initial microstructure represented by 50 columnar grains. (b) Finite-element mesh. (c) Experimental (111) pole figure of as-received electrode-posited nc-nickel from Xiao *et al.* (2001); pole figure projection direction is normal to the plane of the sheet specimen. (d) (111) pole figures corresponding to the grain orientations used in the polycrystal simulation. Traction-separation curves used for nc-nickel: (e) in the normal direction to an interface, and (f) in the tangential direction to an interface.

absorbed in the opposite grain boundaries. With this mechanism in mind, we assume that the slip resistances for the grain interiors s^α are all equal to a *constant* s_0, interpreted as the resistance to emission of partial dislocations from grain-boundaries into the grain interiors. The value of s_0 is estimated as

$$s_0 \approx \frac{Gb}{D}, \tag{1}$$

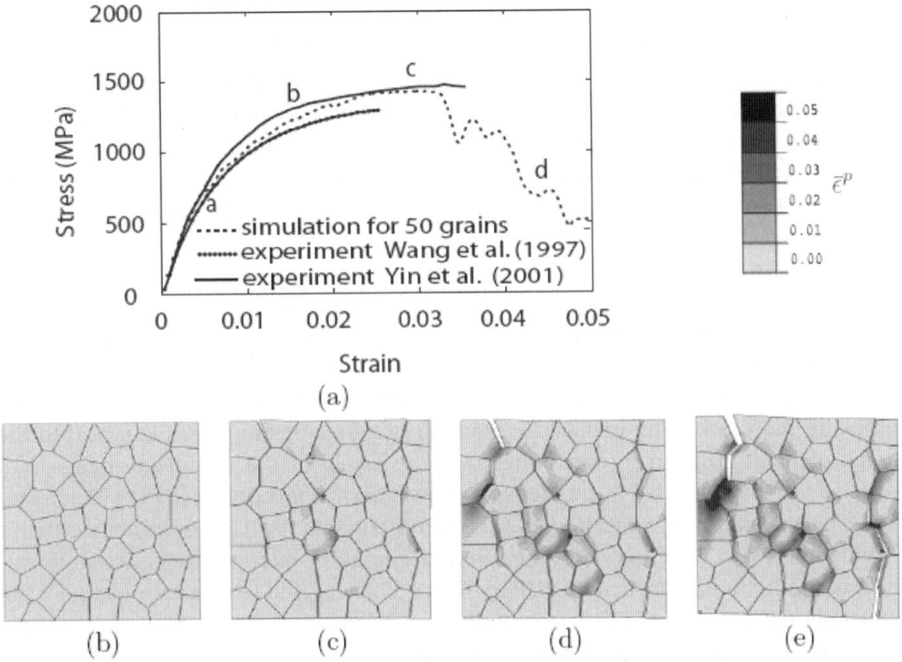

Fig. 2. (a) Stress-strain curves from the simulation and literatures. Contour plots (b) through (e) of the equivalent plastic strain in the grain interiors corresponding to different macroscopic strain level (a–d) keyed to the stress-strain curve.

with $G = \sqrt{(C_{11} - C_{12})C_{44}/2}$ denoting a shear modulus, b the magnitude of the Burgers vector, and D denoting the grain size.[3] For typical values of $G \approx 80$ MPa and $b \approx 0.3$ nm for Ni, and a grain size $D \approx 30$ nm, we obtain a value

$$s_0 \approx 800 \text{ MPa}, \tag{2}$$

for a 30 nm grain-sized nc-nickel.

Using the previously listed assumptions for the grain-interior response, we estimated the grain-boundary traction-separation response by matching our corresponding numerical simulations to the experimental tensile stress-strain curves published in the literature. The estimated traction-separation curves for nc-nickel in the normal direction to an interface, and in the tangential direction to an interface are shown Figures 1e and 1f, respectively. The fitted macroscopic stress-strain curve is shown in Figure 2a.

Figures 2b through 2e show contours of the *equivalent plastic strain* $\bar{\varepsilon}^p$ *in the grains*, corresponding to four different points, labelled a through d, on the stress-strain response, Figure 2a; the tension direction is horizontal in the plane of the

[3] More refined estimates based on stacking fault energies of partial dislocations may be made, but we do not go into such refinements here (Asaro *et al.*, 2003).

paper.[4] *Since Figure 2b shows essentially no plastic strain in the grain interiors, the nonlinearity in the macroscopic stress-strain curve up to point a is due entirely to the nonlinear elastic-plastic interface response.* By point b (Figure 2c) on the stress-strain curve, there is some plastic strain in the vicinity of a few grain-boundary triple junctions; however, the major cause for the macroscopic nonlinear stress-strain response is still the nonlinear interface response; indeed, by this stage a few of the interfaces have visibly failed. By point c (Figure 2d), near the peak of the stress-strain curve, there is a dominant macroscopic crack traversing along several grain boundaries (top left of Figure 2d), and substantial plastic strain within the interiors of the grains at the tip of the crack. By point d (Figure 2e) on the stress-strain curve at least three dominant interface cracks are clearly visible, and the material has lost a large fraction of its stress-carrying capacity. The propagation of grain-boundary cracks causes a substantial amount of plastic deformation in the interior of the grains blocking the crack-path, and in the boundary regions of the grains adjacent to the crack path. In order to further understand the origin of the nonlinear stress-strain response of the nc-nickel during tension, we numerically suppressed plastic deformation in both the grain interiors and the grain boundaries. This gives the response labelled "Elastic" in Figure 3. When in addition grain-interior-plasticity is also allowed, the curve is indicated by GIP in Figure 3. These first two cases, clearly do not resemble the shape of the experimentally-measured macroscopic stress-strain curve of nc-nickel. Next the case when elastic-plastic grain boundary deformation is allowed, and grain interior only deforms elastically is labelled as GB in the figure; the stress-strain curve is close to the experimentally-measured curve. The case when both grain boundary and grain interior plastic deformation is allowed is labelled as GB & GIP, and this gives (as expected) a slightly softer response. *Thus, it is clear that up to $\approx 3\%$ strain, the macroscopic stress-strain is dominated by the nonlinear grain-boundary response.*

4 Concluding Remarks

To model the effects of grain boundaries in polycrystalline materials we have coupled a standard crystal-plasticity model for the grain interiors with a new elastic-plastic grain-boundary interface model. We have used this new computational capability to study the deformation and fracture response of nc-nickel in simple tension. The results from the simulations capture the experimentally-observed stress-strain curves, and the dominant fracture mechanisms in this material (Kumar *et al.*, 2003). The macroscopically-observed nonlinearity in the stress-strain response is mainly due to the inelastic response of the grain-boundaries. Plastic deformation in the interior of the grains prior to the formation of grain-boundary cracks was rarely observed. The stress concentrations at the tips of the distributed grain-boundary cracks, and at grain-boundary triple junctions, cause a limited amount of plastic deformation in the high-strength grain interiors. The competition of grain-boundary deformation

[4] For clarity, the maximum value of the $\bar{\varepsilon}^p$ contours is set to 5% for all four contour plots.

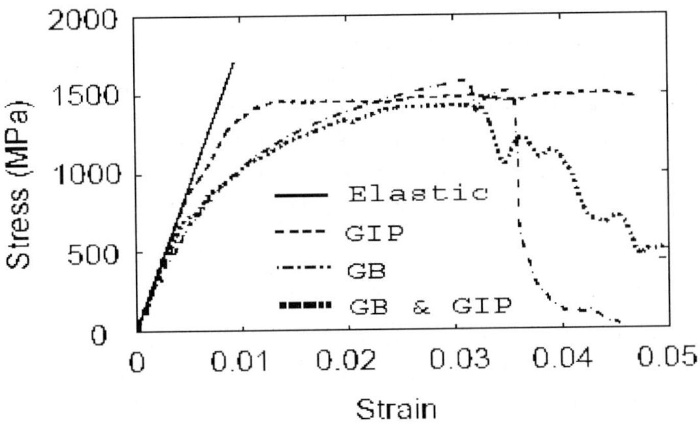

Fig. 3. Comparison of stress-strain curves in tension from several numerical simulation cases: only elastic deformation is allowed for both grain boundary and grain interior (Elastic); grain boundary deformation is suppressed and grain interior may deform elastic-plastically (GIP); elastic-plastic grain boundary deformation is allowed and grain interior only deforms elastically (GB); as well as both grain boundary and grain interior plastic deformation is allowed (GB & GIP).

with that in the grain-interiors determines the observed macroscopic stress-strain response, and the overall ductility. In nc-nickel, the high yield strength of the grain interiors and relatively weaker grain-boundary interfaces account for the low ductility of this material in tension. Full details of our study may be found in Wei and Anand (2004).

Acknowledgements

This work was supported by the Defense University Research Initiative on Nano-Technology (DURINT) on "Damage- and Failure-Resistant Nanostructured and Interfacial Materials" which is funded at the Massachusetts Institute of Technology (MIT) by the Office of Naval Research under grant N00014-01-1-0808.

References

ABAQUS Reference Manuals (2002) Hibbit, Karlsson & Sorenson, Inc., Pawtucket, RI.

Asaro, R.J., Krysl, P. and Kad, B. (2003) Deformation mechanism transitions in nanoscale fcc metals, *Philos. Mag. Lett.*, **83**, 733–743.

Farkas, D., Swygenhoven, H.V. and Derlet, P.M. (2002) Intergranular fracture in nanocrystalline metals, *Phys. Rev. B*, **66**, 060101-1–060101-4.

Gleiter, H. (1989) Nanocrystalline materials, *Progr. Mater. Sci.*, **33**, 223–315.

Kumar, K.S., Suresh, S., Chisolm, M.F., Horton, J.A. and Wang, P. (2003) Deformation of electodeposited nanocrystalline nickel, *Acta Mater.*, **51**, 387–405.

Schiotz, J., Tolla, F.D. Di and Jacobsen, K.W. (1998) Softening of nanocrystalline metals at very small grain sizes, *Nature*, **391**, 561–563.

Simmons, G. and Wang, H. (1971) *Single Crystal Elastic Constants and Calculated Aggregate Properties*, The MIT Press, Cambridge, MA.

Su, C., Wei, Y.J. and Anand, L. (2004) An elastic-plastic interface constitutive model: Application to adhesive joints, *Int. J. Plasticity*, **20**, 2063–2081.

Suryanarayana, C. (1995) Nanocrystalline materials, *Int. Metal. Rev.*, **40**, 41–46.

Swygenhoven, H.V., Derlet, P.M. and Hasnaoui, A. (2002) Atomic mechanism for dislocation emission from nanosized grain boundaries, *Phys. Rev. B*, **66**, 024101-1–024101-8.

Torre, F.D., Swygenhoven, H.V. and Victoria, M. (2002) Nanocrystalline electrode-posited Ni: Microstructure and tensile properties, *Acta Mater.*, **50**, 3957–3970.

Wang, N., Wang, Z.R., Aust, K.T. and Erb, U. (1997) Room temperature creep behavior of nanocrystalline nickel produced by an electrodeposition technique, *Mater. Sci. Engrg. A*, **237**, 150–158.

Wei, Y.J. and Anand, L. (2004) Grain-boundary sliding and separation in polycrystalline metals: Application to nanocrystalline f.c.c. metals, *J. Mech. Phys. Solids*, **52**, 2587–2616.

Wolf, D., Yamakov, V., Phillpot, S.R., Mukherjee, A. and Gleiter, H. (2005) Deformation of nanocrystalline materials by molecular-dynamics simulation: Relationship to experiments?, *Acta Mater.*, **53**, 1–40.

Xiao, C.H., Mirshams, R.A., Whang, S.H. and Yin, W.M. (2001) Tensile behavior and fracture in nickel and carbon doped nanocrystalline nickel, *Mater. Sci. Engrg. A*, **301**, 35–43.

Xu, X.P. and Needleman, A. (1994) Numerical simulations of fast crack growth in brittle solids, *J. Mech. Phys. Solids*, **42**, 1397–1434.

Yamakov, V., Wolf, D., Salazar, M., Phillpot, S.R. and Gleiter, H. (2001) Length-scale effects in the nucleation of extended dislocations in nanocrystalline Al by molecular-dynamics simulation, *Acta Mater.*, **49**, 2713–2722.

Yin, W.M. and Whang, S.H. (2001) Creep in Boron-doped nanocrystalline nickel, *Scripta Mater.*, **44**, 569–574.

Dislocation-Assisted Grain Growth in Nanocrystalline Copper under Large Deformation

X.L. Ma, H.T. Wang and W. Yang*

Department of Engineering Mechanics, Tsinghua University, Beijing 100084, China;
**E-mail: yw-dem@tsinghua.edu.cn*

Abstract. Dislocation plasticity would conventionally induce textures in polycrystalline metals after a large deformation. We used parallel molecular dynamics (MD) to simulate the plastic deformation of nanocrystalline copper to an isochoric stretch up to 100% logarithmic strain. We found that the movements of partial dislocations that dominate the deformation process do not lead to texture formation. The grain size distribution becomes extremely inhomogeneous. By observing the structural evolution, we demonstrate that partial dislocations assisted the grain growth that destroys the texture.

Key words: grain growth, nanocrystals, MD simulation, copper.

1 Introduction

Plastic deformation in polycrystalline metals at room temperature is generally caused by the movements of dislocations. Limited by the number of available slip systems, the grains will change their shapes and rotate towards a limited number of stable configurations, thus create a texture after a large deformation [1]. Nanocrystalline metals may behave differently from their coarse-grained counterparts. Then it is natural to ask whether the same deformation induced textures will develop in nanocrystalline metals. Cold rolling experiments on nanocrystalline copper [2] and palladium [3] showed no trace of textures formation and the grain shape remained equiaxed. Dislocation activities were evidenced during the rolling processes by the increase of the microstrain or the stacking fault density. However, the rather low strain rate suggest that the deformation mechanism was dominated by the grain boundary sliding and the grain rotation assisted by the grain boundary diffusion [2, 3]. The formation of textures was greatly suppressed by the grain boundary activities, similar to what had been observed in fine-grained superplastic alloys and ceramics deformed at high temperatures. The important role of partial dislocations was recognized by MD simulations [4–6] in rapidly deforming the nanocrystalline metals. These simulations are typically performed under high strain rate and low temperatures. Therefore the grain

Yilong Bai et al. (eds), IUTAM Symposium on Mechanical Behavior and Micro-Mechanics of Nanostructured Materials, 11–18.

boundary diffusion was greatly reduced and the deformation process was mainly controlled by the movements of partial dislocations. The logarithmic strains achieved in the early MD simulations were no more than 10%, incapable of the prediction of texture formation. This paper will examine the issue for nanocrystalline copper to the regime of large deformation by MD simulations that naturally down-plays the effect of the grain boundary diffusion.

2 Simulation Scheme

The simulation was carried out by standard MD techniques, with the atomic interactions for copper [7] described by an embedded atom method potential. Three-dimensional Voronoi construction [8] was used to generate the nanocrystalline model for computation. The sample configuration involves 1,048,645 atoms and 64 randomly oriented grains with an average grain size about 7 nm. Periodic side boundaries were maintained throughout the simulation. The isochoric stretching of a nanocrystalline copper box was conducted by parallel molecular dynamics simulations under the Linked-Cell-List algorithm [9]. The initial MD simulation carried 3000 annealing relaxation steps at 300 K, using Verlet leapfrog algorithm [10] with MD time step of 2 fs. The loading was imposed under a quasi-homogeneous manner. A homogeneous affine deformation of 0.1% stretching and 0.05% lateral contraction was enforced. Then 500 relaxing steps were executed at 300 K by maintaining a NVT ensemble. The stretching rate was 10^9 s^{-1} and the total simulation time was 1.006 ns. Figures 1a and 1b showed the atom configurations of 6% and 91% logarithmic strains respectively. The atoms in grain boundaries, stacking faults or inside grains were identified by local crystalline order [11, 12]. Comparing the two configurations in Figure 1, one finds the grain boundaries were widened and the disordered atoms increased after a large deformation. In order to maintain the plastic deformation, more and more partial dislocations emitted from grain boundaries or triple junctions and passed through grains. Partial dislocations were left behind the stacking faults. At the stage of both small (Figure 1a) and large deformation (Figure 1b), only a few slip systems, generally less than three in each grain, were activated. The strain compatibility could not be met along the grain boundaries if the plasticity was only mediated by movements of partial dislocations. The atoms near grain boundaries adjusted their locations locally under the high flow stress, in the form of a large number of small sliding events. These small slipping facilitated the grain rotation process. It was found in the simulation that the grains tended to rotate to the tensile direction, with a grain rotation rate less than 4 degree per 1% logarithmic strain. As the deformation continued, the grain boundaries widened and triple junctions enlarged to accommodate the accumulated misfit strain among grains.

3 Key Results

The flow stress, shown in Figure 2a, was kept about 2.25 GPa after about 5 to 6% logarithmic strain. However, a stable flow stress does not necessarily imply a steady

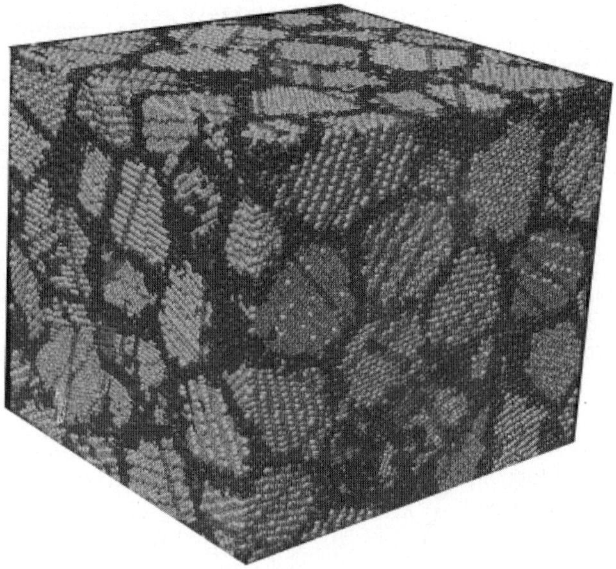

(a)

(b)

Fig. 1. Local crystalline order of nanocrystalline Cu under uniform stretching of (a) 6% and (b) 91% logarithmic strains. The atoms in FCC stacking sequence are referred to as the perfect lattice and colored green; the atoms of HCP stacking sequence are regarded as the stacking fault and colored red, while the atoms with other crystalline orders are thought as grain boundary and painted blue.

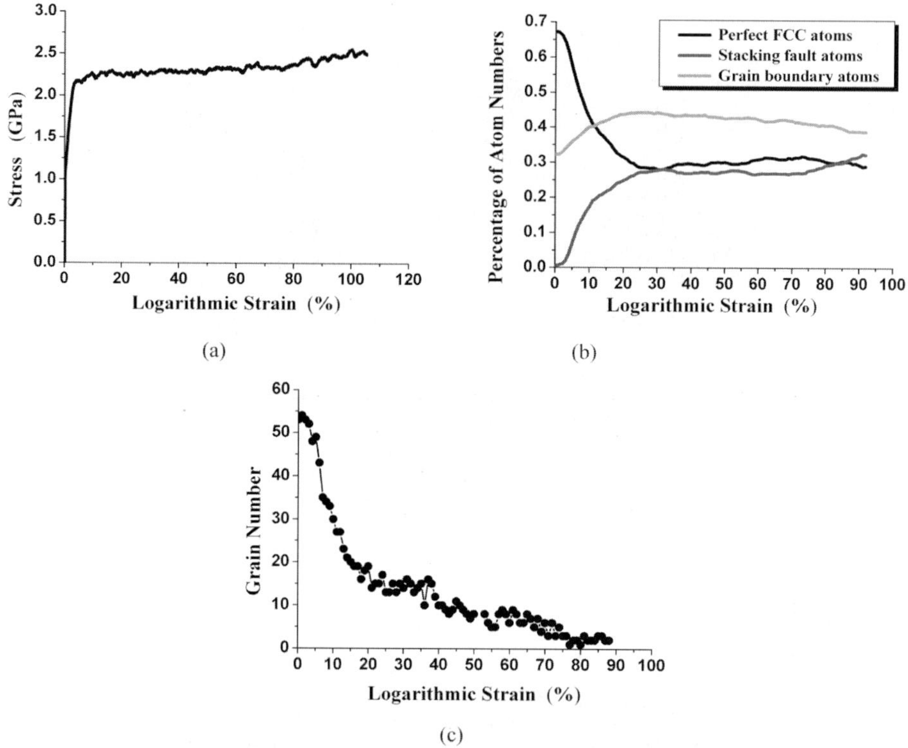

Fig. 2. (a) The curve for the stress versus logarithmic strain; (b) the percentage variations of atoms in perfect FCC, stacking faults and grain boundaries; (c) the variation of the grain number during the deformation process.

state. The atom configuration kept evolving as the deformation proceeded. Figure 2b showed the percentage variations of perfect FCC, stacking fault and grain boundary atoms. The regular FCC atoms steadily decline from the initial percentage of about 70% to merely 30% at 30% logarithmic strain. At the same strain level, the atoms percentage in stacking faults shoots up to nearly 30%. The atom percentage in grain boundaries modulates from the initial percentage of 30% to 40%. After 30% logarithmic strain, these percentages remained steadily with the deformation up to 100% logarithmic strain. The increases on the percentages of the stacking fault and grain boundary atoms indicated the disorder of the nanocrystalline sample was enhanced at large deformation. Hidden behind the steady percentages of perfect FCC, stacking fault, and grain boundary atoms, the changes in the grain configuration were still enormous in the large stretch regime. To reveal this change, one may classify atoms of the sample into three categories: the "grain boundary atoms" unrecognized by the local crystalline order method; the "pseudo-boundary atoms" locating within three layers thickness between the perfect FCC atoms and the grain boundary ones; and the remaining "pure grain atoms". This classification obtains clean isolated grains and

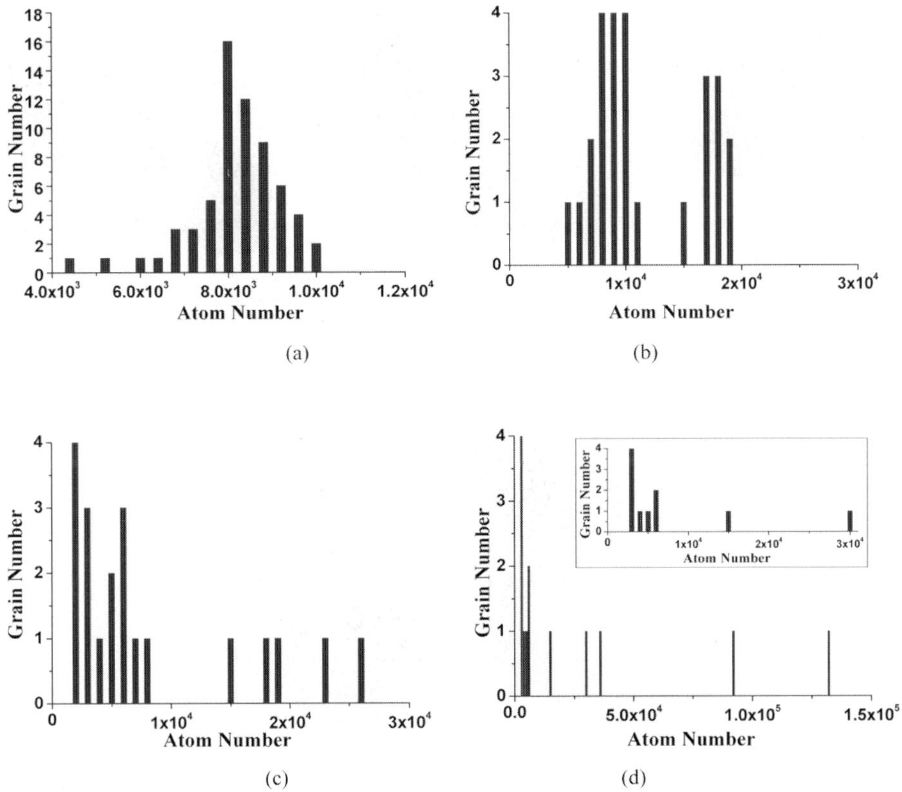

Fig. 3. Size distribution of grains at (a) 0%, (b) 10%, (c) 50% and (d) 91% logarithmic strains. Only the grains containing more than 1000 atoms are considered.

eliminates the scattering clusters of a few atoms which observe the FCC crystalline order. The number of grains which is larger than the average one decreases sharply at the initial 20% logarithmic strain. Then the trend became slow with the stretching strain, as plotted in Figure 2c. At the end of the deformation, most of the grains merged into a few super grains and others were cut into small colonies, containing only several thousands of atoms.

The grain size distribution became inhomogeneous in the regime of large deformation, as delineated in Figure 3. The grain size distribution became bimodal after 10% of deformation (Figure 3b). A few grains grew up obviously. When the deformation reached 50% logarithmic strain, the first peak moved toward a smaller size and the second peak became flat (Figure 3c). Two grains grew to rather large sizes. After 91% logarithmic strain, the size distribution of the grains in the sample became very dispersive (Figure 3d). The largest grain had the number of atoms about 3.8×10^5 and an approximate size of 18 nm. The upper right inset showed that the number of the smaller grains also decreased. The reason might lie in the merging

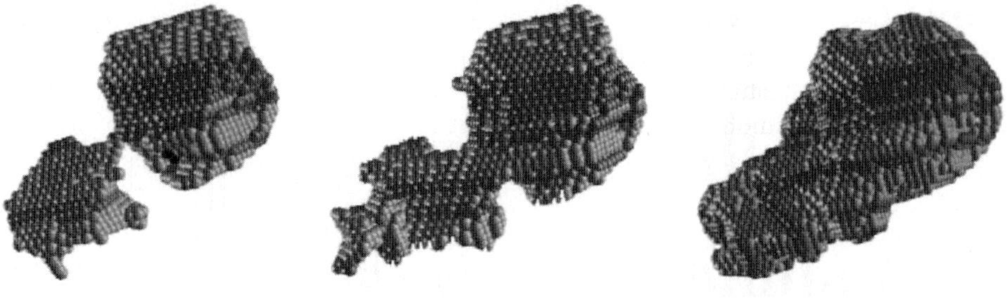

(a)

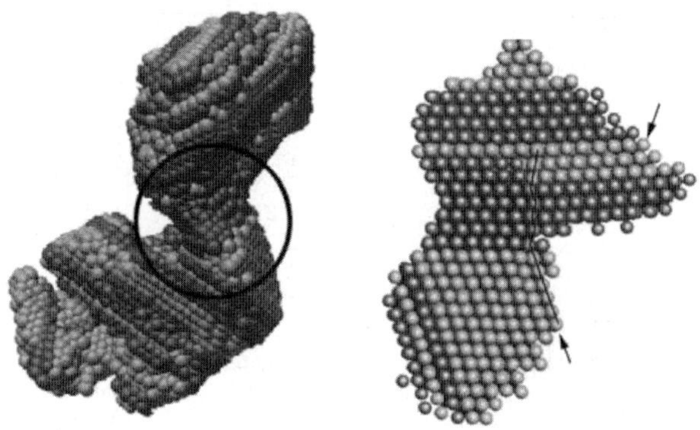

(b)

Fig. 4. (a) Stacking faults vertically pass through small angle grain boundary (from the left to the right). Three stages of the stacking fault penetration are shown: onset of stacking fault passing (left); partial merge of neighboring grains (middle) and complete merge (right). (b) Configuration of a large pure grain under 40% stretching strain. The connection area has a twin-like structure manifested by stacking faults parallel to the previous grain boundary. The cross-section view gives the orthomorphic projection of the area circled in the left.

process among grains. The process of grain refinement is unlikely since the grain boundary sliding, rather than the movements of partial dislocations, would dominate the deformation mechanisms for a small grain size of 4 or 5 nm.

4 Discussions

Two modes of grain combination processes were revealed: penetration of the stacking faults through low angle grain boundaries at small deformation (Figure 4a) and the formation of twin-like structures at large deformation (Figure 4b). The two mechanisms were different from the observations of MD simulation by Hasnaoui et al. [6]

and Haslam et al. [13]. In [6] the sample of the nanocrystalline nickel was deformed at $0.46T_m$ and in [13] that of the nanocrystalline palladium at $0.66T_m$, with T_m being the melting temperature. For the relative high temperature, grain boundaries diffusion gained higher mobility to replace the role of partial dislocations. Both in [6] and [13], low-angle grain boundaries were found to disassociate into partial dislocations, propagate into grains and lead to the grain coalescence. In [13], grain boundary migration assisted by diffusion accounted for the most grain growth. The simulation presented in this paper was undertaken at rather low temperature, i.e. $0.22T_m$. Two modes of grain growth concerned with the kinematics of stacking fault propagation and the grain rotation. In the small deformation regime, grain merge gave priority to those neighboring grains whose mis-orientations are below 10 degrees. The stacking faults burst through the boundary between the neighboring grains vertically, see Figure 4a from the left to the right. The mis-orientation of two neighboring grains would be eliminated gradually, which was presented as the partial merge (the middle graph in Figure 4a) and the final consolidation (the right graph of Figure 4a) processes. The larger grain formed by such mode was stable in the later evolution. Another mode dominated the merge during the large strain regime for the neighboring grains with large mis-orientations. The merge process was featured by parallel (instead of vertical in the small strain regime) passing of stacking faults through the grain boundary. Figure 4b showed the configuration of a large grain under 40% logarithmic strains. The grain was evolved by combining two small ones situated at the top and bottom of the figure. The central connection neck is composed of stacking faults which are parallel to the previous grain boundary. A cross-section view of the orthomorphic projection of the area circled in the left was delineated in the right of Figure 4b. The red neck-like HCP configuration area can be viewed as the twin-type grain boundary layer for the adjacent grains (in green color), as evidenced by the twin-type orientations marked by the arrows. The zigzag path indicates a linking route through the closest atoms to depict the twin-like structure. As suggested by Derlet et al. [14] and Kumar et al. [15], this twin-like structure was formed at large deformations and was caused by emission of a second partial dislocation on an adjacent slip plane. The merged grain was comparatively unstable and might separate again.

Careful observations on the structure evolution showed no textures were formed in our simulation. Initial deformation was dominated by partial dislocations. As the deformation continued, the disordered atoms and the formation of twin-like structures encumbered the activities of partial dislocations. The twin-like structures, formed at large deformations, are effective barriers to the movements of dislocations and thus disturb and hinder the texture evolution among large grains. The grain merge process gives rise to inhomogeneous grain size distribution in the sample. The extremely small grains would facilitate the grain boundary activities that also served for the reduction of textures. This nature of nanocrystalline metals provides a mechanism to maintain their physical properties isotropic even after large deformations.

Acknowledgements

The authors are grateful for the financial supports from the National Science Foundation of China and the State 973 Project.

References

1. W. Yang and W.B. Lee, *Mesoplasticity and Its Applications*, Springer-Verlag, New York (1993), pp. 299–304.
2. L. Lu, M.L. Sui and K. Lu, *Science* **287**, 1463 (2000).
3. J. Markmann et al., *Scripta Mater.* **49**, 637 (2003).
4. J. Schiøtz, F.D. Di Tolla and K.W. Jacobsen, *Nature* **391**, 561 (1998).
5. J. Schiøtz, T. Vegge, F.D. Di Tolla and K.W. Jacobsen, *Phys. Rev. B* **60**, 11971 (1999).
6. A. Hasnaoui, H. Von Swygenhoven and P.M. Derlet, *Phys. Rev. B* **66**, 184112 (2002).
7. R.A. Johnson, *Phys. Rev. B* **39**, 12554 (1989).
8. X.L. Ma and W. Yang, *Nanotechnology* **14**, 1208 (2003).
9. S.J. Plimpton, *J. Comput. Phys.* **117**, 1 (1995).
10. M.P. Allen and D.J. Tildesley, *Computer Simulation of Liquid*, Clarendon, Oxford (1987), pp. 140–181.
11. J.D. Honeycutt and H.C. Andersen, *J. Phys. Chem.* **91**, 4950 (1987).
12. H. Van Swygenhoven, D. Farkas and A. Caro, *Phys. Rev. B* **62**, 831 (2000).
13. A.J. Haslam et al., *Acta Mater.* **51**, 2097 (2003).
14. P. Derlet, A. Hasnaoui and H. Van Swygenhoven, *Scripta Mater.* **49**, 629 (2003).
15. K. Kumar, H. Van Swygenhoven and S. Suresh, *Acta Mater.* **51**, 5743 (2003).

Microstructure and Tensile Strength of Cu with Nano-Scale Twins

Y.F. Shen, X.H. Chen, B. Wu and L. Lu*

Shenyang National Laboratory for Materials Science (SYNL), Institute of Metal Research, Chinese Academy of Sciences, Shenyang 110016, People's Republic of China;
**E-mail: llu@imr.ac.cn.*

Abstract. Tensile test was performed on an electrodeposited pure Cu sample with high density growth-in twins confined in the submicro-sized grains. It is found that the Cu sample has an ultra-high tensile strength (~1.0 GPa) and a good ductility (~13.5%). Microstructure observations indicated that twin boundary provides a strong barrier for the motion of dislocations during plastic deformation.

Key words: nano-scale twins, twin boundary, copper, strength.

1 Introduction

Grain refinement which introduces more grain boundaries (GBs), one of the main barriers to lattice dislocation motion, is effective in strengthening metals. When grains are refined down to an extreme dimension, e.g., into the nanometer regime, where lattice dislocation motion is significantly suppressed by a high density of GBs, an extremely high strength will be expected. Experimental observations showed that most nanocrystalline (nc) metals have significantly high strength (which can be 4–6 times higher than that of coarse-grained (CG) Cu [1–4]). However, they are very brittle typically with an elongation-to-failure less than a few percent in tension [1, 2, 5], even for those ductile metals in CG forms such as Cu and Al. The strength and ductility are trade-off in the metallic materials. The brittleness of nc metals, which greatly limits their technological applications, originates intrinsically from the suppressed dislocation activities by a high density of GBs besides the extrinsic effect from the processing flaws, such as porosities and contaminations, etc. [1, 2].

Twin boundary (TB), a special kind of boundary, may block the motion of dislocations to strengthen materials, analogous to the conventional GBs [6–9]. An early study on an à-brass with twin spacing in the micrometer regime showed that TBs are equivalent to conventional GBs with respect to the H–P relation strengthening [10]. It seems TBs can be used as an alternative practical approach for strengthening materials. In this work, high density of twins with nanoscale spacing was introduced

Yilong Bai et al. (eds), IUTAM Symposium on Mechanical Behavior and Micro-Mechanics of Nanostructured Materials, 19–24.

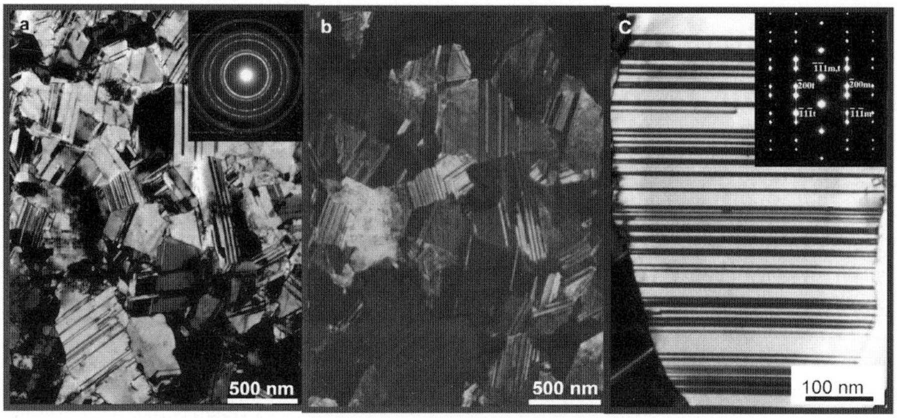

Fig. 1. Typical TEM planar-view microstructure observations ((a) bright field with corresponding SAED pattern and (b) dark field) of the as-deposited sample. (c) A close TEM observation of the twins with a SAED pattern in inset.

in a pure Cu sample by means of pulsed electrodeposition technique. The strength of the Cu sample was investigated by tensile tests at room temperature.

2 Experimental Details

High purity copper foils ($20 \times 10 \times 0.1$ mm^3 in size) with nano-scale growth twin lamellae were synthesized by means of the pulsed electro-deposition technique from a solution of CuSO4. Details of the sample preparation were described in [11]. Chemical analysis indicated that the purity of as-deposited Cu was better than 99.998 at.% with a sulfur content of less than 8 ppm. Contents of oxygen and hydrogen were determined to be less than 20 ppm and 15 ppm, respectively. Density of the as-deposited Cu sample was 8.93 ± 0.03 g/cm^3.

Microhardness measurements were performed on a MVK-H300 hardness testing machine with a load of 5 g and a loading time of 10 s. Both sides of the Cu specimens were mechanically ground and electrolytically polished before the measurements. Uniaxial tensile tests were performed on a Tytron 250 microforce testing system (MTS) at a constant strain rate of 6×10^{-3} s^{-1} at room temperature. The gauge length of the dogbone-shaped specimen was 4 mm with a width of 2 mm. Four Cu samples were tested with a final thickness after electro-polishing of about 16–25 μm.

Microstructures of the as-deposited and the as-deformed Cu samples were characterized by means of transmission electron microscopy (TEM) on a JEM2010 microscope (with an accelerating voltage of 200 kV). Thin foil specimens for TEM observations were prepared by using conventional twin-jet polishing technique at about $-10°$C.

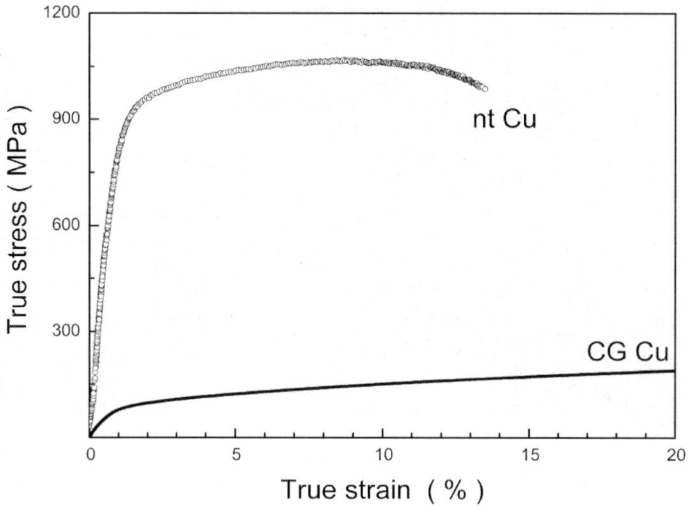

Fig. 2. Tensile true stress–true strain curves for as-deposited nt Cu sample. For comparison, tensile stress–strain curves for a coarse-grained (CG) Cu is also included.

3 Results and Discussion

From the TEM images of the as-deposited nano twin (nt) Cu sample (Figures 1a and 1b), one can see clearly irregular-shaped grains with random orientations, as indicated by the selected area electron diffraction (SAED) pattern. The grain sizes vary within 100 nm to 1 μm with an average value of about 400 nm. As the TEM images show, most grains contain multiple lamellar structures (as shown in Figure 1c). Most of them are of $\{111\}/[112]$ type with twinning elements, e.g., M: $(\bar{1}1\bar{1})/[\bar{1}12]$ and T: $(\bar{1}11)/[1\bar{1}2]$ (as indicated by the corresponding SAED pattern in Figure 1c). Statistic measurements of the twin lamellar spacing show an average value of about 20 nm. Clearly, the high density growth-in coherent TBs subdivided the submicro-sized grains into nanometer-sized twin/matrix lamellar structure. Lattice dislocation can hardly be detected in most thin lamellae, but some dislocations are visible in thick layers. This agrees with the X-ray diffraction result that a negligible atomic level lattice strain was identified. X-ray diffraction patterns show an evident (110) texture in the samples, which is consistent with previous observations in electrodeposited Cu specimens with growth twins [12, 13]. The length of twin lamellar geometry varies from about 100 nm to $\sim$1 μm (depending on the grain diameter).

Figure 2 shows typical true stress vs true strain curves for as-deposited nt Cu sample, in comparison with that for the annealed CG Cu sample (grain size $>$ 100 μm). Obviously, the as-deposited nt Cu sample exhibits much higher strength than that of the CG counterpart. For as-deposited nt Cu sample, The tensile yield strength σ_y (at 0.2% offset) reaches as high as 900 MPa and the ultimate tensile strength (σ_{UTS}) is 1068 MPa, which are about one order of magnitude higher than those of the CG Cu (the yield stress is about 50 MPa and the ultimate tensile

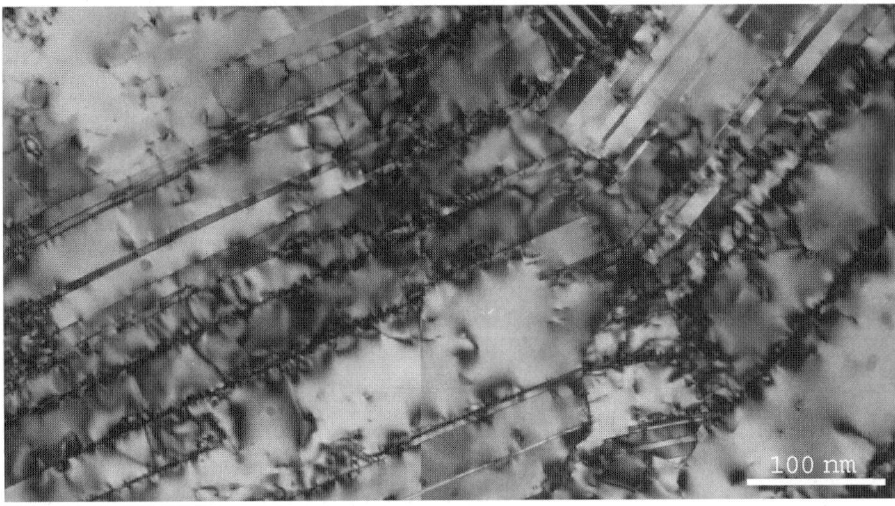

Fig. 3. A TEM image of the microstructure close to the failure surface of as-deposited sample after tensile tests.

strength is about 250 MPa). The microhardness result agrees very well with the tensile data. The values of $H_v/3$ can be approximated as the yield strength in tension ($H_v \approx 2.67 \pm 0.05$ GPa, an average value from 12 tests). In addition, the strength values for the as-deposited nt Cu sample are evidently higher than those reported for polycrystalline pure Cu with 3D grain sizes down into the nanometer scale [2, 5, 14].

In Figure 2, one can also see the nt Cu sample shows a considerable tensile ductility, with an elongation-to-failure value of 13.5%, which is also much larger than that of the previous nanocrystalline Cu specimens with comparable grain size (tens nm) [5]. A slight strain hardening appears in the major plastic deformation stage for sample, indicated that some dislocations accumulated during the plastic straining prior to failure. Obvious necking behaviors were noticed in the deformed samples after tensile tests.

Figure 3 shows a typical TEM image of the strained nt Cu sample (after tensile tests). Comparing with the clear microstructure of the as-deposited nt Cu sample (in Figures 1a–1c), the structure of the after-tensile nt Cu sample is much strained. Most TBs become "dirty" (with stress field induced contrasts) under TEM observations. Some TBs are not straight lines as in the as-deposited sample and a few TBs seem disappeare. Lots of dislocations are detected, but the distributions of dislocations were not uniform; the inhomogeneous dislocation distribution may originate from the layer thickness effect as well as from different grain orientations. Plenty of dislocations (and dislocation tangles) are identified inside the thick lamellae. Most dislocations in the lamellae seem to terminate either at TBs or at grain boundaries. TBs separating thick lamellae become less distinct at which a high density of dislocations locates. Few dislocations are found in thin lamellae and TBs separating thin layers are as distinct as in the as-deposited sample. A close observation of the

TBs revealed that a large number of dislocations locate at these TBs with a spacing ranging from several to several tens nanometers.

Microstructural investigations of the Cu samples after deformation confirmed that the interaction between TBs and dislocations plays a crucial role in the plastic deformation process. The interaction between slip dislocation and coherent TBs in face-centered cubic (fcc) metals has been extensively investigated [6]. It is known that when glide dislocations (which may be generated at GBs or triple junctions upon straining) encounter TBs inside the grain, in most cases the dislocation glide will be inhibited by the high density of TBs. However, these dislocations could also pile-up and propagate across the twins if they were to undergo dislocation dissociation reactions, which require stress concentrations at twin-slip band intersection, thus leading to strengthening. In this case, coherent TBs behave more or less like GBs in acting as obstacles to plastic deformation [9, 15, 16]. When the twin lamellae are thick, dislocation pile-ups form and produce a certain stress concentration at the TBs; in this case, a lower applied stress is needed to activate the slip transmission across TBs. The thinner the twin lamellae, the higher the external stress required for the dislocations to cross the TBs. If the twin spacing is too thin to have a dislocation pile-up, only a single dislocation penetrates the TB. This is the extreme case and a very high stress is required. Recent molecular dynamic (MD) simulation results on the role of TB in blocking a single dislocation transmission in fcc metals strongly supports the observed experimental phenomenon [17]. It was found for the TBs in Ni, the critical stress can be as high as 1.77 GPa for a single dislocation penetrating a coherent TB.

The observed dislocation configurations in the deformed sample might also account for the measured plasticity. For the thick lamellae, plastic deformation will be carried mainly by lattice dislocations inside the layers. While for thin lamellae, larger plastic strains can be accommodated by a high density of dislocations locating at numerous TBs. In addition, TB migration via movements of dislocation at TBs will also be responsible for carrying plastic strains. Therefore, a higher plasticity is obtained in the sample with higher density of twin lamellae. This observation is consistent with a recent MD simulations [17] that have also indicated that when a perfect glide dislocation with $b = 1/2\ [101]$ crosses a symmetric (111) TB, a Shockley partial (with $b = 1/6\ [-1-12]$) is left behind at the TB. The capacity of dislocation accumulation at the TBs is enhanced by a high TB density, which consequently results in an enhanced ductility during plastic deformation.

4 Summary

Presence of nano-scale twins in a pure ultrafine-grained Cu sample enhances the tensile strength up to 1 GPa with an elongation-to-failure of 13.5%. The experimental investigation demonstrated that TBs may act as effective obstacles to dislocation motion and a high density of TBs with twin lamellae thickness in the nanometer scale may significantly strengthen metals.

Acknowledgements

The authors thank NSFC (Grant No. 50201017 and No. 90206044), MOST of China (G1999064505) and MPG of Germany for financial supports, X. Si and H.B. Ma for sample preparation.

References

1. Koch, C.C., Morris, D.G., Lu, K. and Inoue, A., Ductility of nanostructured materialsm, *MRS Bull.* **24**, 1999, 54–58.
2. Sanders, P.G., Eastman, J.A. and Weertman, J.R., Elastic and tensile behavior of nanocrystalline copper and palladium, *Acta Mater.* **45**, 1997, 4019–4025.
3. Wang, Y.M., Wang, K., Pan, D., Lu, K., Hemker, K.J. and Ma, E., Microsample tensile testing of nanocrystalline copper, *Scripta Mater.* **48**, 2003, 1581–1586.
4. Weertman, J.R., Farkas, D., Hemker, K., Kung, H., Mayo, M., Mitra, R. and Van Swygenhoven, H., Structure and mechanical behavior of bulk nanocrystalline materials, *MRS Bull.* **24**, 1999, 44–50.
5. Wang, N., Wang, Z.R., Aust, K.T. and Erb, U., Room temperature creep behavior of nanocrystalline nickel produced by an electrodeposition technique, *Mater. Sci. Eng. A-Struct. Mater. Prop. Microstruct. Process.* **237**, 1997, 150–158.
6. Mahajan, S. and Chin, G.Y., The interaction of twins with existing substructure and twins in cobalt-iron alloys, *Acta Metallurgica* **22**, 1974, 1113–1119.
7. Lim, L.C., Slip-twin interactions in nickel at 573K at large strains, *Scripta Metallurgica* **18**, 1984, 1139–1142.
8. Mayers, M.A. and Chawla, K.K., *Mechanical Behavior of Materials*, Prentice Hall, Upper Saddle River, NJ, 1999.
9. Andrade, U., Meyers, M.A., Vecchio, K.S. and Chokshi, A.H., Dynamic recrystallization in high-strain, high-strain-rate plastic-deformation of copper, *Acta Metall. Mater.* **42**, 1994, 3183–3195.
10. Babyak, W.J. and Rhines, F.N., *Trans. Metall. Soc. AIME* **218**, 1960, 21.
11. Lu, L., Shen, Y.F., Chen, X.H., Qian, L.H. and Lu, K., Ultrahigh strength and high electrical conductivity in copper. *Science* **304**, 2004, 422–426.
12. Dahlgren, S.D., Nicholson, W.L., Merz, M.D., Bollmann, W., F. D.J. and Wang, R., Microstructural analysis and tensile properties of thick copper and nickel sputter deposits. *Thin Solid Films* **40**, 1977, 345–353.
13. Rasmussen, A.A., Jensen, J.A.D., Horsewell, A. and Somers, M.A.J., Microstructure in electro deposited copper layers; the role of the substrate. *Electrochim. Acta* **47**, 2001, 67–74.
14. Champion, Y., Langlois, C., Guerin-Mailly, S., Langlois, P., Bonnentien, J.L. and Hytch, M.J., Near-perfect elastoplasticity in pure nanocrystalline copper. *Science* **300**, 2003, 310–311.
15. Christian, J.W. and Mahajan, S., Deformation twinning. *Prog. Mater. Sci.* **39**, 1995, 1–157.
16. Valiev, R.Z., Structure and mechanical properties of ultrafine-grained metals. *Mater. Sci. Eng. A-Struct. Mater. Prop. Microstruct. Process.* **234**, 1997, 59–66.
17. Zhang, X., Misra, A., Wang, H., Nastasi, M., Embury, J.D., Mitchell, T.E., Hoagland, R.G. and Hirth, J.P., Nanoscale-twinning-induced strengthening in austenitic stainless steel thin films. *Appl. Phys. Lett.* **84**, 2004, 1096–1098.

Microstructural Evolution in Crystalline Metal Induced by Plastic Deformation

Atomistic Study of Grainsubdivision in Tension, Torsion and Rolling

A. Nakatani[1] and T. Shimokawa[2]

[1]*Department of Adaptive Machine Systems, Osaka University, 2-1 Yamadaoka, Suita, Osaka 565-0871, Japan; E-mail: nakatani@ams.eng.osaka-u.ac.jp*
[2]*Division of Mechanical Science and Engineering, Kanazawa University. Kakuma-machi Kanazawa, Ishikawa 920-1192, Japan; E-mail: simokawa@t.kanazawa-u.ac.jp*

Abstract. Atomistic studies of deformation-induced microstructural evolution in single crystalline metals are carried out to elucidate the mechanism of grain subdivision in a perfect crystal. We review our simulations of tensile and torsion of miniaturized models. Then, rolling of the iron nanoplate process is performed. Remarkable polycrystallization is observed in the twisting and rolling problems, compared to the tension problem. According to the results, it is concluded that the strain gradient plays an important role in grain subdivision.

Key words: plasticity, molecular dynamics, polycrystalline metal, grain subdivision.

1 Introduction

Improvement of a material's properties by controlling its microstructure, such as grain re-finement, attracts a great deal attention because it has the advantage of recycling over other methods, such as using chemical composition or composite materials. In particular, severe plastic deformation (SPD) processes, e.g., high-pressure torsion (HTP), equal-channel angular pressing/extrusion (ECAP/ECPE) and accumulative roll-bonding (ARB) (Saito et al., 1999) can produce an ultrafine granular structure. The microstructure obtained by inducing a large strain has a high defect density. Studies of the internal structure obtained by SPD have been performed and the relationship between texture and mechanical properties has been clarified in recent research. For example, one study of the SPD process is examined using crystal plasticity (Sivakumar and Ortiz, 2004) and another is applied the theory of higher-order lattice defect (Seefeldt et al., 2001) to investigate grain subdivision (Hansen, 2001). However, many of the details of the mechanism of grain refinement or grain subdivision are unknown. In this paper, atomistic studies of deformation-induced microstructural evolution in single crystalline metals are described with the aim of

Yilong Bai et al. (eds), IUTAM Symposium on Mechanical Behavior and Micro-Mechanics of Nanostructured Materials, 25–35.

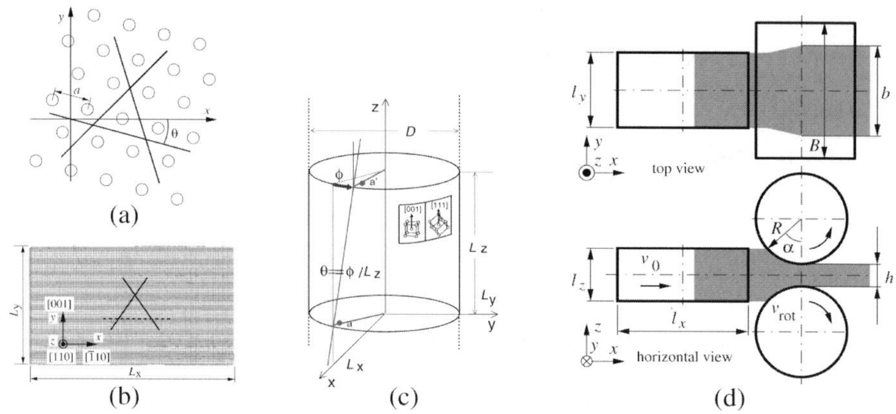

Fig. 1. Analysis models. (a) Triangular two-dimensional lattice with a rotation of θ. (b) Fcc single crystal model and active slip systems. (c) Torsion model. (d) Rolling model.

investigating the mechanism of grain subdivision in a perfect crystal. First, tensile deformation of a two-dimensional triangular lattice system of iron, and tensile deformation of an aluminum single crystal in plane strain are studied. Second, we review the torsion test of a nanowire that has been performed previously by the authors and coworkers. Third, a nanoplate is rolled to investigate the grain subdivision of a perfect crystal.

2 Models and Methodologies

We produce four different simulations to study the mechanism of the grain subdivision process in a single crystal under plastic deformation. Figure 1 is a schematic of these models.

2.1 Tension of the Two-Dimensional Triangular Lattice System of Iron

Figure 1a shows a two-dimensional triangular lattice system of iron atoms that we use for the tensile deformation simulation. In the figure, a is the lattice constant and the lattice is rotated θ clockwise. The Morse potential is adopted as the interatomic potential,

$$\phi(r) = A\{\exp[-2\alpha(r - r_0)] - 2\exp[-\alpha(r - r_0)]\}, \tag{1}$$

where $A = 8.160 \times 10^{-20}$ J, $r_0 = 0.258$ nm, and $\alpha = 14.57364$ [1/nm]. The shifted-force potential $\varphi_s(r)$, which is defined by

$$\phi_s(r) = \begin{cases} \phi(r) - \phi(r_c) - [r - r_c]\left(\frac{d\phi}{dr}\right)_{r_c}, & r \leq r_c \\ 0, & r \succ r_c \end{cases} \tag{2}$$

is used, where $r_c = 0.6$ nm is the cut-off distance. The lattice constant is $a_0 = 2.4935$ Å and the cohesive energy is $E_0 = -1.573$ eV in the static equilibrium state.We define the thickness of this 2D model as a reference length a_0 (lattice constant). The elastic moduli are $C_{11} = 203.4$ GPa and $C_{12} = C_{44} = 67.8$ GPa.

From a simple preliminary calculation, the melting temperature of this system is estimated to be greater than 3000 K but less than 3300 K. The analysis model consists of the arrangement of N_{atom} atoms in a rectangular region, and the dimensions in the x- and y-directions are L_x and L_y, respectively. Periodic boundary conditions are used in the x- and y-directions. First, a relaxation calculation is used to obtain the thermal equilibrium state at 500 K. The pressure of the system is controlled to be zero by the Parrinello–Rahman method. The tensile deformation is incorporated into the elongation of the unit cell length L_y. The increment, 0.001 $:_y$, is added to the cell length every 0.5 ps. Therefore, the nominal strain rate is 2×10^9 [1/s]. The stress σ_{xx} normal to the x-axis is controlled to be zero and the momentum of every atom is rescaled to be equivalent to 500 K. The velocity Verlet integrator is used for integration with respect to time. Three models (Models 1, 2, and 3) are considered to examine the dependence of symmetry and specimen size. The parameters (θ, L_x, L_y, N_{atoms}) of Models 1, 2, and 3 are (0°, 100.4 nm, 100.7 nm, 184338), (15°, 113.8 nm, 109.5 nm, 226890), (0°, 403.7 nm, 403.6 nm, 2970608), respectively.

2.2 Tension of an Aluminum Single Crystal in Plane Strain

Figure 1b shows a model of an aluminum crystal. The embedded atom method potential proposed by Mishin et al. (1999) is used as the interatomic potential for the aluminum single crystal. The periodic boundary condition is applied for all of the three Cartesian axes. The mirror indices of x-, y-, and z-axes correspond to $[\bar{1}10]$, [001], and [110], respectively. The dimensions of the specimen are set to $(L_x, L_y, L_z) = (33.37$ nm, 19.55 nm, 2.58 nm). Since the unit cell length in the z-direction L_z is quite small, the periodic boundary condition causes an artifact, i.e., active slip systems are limited to only two 60° dislocations on A $(11\bar{1})$ and B (111) planes and one zigzag lip (Shimokawa et al., 2005).

After a relaxation calculation, the tensile deformation is included with a nominal strain increment of 0.001 which is given to a unit cell length in the y-direction L_y every 0.4 ps, then the atomic system is relaxed under the fixed L_y condition. Consequently, the nominal strain rate is 2.5×10^9 s^{-1}, and the unit cell length in the x-direction L_x is changed under a pressure-free condition during the relaxation. The temperature of the entire atomic system is controlled at room temperature, at 300 K.

2.3 Torsion of Iron Nanowire

The material models for the torsion problem of a nanowire are single-crystalline iron in which interatomic interaction is introduced by a Finnis-Sinclair-type potential (Finnis and Sinclair, 1984, 1986). Solid cylindrical wire with radius R is modeled, as shown in Figure 1c. A twisted periodic boundary condition (TPBC) (Nakatani et al., 2004) is used. The lateral dimensions L_x and L_y of the unit cell are assumed to be sufficiently larger than the diameter $2R$ of the wire, so that the periodic boundary condition is assumed only in the z-direction, with dimension L_z, of the unit cell. In the present study, the specimen having dimensions $R = 10$ nm and $L_z = 20$ nm is considered. Two single crystal models are examined, in which $\langle 001 \rangle$ and $\langle 111 \rangle$ respectively are set as the twist axis (z-direction). The wire is twisted monotonically by the increment of a specific twist angle $\Delta\theta = 1.25 \times 10^6$ [deg/m], which is added to the specific twist angle θ every 0.5 ps. The nominal surface shear strain $\gamma_0 = R_\theta$ is used to express the strain level. The system temperature is set to 300K during the calculation.

2.4 Rolling of Iron Nanoplate

The basic computational procedures and material model are the same as for the torsion problem, i.e., the materials of the nanoplate for the rolling problem are a single crystalline iron in which interatomic interaction is introduced by a Finnis–Sinclair-type potential. The model for nanoplate rolling is shown in Figure 1d. The specimen consisted of a body center cubic lattice of N_{atoms} α-iron atoms with a lattice constant of $a = 0.28665$ nm. Rolling plane and rolling direction are (111) and $[\bar{1}\,\bar{1}2]$, respectively. The dimensions of a unit box, which is a rectangular parallelepiped including six atoms, are $\Delta x = \sqrt{6}\,a$, $\Delta y = \sqrt{2}\,a$, and $\Delta z = \sqrt{3}\,a/2$ in the x-, y-, and z-directions, respectively. The specimen is composed of an arrangement of $n_x n_y n_z$ units, with dimensions $l_x = n_x \Delta_x$, $l_y = n_y \Delta_y$, $l_z = n_z \Delta_z$. Rollers with a radius of $R = 10$ nm are modeled as rotating rigid bodies with an atomic arrangement on their (111) surface. The atomic interaction between roller and specimen is assumed to be a contribution to the pair potential term of the FS potential. The distance between the upper and lower rollers h is adjusted for each specimen. The circumferential velocity of the rollers is $v_{\text{rot}} = 1000$ [m/s]. The parameters for Cases 1, 2, 3 and 4 are $(n_x, n_y, n_z, h) = (20, 10, 10, 1.6\,\text{nm})$, $(20, 10, 20, 3.2\,\text{nm})$, $(20, 20, 40, 6.4\,\text{nm})$ and $(80, 40, 40, 6.4\,\text{nm})$, respectively. The initial translational velocity v_0 of the specimen is set to v_{rot}. The atomic velocity reduced by the average translational velocity is rescaled to be equivalent to 300 K.

3 Results and Discussion

3.1 Tension of a Two-Dimensional Triangular Lattice System

The lattice constant obtained by the relaxation calculation at 500 K is $1.0095a_0$. Figure 2a shows the relationship between tensile stress and nominal strain for each model. Figures 2b, 2c and 2d show the deformed atomic arrangement at a nominal strain of $\varepsilon = 0.5$ for Model 1, Model 2, and Model 3, respectively. In these figures, the grayscale of atoms denotes the lattice rotation angle, α , which is calculated using the relative coordinates to neighboring atoms. White atoms and black atoms mean $\alpha \leq -5°$ and $\alpha \geq 5°$, respectively. Stress increases linearly up to the first peak, which can be defined as a yield stress, for each model (see Figure 2a). After

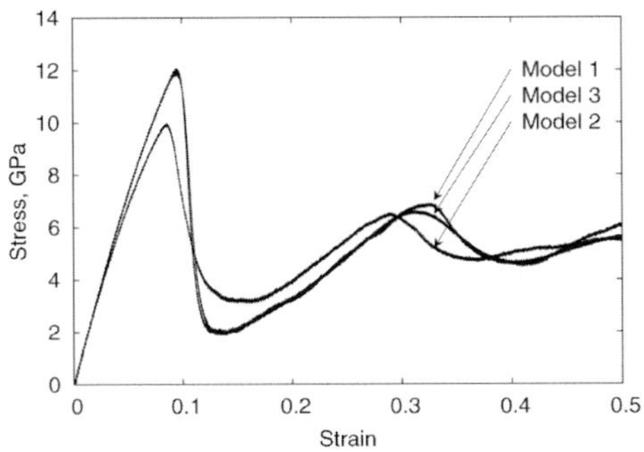

(a)

(b) (c) (d)

Fig. 2. (a) Stress–strain curves. (b), (c) and (d) Distribution of lattice rotation angle α between initial state and $\varepsilon = 0.5$: (b) Model 1, 100.4×100.7 nm, 184,338 atoms, $\theta = 0$. (c) Model 2, 113.8×109.5 nm, 226,890 atoms, $\theta = 15$. (d) Model 3, 403.7×403.6 nm, 2970.608 atoms, $\theta = 0$. White and black represent $\alpha \leq -5$ and $\alpha \geq 5$, respectively.

that, the stress decreases suddenly. During this decreasing stress stage, many pairs of dislocations are nucleated almost simultaneously. Then, the dislocations move in the model. The peak stress of Model 1 is almost the same as that of Model 3, which has the same crystallographic orientation as Model 1. These model a bi-slip system, which initially has two equivalent slip systems. On the other hand, the peak stress of Model 2, which has a nonsymmetric configuration is smaller than that of Model 1 and Model 3. The oscillating shape of the stress–strain curves is qualitatively similar in each model and the curve of Model 1 quantitatively corresponds to that of Model 3. The temporal period, 120~130 ps, is relatively large and this corresponds to the change of mesoscopic structure of the dislocations caused by dislocation–dislocation interactions, i.e., the fixation, separation, nucleation, annihilation, and mixture of

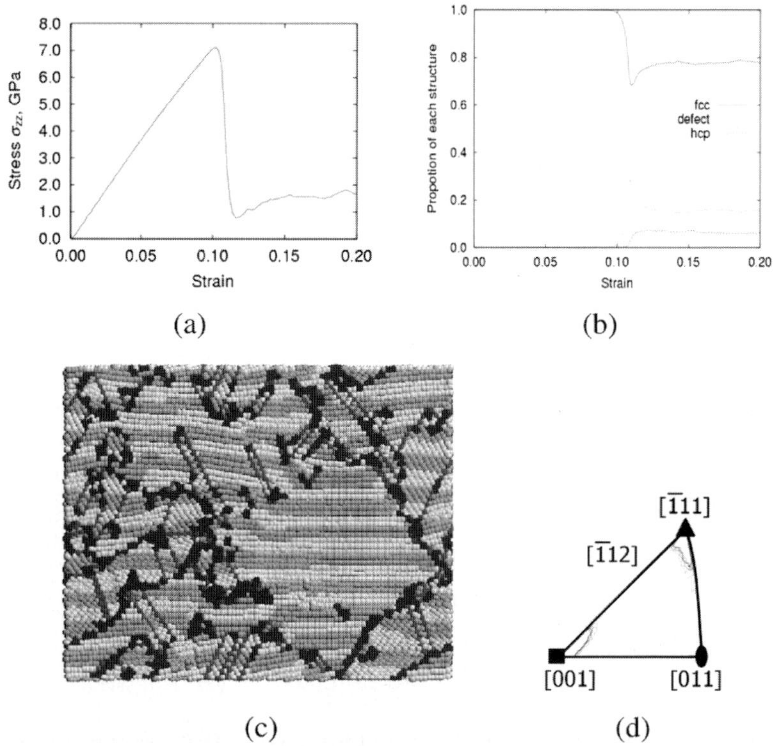

Fig. 3. (a) Stress–strain curves. (b) The number of defect atoms. Crystallographic orientation of local fcc structures. (c) 80 ps at 300 K. (d) Relaxation of 20 ps at 900 K.

dislocations. There appear to be some domains that have the same rotation angle but are not stabilized. There is no structure similar to a grain boundary. Consequently, we conclude that no grain subdivision phenomena occur in such two-dimensional triangular lattice systems.

3.2 Tension of Aluminum in Plane Strain

The stress–strain curve, the fraction of atoms in a perfect crystal or in the defects (hexagonal closed-packing, hcp; body-center-cubic, bcc; and others), the atomic arrangement at $\varepsilon = 0.2$, and the distribution of crystallographic orientation of tensile axis on the standard stereographic triangle are shown in Figures 3a, 3b, 3c, and 3d, respectively. The defect structure is detected by common neighbor analysis (CNA) (Honeycutt and Andersen, 1987). In Figure 3b, dark-colored atoms means that they are located in the defect.

After tensile deformation, the models are divided into many domains, surrounded by the boundary of defect atoms (see Figure 3c). At an initial state, light-gray stripes are marked horizontally (see Figure 1b), but these incline in some domains

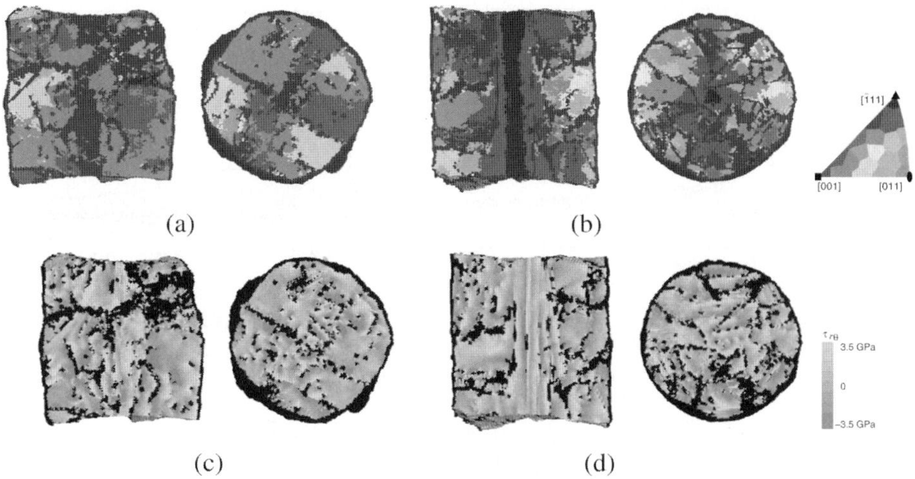

Fig. 4. Cross-sectional views after unloading. Defect atoms are colored black, and other colors correspond to the crystallographic orientation of the twist axis in the unit triangle of the reverse stereographic pole figure. (a) Single crystal in ⟨001⟩ torsion. (b) Single crystal in ⟨111⟩ torsion. Cross-sectional views after unloading. Defect atoms are colored black, and the gray scale corresponds to the atomic shear stress value. (c) Single crystal in ⟨001⟩ torsion. (d) Single crystal in ⟨111⟩ torsion.

after the deformation. This means that crystallographic rotation occurs, and each domain separated by defect atoms corresponds to grains separated by a grain boundary. However, investigating the atomic structure in detail shows that most grain boundaries consist of an HCP atomic structure. Figure 3d shows that two dense distributions of tensile axis appear near the [001] structure and [$\bar{1}$11]. These show that the polycrystalline structure obtained consists of twin systems.

3.3 Grain Refinement of Single Crystal by Torsion

To investigate the deformed internal structure in detail, an unloading process is simulated in which the starting point of the unloading is $\gamma_0 = 1$, and the process is realized by the reverse twist to the torque-free state. Figures 4a and 4b show cross-sectional views after the unloading. CNA is used and atomic sites are regarded as belonging to bcc when more than 10 bcc-characteristic atoms are found among the eight first neighbors and the six second neighbors. In these figures, undetermined orientation atoms and lattice defect atoms, except for the bcc atoms, are indicated by dark coloring, and the coloring of the bcc atoms shows the crystallographic orientation of the twist axis, which is determined using the second neighbor atoms in the unit triangle of the reverse stereographic pole figure. In Figures 4c and 4d, lattice defect atoms are shown as dark, and the coloring for bcc atoms shows the distribution of a cylindrical shear component $\tau_{\theta z}$ of residual stress. It is observed that a polycrystal structure develops near the surface in the cases of a twisted single crys-

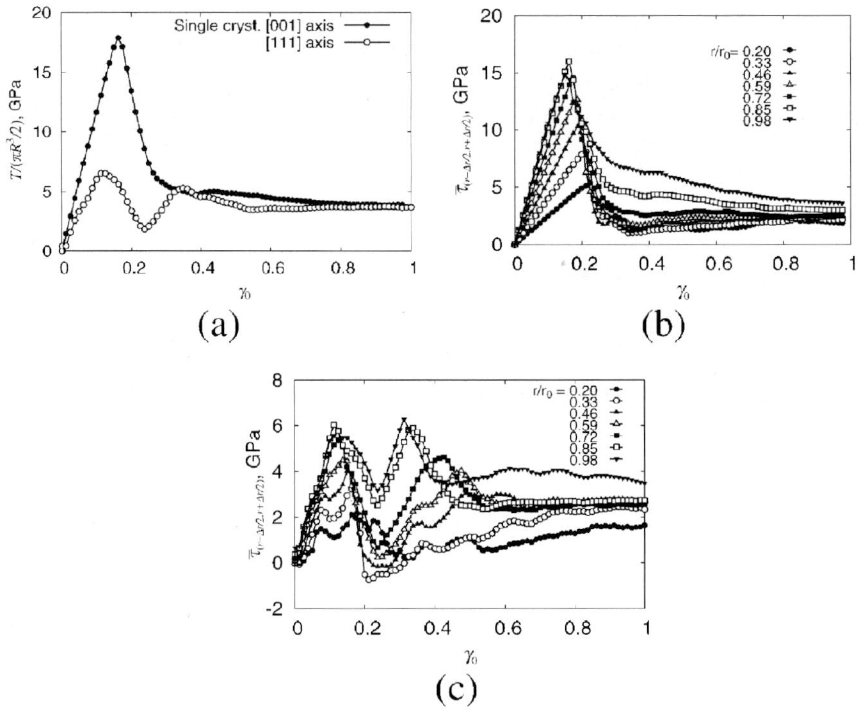

Fig. 5. Torque versus twist curves for single crystals. Torque value is converted to a representative value having the dimension of stress multiplied by a factor of $2/\pi R^3$, and twist is expressed by nominal surface strain $\gamma_0 = R\theta$. Shear stress versus twist curves for (a) single crystal in $\langle 001 \rangle$ torsion, and (b) single crystal in $\langle 111 \rangle$ torsion.

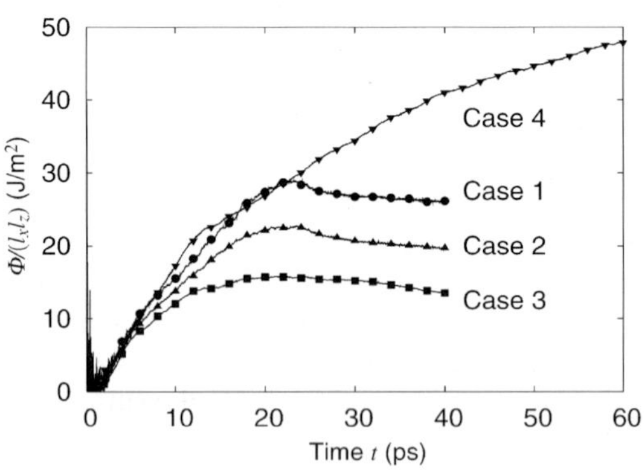

Fig. 6. Temporal change of potential energy.

tal (Figures 4a and 4b). Defect atoms are highly concentrated around the center of the twist axis in the case of ⟨001⟩ torsion, but the elastic core, which drags in some dislocations, remains around the twist axis in the case of ⟨111⟩ torsion. Twin boundaries and low-angle grain boundaries, which consist of a grain boundary dislocation array, are also observed. A large stress gradient is observed around the dislocations (see Figures 4c and 4d). Figure 5a shows the torque versus twist curves. The torque is calculated as the resultant moment of the transverse shear component of atomic stress $(\tau^{\alpha})_{\theta z}$ around the torsion axis. Anisotropy in the maximum torque is significant in the cases of a single crystal, and the maximum torque in ⟨001⟩ torsion is two and a half times larger than in ⟨111⟩ torsion. The torque in ⟨001⟩ torsion monotonically decreases after $\gamma_0 = 0.2$, but an oscillation is observed in ⟨111⟩ torsion. In both cases, the torque is almost constant within the range $\gamma_0 \succ 0.5$, and the value of torque standardized by a factor of $\pi R^3/2$ is approximately 4 GPa.

The average shear stress $\tau_{\theta z}$ in an annular ring is shown as a function of twist in Figures 5b and 5c. In the elastic deformation range, the shear stress monotonically increases as the radial coordinates r increase, except for the value in a few layers near the outer surface, in which the stress value is almost independent of r. Since inelastic deformation occurs near the surface and it decreases toward the interior, the peak stress of the interior annular ring is smaller than for the outer ring. The average stress distribution in the annular rings with different radial coordinates r becomes uniform, and the value is almost constant at 3 GPa during the deformation after $r_0 = 0.4$, in every model.

3.4 Rolling of Iron Nanoplate

Figure 6 shows the temporal change of potential energy. The reduction ratio is approximately 50% in each case. Therefore, the value of potential energy divided by the initia cross-sectional area $l_x l_z$ is shown. Figure 7 shows a snapshot of the atomic arrangement, which is evaluated by averaging rom 0.4 ps before the indicated time to the indicated time. The atoms that are determined to have a bcc structure by CNA are colored with the color that corresponds to the location of the normal vector of the plate surface on the stereographic triangle, and the atoms that are detected as defects are shown in black with a smaller radius. For Case 1, homogeneous deformation is realized, but the thickness of the plate is too small to retain the crystallographic structure (see Figure 7a). The material obtained is amorphous and there is no bcc structure after the deformation. On the other hand, for the grains that have different orientations from that of the initial specimen, polycrystallization occurs in Cases 2, 3, and 4. In the initial stage, the deformation near the surfaces precede to the center, and two bump profiles are observed in the side view. In the top view, an enlargement of the width of the plate is observed. For this orientation of specimen, the ratio $(b - l_y)/l_y$ is relatively large, approximately 20% in the experiment. The elongation of this simulation is the same as in the experiment. A comparison of Figures 6 and 7, shows that the curves for the stage in which the specimens are inserted symmetrically are identical. For Cases 1, 2, and 3, in which the rolling process is complete, the potential energy decreases slightly but is maintained constant after rolling. Ex-

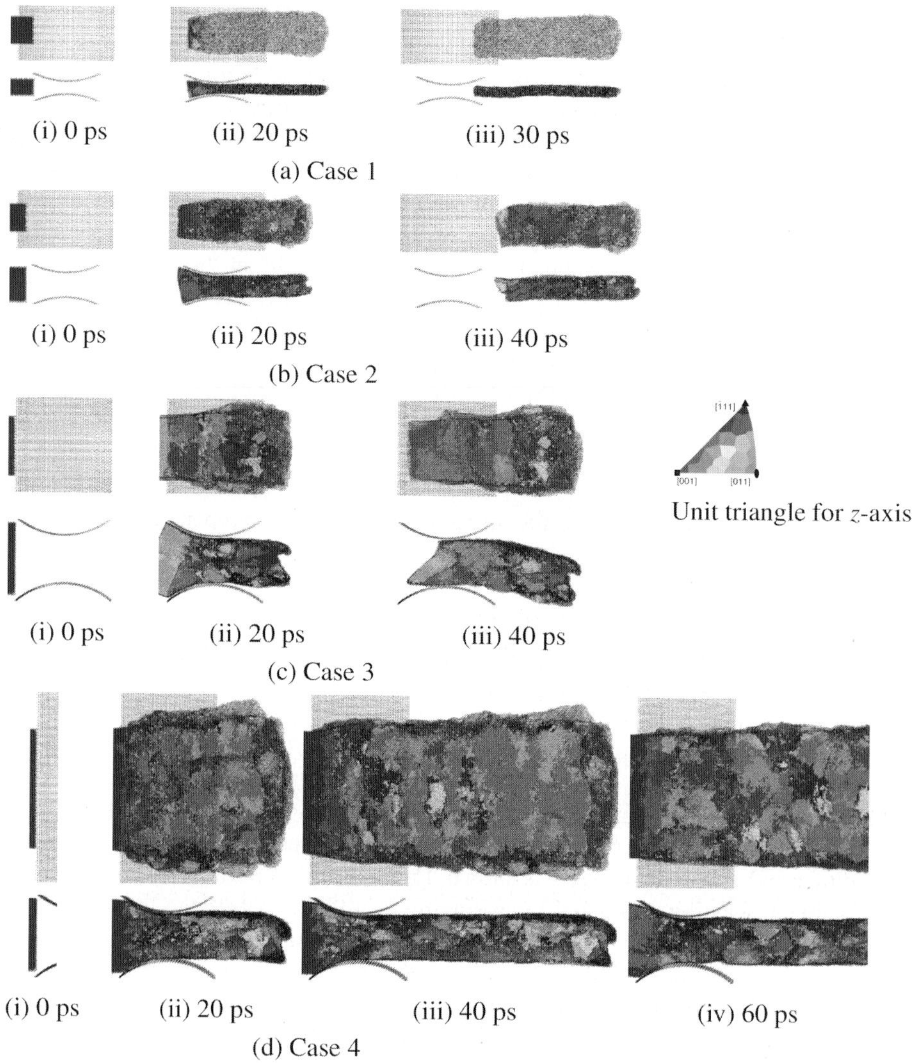

Fig. 7. Atomic arrangement and crystallographic orientation of grains.

cept for the final stages, when the rotation due to asymmetric shear deformation is significant, the rolling process is steady and the thickness of rolled plate is almost constant. There is no significant recrystallization after rolling. Except for the final stages, when the rotation due to asymmetric shear deformation is significant, therolling process is steady and the thickness of rolled plate is almost constant. There is no significant recrystallization after rolling.

4 Conclusions

Remarkable polycrystallization is observed in the twisting and rolling problems, compared to the tension problem. From the results of a series of analyses it is concluded that the strain gradient plays an important role in grain subdivision. The results of the polycrystallization and reorganization processes in such miniaturized models should not be compared directly with macroscopic phenomena. However, we can study the deformation mechanism under nonuniform stress or strain distribution using an analogous scaled-down model of an actual system.

Acknowledgments

The authors are pleased to acknowledge fruitful discussions with Dr. Nobuhiro Tsuji who is one of the inventors of ARB. The authors also thank former students Mr. Syunsuke Igawa and Mr. Hirokazu Mino for their calculations.

References

Finnis, M.W. and Sinclair, J.E. (1984) A simple empirical N-body potential for transition metals. *Philos. Mag., A* **50**(653), 45–55.

Finnis, M.W. and Sinclair, J.E. (1986) Erratum. *Philos. Mag., A* **53**(1), 161.

Hansen, N. (2001) New discoveries in deformed metals. *Metall. Mater. Trans. A* **32A**, 2917–2935.

Honeycutt, J.D. and Andersen, H.C. (1987) Molecular dynamics study of melting and freezing of small Lennard–Jones clusters. *J. Phys. Chem.* **91**(19), 4950–4963.

Mishin, Y., Farkas, D., Mehl, M.J. and Papaconstantopoulos, D.A. (1999) Interatomic potentials for monoatomic metals from experimental data and ab initio calculations. *Phys. Rev. B* **59**(5), 3393–3407.

Nakatani, A., Shimokawa, T., Matsumoto, R. and Kitagawa, H. (2004) Atomistic study on ideal strength of nanocrystal and deformation induced nanostructures. *Solid Mechanics and Its Applications* **115**, 365–380.

Saito, Y., Utsunomiya, H., Tsuji, N. and Sakai, T. (1999) Novel ultra-high straining process for bulk materials development of the accumulative roll-bonding (ARB) process —. *Acta Mater.* **47**, 579–583.

Seefeldt, M., Delannay, L., Peeters, B., Kalidindi, S.R. and Van Houtte, P. (2001) A Disclination-based model for grain subdivision. *Mat. Sci. Engng. A* **319–321**, 192–196.

Shimokawa, T., Nakatani, A. and Kitagawa, H. (2005) Grain-size dependence of the relationship between intergranular and intragranular deformation of nanocrystalline Al by molecular dynamics simulations. *Phys. Rev., B* **71**, 224110 (1–8).

Sivakumar, S.M. and Ortiz, M. (2004) Microstructure evolution in the equal channel angular extrusion process. *Comput. Methods Appl. Mech. Engng.* **193**, 5177–5194.

Part 2

Super-Strength and Ductility of Nano-Thin Films

The Origin of Superhardness in Nanocomposite Coatings: Analysis of Nanoindentation and Scratch Tests

Chunsheng Lu[1,*], Yiu-Wing Mai[1] and Yao-Gen Shen[2]

[1] *Centre for Advanced Materials Technology (CAMT), School of Aerospace, Mechanical and Mechatronic Engineering J07, The University of Sydney, Sydney, NSW 2006, Australia;*
**E-mail: chunsheng.lu@aeromech.usyd.edu.au*
[2] *Department of Manufacturing Engineering and Engineering Management (MEEM), City University of Hong Kong, Kowloon, Hong Kong, China*

Abstract. Nanocomposite coatings characterized by superhardness exhibit unique microstructures and deformation mechanisms at the nanometre scale. Based on the studies of nanoindentation size effect, we show that dislocation-based deformation is gradually replaced by grain-boundary mediated deformation as the hardness of coatings increases, especially for coatings approaching superhardness. The measured hardness of a superhard coating exponentially increases with the decrease of indentation depth, and its intrinsic hardness can be exactly determined by the analysis of indentation size effect. The optimal conditions extracted from acoustic emission signals from scratching superhard nanocomposite coatings are in good agreement with nanoindentation and drilling tests. In the optimal regime, it is the competition between different deformation mechanisms that results in the origin of superhardness in self-organized microstructures of nanocomposite coatings.

Key words: superhard coatings, nanoindentation, size effect, acoustic emission

1 Introduction

With the rapid development of synthesis, fabrication and testing techniques, there has been much interest in the search for nanostructured materials, especially for novel, superhard coatings defined as those with Vickers hardness $H \geq 40$ GPa. This is driven not only by scientific curiosity, but also by their wide engineering applications, from cutting and polishing tools to wear-resistant coatings. Compared to hard materials, there are only a few superhard materials, such as cubic boron nitride and diamond with hardness $H \approx 70$–100 GPa that are believed to be the hardest known materials available in nature. Unfortunately, even diamond has limitations, which is not effective for cutting steels and some alloys owing to the high solubility of carbon. Hence, it is not surprising that many efforts have been devoted to develop superhard nanocomposite coatings in recent years [1–5].

Yilong Bai et al. (eds), IUTAM Symposium on Mechanical Behavior and Micro-Mechanics of Nanostructured Materials, 39–49.

Superhard nanocomposite coatings comprise two or more phases that are either a nanostructured phase embedded in another amorphous phase, such as nanocrystalline (nc-)TiN and amorphous (a-)Si_3N_4 (nc-TiN/a-Si_3N_4), or two nanocrystalline phases such as nc-TiN/BN. Recent investigations reveal that, in the optimal deposition conditions, these coatings possess an unusual combination of mechanical properties, and their hardness can reach 40 to 100 GPa. However, it was criticized and even doubted that the high hardness values reported might be caused by possible artefacts in synthesis and/or measurements, for example, biaxial compressive stresses caused by physical vapour deposition, indentation size effect, etc. [6–8]. The reason for debate is partially attributed to the lack of a better basic understanding of the implications of hardness and intrinsic superhardening mechanisms of these nanocomposite coatings.

Unlike the elastic modulus and strength, hardness does not have an unambiguous definition. Broadly speaking, hardness is a measure of the capability of a material against imprinting or scratching with another hard material, such as diamond, which microscopically corresponds to the resistance against local plastic deformation due to the creation and movement of dislocations. The debate caused by this relative and ambiguous definition for hardness becomes more serious in nanocomposite coatings possessing superhardness. To our best knowledge, there is still a lack of systematic study. Thus, to design new superhard materials, it is imperative to understand their underlying deformation mechanisms and the origin of superhardness, that is, what makes superhard nanocomposite coatings unique [7]. In this paper, studies are made by using nanoindentation and scratch tests of several hard and superhard nanocomposite coatings such as AlN, Ti-B, and nc-TiN/a-Si_3N_4. Based on the analysis of indentation size effect and acoustic emission signals due to scratching, plausible superhardening mechanisms in these coatings are discussed.

2 Nanoindentation Size Effect

From a well-accepted definition, the indentation hardness of a material is given by

$$H = \frac{P_{\max}}{A}, \tag{1}$$

where $P_{\max}$ is the maximum load and A is the surface or projected area of a remaining impression. The nanoindentation method for measuring hardness and elastic modulus, introduced by Oliver and Pharr [9], has been widely used in the characterization of mechanical properties of materials at small scales. Without the need to image the indent impression, hardness and elastic modulus can be determined directly from indentation load and displacement curves, as illustrated in Figure 1. In nanoindentation tests, three important quantities are measured from the load-displacement curves: maximum load $P_{\max}$, maximum displacement $h_{\max}$, and elastic (or unloading) contact stiffness $S = \mathrm{d}P/\mathrm{d}h$, see the inset of Figure 1. Once the contact area is determined, the hardness can be easily estimated from Equation (1). Here, the so-called indenter shape function, $A = F(h_c)$, must be carefully calibrated by independent measurements using standard samples such as fused silica ($E = 72.0$ GPa

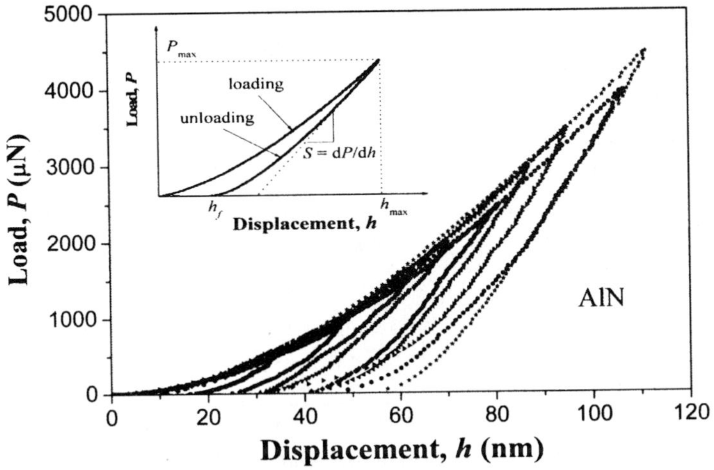

Fig. 1. Nanoindentation load-displacement curves for AlN coatings, where the inset is a schematic illustration of indentation load-displacement data with three important measured parameters.

and $\nu = 0.17$), where $h_c = (h_{\max} - \varepsilon P_{\max}/S)$ and ε is a constant relevant to the indenter. Modulus follows the relationship

$$E_r = \frac{\sqrt{\pi}\, S}{2\beta\sqrt{A}}, \tag{2}$$

where $\beta = 1.05$ is a dimensionless correct parameter and E_r is the reduced or effective elastic modulus defined as $1/E_r = (1-\nu^2)/E + (1-\nu_i^2)/E_i$, with Young's modulus E and Poisson's ratio ν for the specimen and elastic constants $E_i = 1141$ GPa and $\nu_i = 0.07$ for the diamond indenter [9, 10].

Nanoindentation tests largely facilitate the measurement of mechanical properties at small scales. It seems, however, not so easy to obtain the intrinsic hardness and elastic modulus of a nanocomposite coating. Many factors may affect the measured results of hardness and elastic modulus, which include external factors such as indenter geometry, tip rounding, machine compliance, etc.; and material-related factors such as the surface roughness, indentation size effect, residual stress, coating thickness, substrate, etc. [6–10]. To avoid the effect of a substrate on the hardness of a coating, the maximum indentation depth should not exceed ~10% of the coating thickness according to a rule-of-thumb criterion [4]. Therefore, the indentation size effect, that is, an increase in hardness with decreasing indentation depth, must be considered carefully when measuring the intrinsic hardness of a nanocomposite coating that is a few micrometers thick.

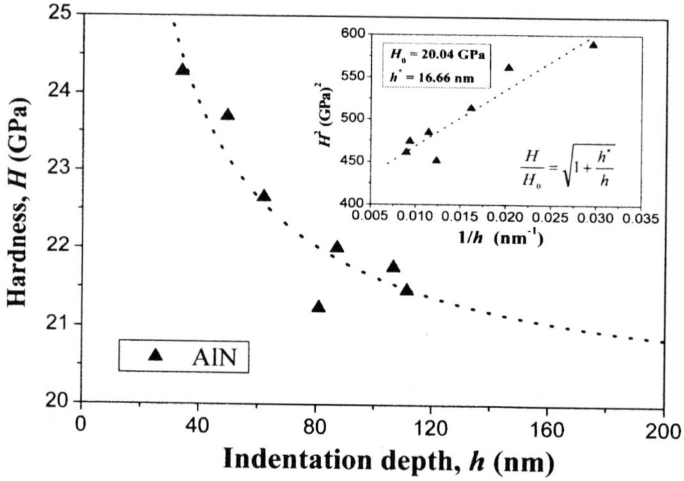

Fig. 2. Hardness versus indentation depth for AlN coatings. The dotted line is the fitted result by using Equation (3) and the inset displays the data as a plot of H^2 versus $1/h$.

2.1 AlN Coating

The interest in nano-structured aluminium nitride (AlN) as thin films or coatings in microelectronic devices and optical sensors originates in their specific properties such as high melting point, chemical stability, high thermal expansion, etc. [11]. By using reactive close-field unbalanced magnetron sputtering, nano-structured AlN coatings ($\sim$1 μm thick) were deposited onto silicon (111) or M42 high-speed-steel substrates at room temperature [12]. Figure 1 shows typical indentation load-displacement curves subject to different maximum loads from 500 to 4500 μN. According to the Oliver–Pharr method, the measured hardness H versus indentation depth h is shown in Figure 2.

It is noticed that large strain gradient inherent in small indentation leads to geometrically necessary dislocations [13]. Using this concept, Nix and Gao [14] have indicated that the indentation size effect of crystalline materials can be described by the following expression for the depth (h) dependence of the hardness H:

$$\frac{H}{H_0} = \sqrt{1 + \frac{h^*}{h}}, \tag{3}$$

where H_0 is the hardness in the limit of infinite depth and h^* is a characteristic length that depends on the shape of the indenter, the shear modulus and H_0. As displayed in the inset of Figure 2, the indentation size effect of hardness of AlN coatings is well fitted by Equation (3). The intrinsic hardness value $H_0 = 20.04$ GPa, which agrees well with the value reported for single-crystal AlN [11], and the characteristic length $h^* = 16.66$ nm, which is a little bigger than the grain size ($\sim$10 nm) of AlN coatings. Similar to TiN coatings under the same testing conditions, AlN coatings have high elastic recovery (about 80% or more), and their elastic modulus, $E_r = 245$ GPa ($E = 327$ GPa), were estimated by Equation (2). The high elastic response of AlN coatings makes the measurement of hardness difficult; however, their intrinsic hardness can be well predicted by the analysis of indentation size effect for crystalline materials based on the concept of geometrically necessary dislocations.

2.2 Ti-B Coating

It is obvious that, based on Equation (3), the large value of intrinsic hardness H_0 would cause h^* to be very small, and the hardness of a coating depends weakly on the depth at a given level of indentation. Recent nanoindentation tests of Ti-Al-N thin films ($H_0 \approx 35$ GPa) showed that, however, the indentation size effect of hardness clearly exists, which cannot be described or fitted by Equation (3). As the intrinsic hardness of coatings increases, the indentation size effect becomes more serious. As shown in Figure 3, the measured hardness of Ti-B coatings increases exponentially as the indentation depth decreases, this can be well fitted by

$$H = H_0 + H' \exp(-h/h^*), \tag{4}$$

where H_0 and h^* have been earlier defined and H' is a constant. For Ti-B coatings, the fitted parameters are: $H_0 = 70.07$ GPa, $h^* = 116.28$ nm, and $H' = 115.94$ GPa. The estimated value of intrinsic hardness is consistent with that of the depth-independent hardness. It is also observed that the influence of a substrate on the hardness of Ti-B coatings exists even if the maximum indentation depth is less than 10% of the thickness of the coatings. Similar behaviour was also discovered in the tests of Ti-Al-N thin films [15].

Further study on nanoindentation of Ti-B and Ti-Al-N coatings excluded that the indentation size effect in Figure 3 is due to the influence of surface roughness, which can be described by a formula like: $H = H_0 + C/h$, where C is a constant [16]. Thus, there exist some new deformation mechanisms rather than dislocation pile-ups, which are responsible for the deformation processes of superhard nanocomposite coatings.

3 Statistical Analysis of Acoustic Emission Signals

In addition to nanoindentation, scratch is another simple and convenient method to evaluate the integrity of superhard nanocomposite coatings. It is well-known that the tribological properties of coatings are rather complicated, and influenced by many factors over multi-scales, such as dislocation pile-ups, cracking, delamination, buckling, etc. [17]. However, the influence of these factors can be directly monitored by acoustic emission (AE) sensors because the creation and movement of dislocations or cracks result in a sudden change in stress or displacement within a solid in the form of elastic waves with ultrasonic frequency.

Standard scratch tests of superhard nc-TiN/a-Si_3N_4 coatings deposited onto M42 high-speed-steel substrate were performed. A Rockwell C diamond indenter

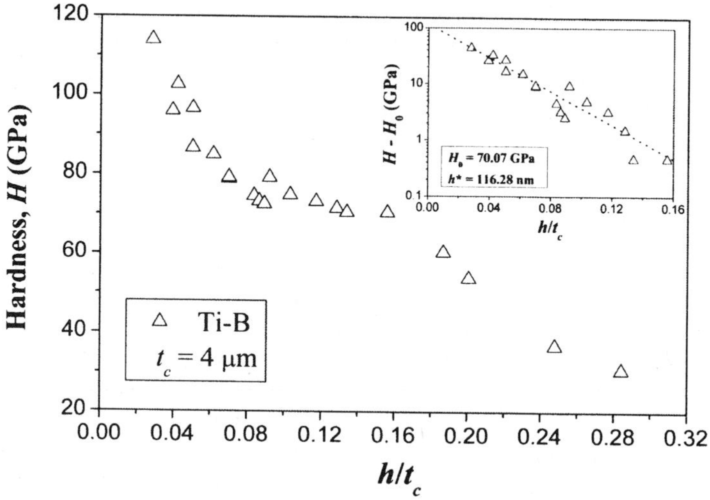

Fig. 3. Hardness versus the ratio h/t_c of indentation depth h and thickness $t_c = 4\ \mu$m for Ti-B coatings, data taken from [7]. The inset displays the data as a plot of $\log_{10}(H - H_0)$ versus h/t_c, where the dotted line is the fitted result by using Equation (4).

(200 μm in tip radius) was drawn across the coating surface at a constant velocity of 10 mm min^{-1} while increasing the load from 10 to 70 N at a constant rate of 100 N min^{-1}. Experiments showed that, with time the fluctuation of friction force increases and the first derivative of AE becomes rougher. There are two time scales in this process: the characteristic relaxation time τ and the measuring time T. Their ratio, i.e. the Deborah number De $= \tau/T \ll 1$, implies that the coating subjected to scratching exhibits a "liquid-like" behaviour. This can be easily described by a stick-slip process that occurs in most sliding friction phenomena from macroscopic to atomic scales. For simplicity, it is assumed that the stress level in a coating can be described by a scalar value, $\sigma(t)$, which increases deterministically between AE events and which is released stochastically as a Markov process. Thus, the evolution of stresses is given by

$$\sigma(t) = \sigma(0) + \rho t - S(t), \tag{5}$$

where $\sigma(0)$ is initial stress level, ρ is external loading rate, and $S(t) = \Sigma S_i(t \mid t_i < t)$ is accumulated stress release from events within the region over the time period $(0, t)$, where t_i, S_i are the origin time and stress release associated with the i-th AE event, respectively [18]. The value of stress release during an AE event can be estimated from its magnitude m, in the unit of decibel (dB), in terms of the empirical formula

$$m = 10 \log_{10} A + m_0, \tag{6}$$

where A is relative strength of an AE signal, such as power, voltage, etc., m_0 is a reference magnitude. Here, the stress drop is roughly considered to be proportional to the power, i.e., $S \sim A$, and $m_0 = 60$ dB (the noise level of normal conversation) was

Table 1. Calculated AIC values and the fitted parameters.

Samples	N	Δ AIC	Δ AIC$/N$	t_0	S_0	η
Without heating	156	−1.703	−0.0109	0.63	0.58	0.19
400°C	81	2.301	0.0284	0.71	1.96	0.99
500°C	130	0.035	0.0003	0.60	1.00	0.27
600°C	161	0.413	0.0026	0.43	0.62	0.19

used in subsequent analysis. The substance of results is not sensitive to the choice of m_0.

The intensity of an AE event occurrence is controlled by a risk function $\psi(\sigma)$, which increases nonlinearly with the stress level. The simplest choice of $\psi(\sigma)$ is taken as an exponential function, $\psi(\sigma) = \exp(\mu + \nu\sigma)$, where μ and ν represent the background and the sensitivity to risk, respectively. This is a compromise between regular (time-predictable) and pure random (Poisson) processes [18–20]. Let us further assume that the size distribution of AE events is independent of the stress level, and then the AE signals in scratch tests can be treated as a marked point process in time-stress space with the conditional intensity function, $\lambda(t) = \psi[\sigma(t)] = \exp\{\alpha + \nu[\rho t - S(t)]\}$, where α $[= \mu + \nu\sigma(0)]$, ν, and ρ are the parameters to be fitted. To remove the influence of sample sizes (the number of AE events), a scale-free sensitivity to risk can be rewritten as

$$\bar{\lambda}(t) = \lambda(t)\exp(-\alpha) = \exp\{\eta[t/t_0 - S(t)/S_0]\}, \tag{7}$$

where $\eta = \nu S_0$, $t_0 = S_0/\rho$, $S_0 = \Sigma S_i/N$ and N is the number of AE events over the interval (t_1, t_2). Estimates of these parameters are found by maximizing its log-likelihood. The comparison between two models is based on the Akaike information criterion (AIC), that is, AIC $= -2\ln\hat{L} + 2k$, where $\ln\hat{L}$ and k are the maximum log-likelihood and the number of parameters for a given model [21]. This represents an approximate way of compensating for the effect of adding parameters, and is a useful heuristic measure of the relative effectiveness of different models. For example, in comparing the stick-slip model with three parameters against the Poisson model with only one ($\nu = \rho = 0$), the former must demonstrate a significantly better fit to justify additional parameters. The relative effectiveness of two models in fitting AE data can be identified if the difference of their AIC values (i.e. Δ AIC $=$ AIC$_p$ $-$ AIC$_s$) is larger than 2, where AIC$_p$ and AIC$_s$ are the AIC values of the Poisson model and stick-slip model [19–21].

As listed in Table 1, let us take the Poisson model (random process) as a reference, it is obvious that the stick-slip model fits AE data better than the Poisson model at 400°C deposition temperature since the difference of AIC values is substantial, i.e. Δ AIC > 2. However, such a direct comparison might be misleading, as coatings deposited at different temperatures yield different numbers of AE events (see Table 1). To allow for this effect, we use the indicator, Δ AIC$/N$, as a measure of the improvement in performance which is approximately independent of sample sizes. The same nonlinear changing order (400°C, 600°C, 500°C, and without heating) is obtained, and the best fitting is still the case of 400°C. Here, the larger the ratio of Δ AIC$/N$,

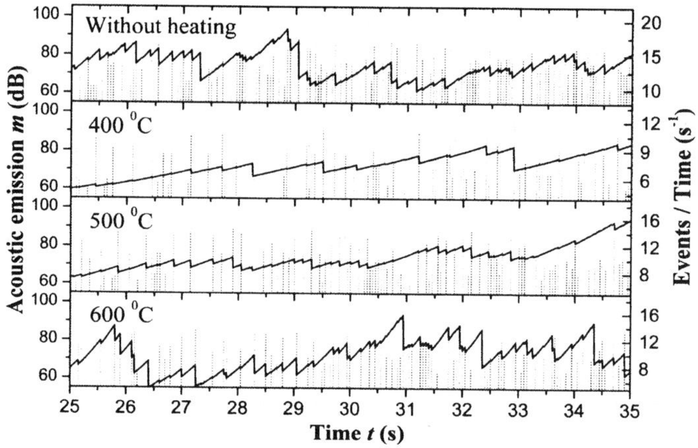

Fig. 4. Risk level (the number of events per second) versus time calculated by the fitted parameters in Table 1. For comparison, AE events with magnitude $m \geq 60$ dB are also plotted [22].

the better the AE data is fitted by the stick-slip model relative to the Poisson model, and the more stable or optimal the coating. The fitted parameters in Equation (7) are also set out in Table 1, and the intensity function versus time for each deposition temperature is shown in Figure 4. For a given stress release S_0, the larger the value η, the higher the sensitivity to risk ν ($\eta = \nu S_0$). The results also indicate that the coating deposited at 400°C is more stable than those in the other three cases.

Some caution should be noted because this method is, after all, a post-mortem examination. We cannot overestimate its role and rush to conclusions before experimental calibrations. Nanoindentation experiments showed that the coating (8.6 at. % Si) deposited at 400°C possessed the hardness of 50.9 GPa, which indeed was larger than those with the same composition but deposited at different temperatures. Furthermore, wear testing on these coatings was assessed. The drills coated with nc- TiN/a-Si_3N_4 deposited at 25°C (without heating), 400°C, 500°C, and 600°C drilled 78, 352, 120, and 225 holes, respectively. For comparison, the drill coated with TiN ($H \approx 20$ GPa), a commonly used hard coating, drilled 107 holes. Here, it is of interest to scrutinize the AE data at 400°C, the optimal deposition temperature obtained by both statistical analysis and real tests. The results show that the magnitude-frequency distribution of AE events follows the formula, $\log_{10} N(> m) = -km + \text{const.}$ with $k = 0.026 \pm 0.001$. Base on the magnitude definition of AE in Equation (6) and the assumption of the energy released in an AE event $E \sim A$, a power-law distribution, $P(> E) \sim E^{-10k}$ is obtained. However, as deposition temperatures increase or decrease relative to this optimal value, the energy frequency of AE events rapidly deviates from the power-law distribution [22].

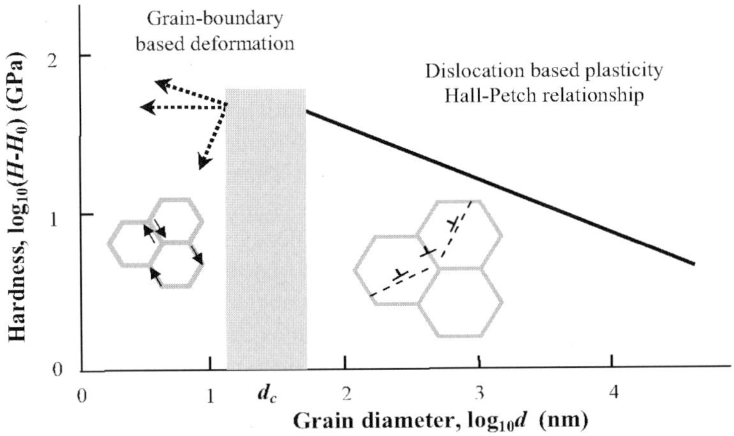

Fig. 5. Illustration of hardness (or yield strength) versus grain size, where deformation is mediated by dislocation pile-ups (coarse-grains) and grain boundaries (nano-grains), respectively. The Hall–Petch relationship works well for coarse-grained materials, where d_c is a characteristic length.

4 Discussion and Conclusions

Computer simulations show that plastic deformation in nanocrystalline materials can be accommodated by grain-boundary mechanisms such as sliding when the size of grains is below ~10 nm. As grains are reduced to the nanometre scale, the percentage of grain boundary atoms increases rapidly (about 10% of atoms are located at grain boundaries for a sample with grain diameters of 20 nm) [23, 24], and bulk nanocrystalline metal has softening behaviour. Compared to coarse-grained crystalline metals, dislocation sources and pile-up are hardly expected to exist in such a material and deformation is believed to be carried mostly by grain boundaries via plastic sliding (see Figure 5). Thus, it is not surprising that there exhibit different indentation size effects. In fact, Equation (3) based on geometrically necessary dislocations is a typical kind of complex scaling, which is consistent with the long-range stress field of dislocations. The exponential increase of hardness as the decrease of indentation depth in coatings such as Ti-B, as indicated in Equation (4), just identifies this transition of deformation mechanisms. Thus, the intrinsic hardness of hard and superhard coatings can be exactly determined by the analysis of their indentation size effects.

In superhard nanocomposite coatings like nc-TiN/a-Si_3N_4, the nc-TiN phase is sufficiently hard to bear the load whilst the a-Si_3N_4 phase provides structural flexibility, in which further increase in hardness requires blocking of grain-boundary sliding by optimal design of their microstructures [25]. In fact, it is due to the competition between different deformation mechanisms in nanocrystalline grains and along amorphous boundary (stick-slip) that a steady self-organized nanostructure with peculiar properties (superhardness) can be formed. This may be a possible reason why AE events due to scratching occur over many scales when the nanostructure of coatings lies in an optimal state. In the self-organized nanostructure for superhard coatings, three correlative conditions should be met: higher bond density or electronic density, shorter bond length, and greater degree of covalent bonding [26].

In summary, the vast range of influence factors in nanocomposite coatings precludes a purely Edisonian (or cooking) approach to identifying promising compositions, it is simply too time-consuming and costly. The most rapid progress will probably be made by using systematic investigations on the superhardening mechanisms of nanocomposite coatings. The power-law frequency-magnitude distribution of AE energy at optical deposition conditions implies that the understanding of its behaviour relies on only a few emergent material parameters as being done in the model. As we have seen that, optimal information, such as Si contents, deposition temperatures, etc., in nc-TiN/a-Si_3N_4 coatings can be extracted from crackling noise by a simple stick-slip model. The results provide us some valuable clues for the understanding of superhardening mechanisms of nanocomposite coatings, and further for the tailoring of new nanostructural materials.

Acknowledgements

This work was supported by the Australian Research Council (ARC) and the Hong Kong SAR Research Grant Council (Project No. CityU 1132/03E). YWM and CL are, respectively, Australian Federation Fellow and ARC Senior Research Associate.

References

1. Veprek, S., The search for novel, superhard materials, *J. Vac. Sci. Technol. A* **17** (1999), 2401–2420.
2. Musil, J., Hard and superhard nanocomposite coatings, *Surf. Coat. Technol.* **125** (2000), 322–330.
3. Tjong, S.C. and Chen, H., Nanocrystalline materials and coatings, *Mater. Sci. Eng. R-Rep.* **45** (2004), 1–88.
4. Veprek, S., Veprek-Heijman, M.G.J., Karvankova, P. and Prochazka, J., Different approaches to superhard coatings and nanocomposites, *Thin Solid Films* **476** (2005), 1–29.
5. Kaner, R.B., Gilman, J.J. and Tolbert, S.H., Designing superhard materials, *Science* **308** (2005) 1268–1269.
6. Lu, C., Mai, Y.-W. and Shen, Y.G., Recent advances on understanding the origin of superhardness in nanocomposite coatings: A critical review, *J. Mater. Sci.* **41** (2006), 937–950.
7. Musil, J., Zeman, H., Kunc, F. and Vlček, J., Measurement of hardness of superhard films by microindentation, *Mater. Sci. Eng. A* **340** (2003), 281–285.
8. Veprek, S. and Argon, A.S., Towards the understanding of mechanical properties of super- and ultrahard nanocomposites, *J. Vac. Sci. Technol. B* **20** (2002), 650–664.
9. Oliver, W.C. and Pharr, G.M., An improved technique for determining hardness and elastic-modulus using load and displacement sensing indentation experiments, *J. Mater. Res.* **7** (1992), 1564–1583.
10. Cheng, Y.T. and Cheng, C.M., Scaling, dimensional analysis, and indentation measurements, *Mater. Sci. Eng. R-Rep.* **44** (2004), 91–149.

11. Yonenaga, I., Shima, T. and Slutter, H.F., Nano-indentation hardness and elastic moduli of bulk single-crystal AlN, *Jpn. J. Appl. Phys.* **41** (2002), 4620–4621.
12. Wong, D.K.L., Jiang, N. and Shen, Y.G., Deposition and tribological behaviour of nano-structured aluminium nitride and titanium nitride thin films, in G.W. Stachowiak (ed.), *Proceeding of the 6th International Tribology Conference*, Perth, 2002, pp. 669–676.
13. Fleck, N.A., Muller, G.M., Ashby, M.F. and Hutchinson, J.W., Strain gradient plasticity – Theory and experiment, *Acta Metall. Mater.* **42** (1994), 475–487.
14. Nix, W.D. and Gao, H.J., Indentation size effects in crystalline materials: A law for strain gradient plasticity, *J. Mech. Phys. Solids* **46** (1998), 411–425.
15. Jeong, S.M., Shum, P.W., Shen, Y.G., Li, K.Y., Mai, Y.-W. and Lee, H.L., Determination of effective nanoindentation range: Case study for hard (Ti, Al)N thin film, *J. Mater. Sci.* (2005), submitted.
16. Zhang, T.Y., Xu, W.H. and Zhao, M.H., The role of plastic deformation of rough surfaces in the size-dependent hardness, *Acta Mater.* **52** (2004) 57–68.
17. Bhushan, B., Israelachvili, J.N. and Landman, U., Nanotribology: Friction, wear and lubrication at the atomic scale, *Nature* **374** (1995) 607–616.
18. Vere-Jones, D., Earthquake prediction: A statistician's view, *J. Phys. Earth* **26** (1978), 129–146.
19. Lu, C., Vere-Jones, D. and Takayasu, H., Avalanche behaviour and statistical properties in a microcrack coalescence process, *Phys. Rev. Lett.* **82** (1999), 347–350.
20. Lu, C., Danzer, R. and Fischer, F.D., Fracture statistics of brittle materials: Weibull or normal distribution, *Phys. Rev. E* **65** (2002), 067102.
21. Akaike, H., New look at statistical-model identification, *IEEE Trans. Autom. Control* **19** (1974), 716–723.
22. Lu, C., Mai, Y.-W. and Shen, Y.G., Optimum information in crackling noise, *Phys. Rev. E* **72** (2005), 027101.
23. Gleiter, H., Nanostructured materials: Basic concepts and microstructure, *Acta Mater.* **48** (2000), 1–29.
24. Ovid'ko, I.A., Deformation of nanostructures, *Science* **295** (2002), 2386–2386.
25. Kauffmann, F., Ji, B., Dehm, G., Gao, H. and Arzt, E., A quantitative study of the hardness of a superhard nanocrystalline titanium nitride/silicon nitride coating, *Scr. Mater.* **52** (2005), 1269–1274.
26. Gao, F.M., He, J.L., Wu, E.D., Liu, S.M., Yu, D.L., Li, D.C., Zhang, S.Y. and Tian, Y.J., Hardness of covalent crystals, *Phys. Rev. Lett.* **91** (2003), 015502.

Micromechanics of Nanocomposites with Interface Energy Effect

Z.P. Huang* and J. Wang

*LTCS and Department of Mechanics and Engineering Science, Peking University, Beijing 100871, P.R. China; *E-mail: huangzp@pku.edu.cn*

Abstract. In this paper, a fundamental framework of micromechanics for predicting the effective properties of a composite is generalized to include the interface energy effect. In this framework, both the interface constitutive relations for multi-phase hyperelastic solids at finite deformation and the Lagrangian and Eulerian descriptions of the generalized Young–Laplace equations are presented. Then, by taking into account the change of the "residual" elastic field due to the change of configuration, the difference of the governing equations across the interface is derived. A discussion of the infinitesimal deformation approximation of these governing equations is also given, and analytical expressions of the size-dependent effective moduli of a particle-filled nanocomposite are obtained. It is shown that the liquid-like interface tension influences the effective properties of the nanocomposite. Thus some misunderstandings of the interface energy effect in the existing literature are clarified.

Key words: micromechanics, interface effects, size-dependent effective properties, nanocomposites.

1 Introduction

The effect of surface/interface energy on the elastic fields and the mechanical properties of nano-size structures and nanocomposites becomes important due to their large ratio of surface/interface to volume. Therefore, the study of this effect has attracted considerable attention of researchers in materials science and mechanics for many years (e.g. Shuttleworth, 1950; Gurtin and Murdoch, 1975; Gumbsch and Daw, 1991; Nix and Gao, 1998; Müller and Saúl, 2004; Sun et al., 2004). The interest in this subject has intensified recently (e.g. Cuenot et al., 2004; Duan et al., 2005; Shenoy, 2005; Dingreville et al., 2005). However, it should be noted that the interface models can be classified into two categories. The first one can be viewed as an approximation of a thin interphase when the interphase thickness approaches zero. It was shown that a thin and stiff interphase can be approximated by the equivalent interface stress model (Benveniste and Miloh, 2001; Hashin, 2002; Wang et al., 2005). The second one considers the difference of the atomistic microstructures between the

Yilong Bai et al. (eds), IUTAM Symposium on Mechanical Behavior and Micro-Mechanics of Nanostructured Materials, 51–59.

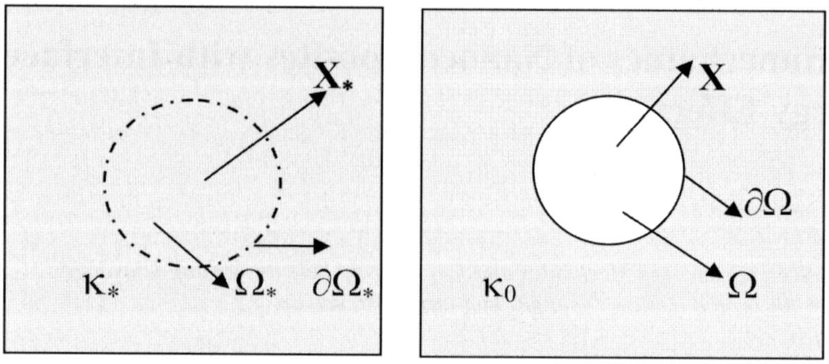

Fig. 1. Stress-free configuration κ_* and self-equilibrium configuration κ_0.

surface/interface and the bulk material. Hence there is excess surface/interface free energy. This can be described by an interface energy model. Although in both the interface stress model and the interface energy model, the tractions across the interface are discontinuous, and the constitutive relations of the interface are needed, yet there is no interface-induced elastic field in the former, whereas there is an interface-induced elastic field in the latter. In this paper, only the interface energy model will be discussed.

Generally, creating a free surface or an interface by bonding two different media together will generate excess surface/interface free energy and a surface/interface stress. Thus, for a multi-phase material, there should exist an intrinsic elastic field (or "residual" elastic field) induced by the surface/interface energy even when it is not subjected to any external body force and boundary traction. This renders the mechanical response of such a material under external loading inevitably dependent upon the deformation generated both by the intrinsic elastic field and by the applied external loading. In this paper, a fictitious stress-free configuration is introduced to facilitate a better understanding and to solve elastic problems involving surface/interface energy effect. Then it can be shown that all the field equations, including the generalized Young–Laplace equations, can be obtained from the stationary condition of a new functional. Unlike previous studies in the literature, the hyperelastic potential in the functional depends upon not only the deformation gradient caused by the external loading, but also the intrinsic elastic field. The latter reflects an intrinsic physical attribute of a multi-phase material containing surfaces/interfaces.

2 Basic Concepts in Interface Energy Models

2.1 Surface/Interface Energy-Induced Elastic Field

In order to study the effect of surface/interface energy, we consider an infinite elastic medium as shown in Figure 1a. The position vector of a material point in the medium is represented by $\mathbf{X}_*$. The medium is stress-free, when there is no external loading.

This stress-free configuration is denoted by κ_*. Suppose that there is a spherical region Ω_* with a boundary $\partial\Omega_*$ and radius R_*. If this region is removed from the medium, then the original spherical region Ω_* will become a void, denoted by Ω, with a radius R and boundary $\partial\Omega$. (Here, we assume that the material is homogeneous and isotropic from a continuum point of view.) The material point $\mathbf{X}_*$ will move to $\mathbf{X}$. As the microstructure (at the atomic-molecular level) of the material at $\partial\Omega$ will be different from that of the interior region away from the boundary, there will be surface energy and surface stress γ_0^* at $\partial\Omega$. In terms of the Young–Laplace equation, the normal traction at the boundary of the void is $\sigma_r = 2\gamma_0^*/R$. Therefore, the stresses in the matrix are no longer zero. This self-equilibrium state is denoted by configuration κ_0.

Therefore, due to the existence of the surface/interface, there are not only the surface/interface energy but also the elastic field induced by the surface/interface energy, even there is no external loading. For the sake of expediency, we call this elastic field the "residual field". Obviously, it is different from the residual stress field in plasticity and in continuum theory of distributed dislocations where the strain field is incompatible. In addition, we will call the configuration κ_* the *fictitious stress-free configuration*, and specifically take the configuration κ_0 as the *reference configuration*. Although, in continuum mechanics, any configuration can be chosen as the reference configuration, yet in this paper, the reference configuration will specifically refer to κ_0 with no external loading, but with the elastic field induced by the surface/interface energy.

2.2 Fictitious Stress Free Configuration

The above elastic field induced by the surface/interface energy can be depicted in terms of the deformation gradient $\mathbf{F}^* = \partial\mathbf{X}/\partial\mathbf{X}_*$ from the stress-free configuration κ_* to the reference configuration κ_0. The reason why the configuration κ_* is referred to as the fictitious stress-free configuration is that the atomistic microstructure of the material at and adjacent to the surface/interface is required to be the same as that of the interior of the material. In general, this condition cannot be fulfilled in reality; thus it is only a fictitious one. Suppose that there is an interface dividing materials 1 and 2. If the two materials are separated at the interface, the two materials will form two new surfaces $A_0^{(1)}$ and $A_0^{(2)}$, respectively, and these two new surfaces will in turn have surface energies. This is what happens in reality, and the configuration corresponding to the separated materials is *not* the fictitious stress-free configuration. The fictitious stress-free configuration is one where the atoms at the surfaces $A_0^{(1)}$ and $A_0^{(2)}$ would have the same environments as their respective counterparts in the interior regions of the two materials. Then the two fictitious surfaces so obtained (imaginarily) are denoted by $A_*^{(1)}$ and $A_*^{(2)}$, respectively, and the corresponding configuration is referred to as the fictitious stress-free configuration.

2.3 Calculation of Total Free Energy

Now we consider a multi-phase hyperelastic medium. The hyperelastic potential is denoted by $\psi(\mathbf{X};\mathbf{C})$, where $\mathbf{X}$ is the position vector and $\mathbf{C}$ is the right Cauchy–Green tensor. Moreover, it can be assumed that $\psi = 0$ when $\mathbf{C} = \mathbf{I}$, if there is no interface energy. For an isothermal process, ψ is actually the Helmholtz free energy. Next, we consider the case with the interface energy. If, in the deformed (current) configuration, the free energy per unit area is γ, then what is the total free energy of this multi-phase medium? In the literature, the total free energy is usually expressed as $\int_{A_0} J_2\gamma\,\mathrm{dA}_0 + \int_{V_0} \rho_0\psi(\mathbf{X},\mathbf{C})\,\mathrm{dV}_0$. However, as discussed in the above expositions, even there is no external loading, that is, $\mathbf{C} = \mathbf{I}$, there may exist an elastic field induced by the interface energy such that $\psi \neq 0$. Therefore, if we take into account the residual elastic field, the total free energy should be expressed as $\int_{A_0} J_2\gamma\,\mathrm{dA}_0 + \int_{V_0} \rho_0\psi(\mathbf{X},\tilde{\mathbf{C}})\,\mathrm{dV}_0$, where $\tilde{\mathbf{C}} = \mathbf{F}^{*T}\cdot\mathbf{C}\cdot\mathbf{F}^*$ is not only dependent upon $\mathbf{C}$, but also upon the deformation gradient $\mathbf{F}^*$ from the fictitious stress-free configuration κ_* to the reference configuration κ_0. Thus, the necessity and importance of the introduction of the fictitious stress-free configuration are explained.

2.4 Effect of Surface/Interface Energy on Mechanical Properties of Composites

During the deformation process of a multi-phase hyperelastic medium, the size and shape of the interface(s) will change, and so do the interface stress and the curvature tensor of the interface(s). Thus the interface energy-induced elastic field will also change. Generally, we are not interested in the elastic field in a particular configuration; instead, we are usually concerned with the change of the elastic field from one configuration to another. The change of configuration is a problem of finite deformation. Therefore, the governing equations taking into account the surface/interface effect should be established within the framework of finite deformation in the first place, especially the constitutive relations of the interface and the generalized Young–Laplace equations. Then the infinitesimal deformation analysis can be derived. In this way, the surface/interface effect can be better considered. The analyses in the literature that are based upon one configuration or adopt the infinitesimal deformation approximation in the first place cannot take into account the effect of the residual surface/interface stress. For example, from these analyses, it is believed that the liquid-like constant surface tension does not affect the effective elastic constants of heterogeneous materials. However, as will be shown below, this constant surface tension will influence the effective properties when the governing equations are established within the framework of finite deformation and then, these equations are reduced to the infinitesimal deformation case.

3 Constitutive Relations of Interface at Finite Deformation and Generalized Young–Laplace Equations

The mechanical response of the interface from the reference configuration κ_0 to the current configuration κ can be described by the constitutive relations of the interface. Suppose that the covariant base vectors in the tangent plane of the interface at a representative point $\mathbf{X}$ in κ_0 are denoted by $\mathbf{A}_\alpha$ $(\alpha = 1, 2)$. After deformation, the point $\mathbf{X}$ moves to $\mathbf{x}$, and these base vectors become $\mathbf{a}_\alpha$ $(\alpha = 1, 2)$ so that the deformation gradient of the interface can be defined as $\mathbf{F}_S = \mathbf{a}_\alpha \otimes \mathbf{A}^\alpha$, where $\mathbf{A}^\alpha$ $(\alpha = 1, 2)$ are the contravariant base vectors in the tangent plane in κ_0. Accordingly, we can define the right Cauchy–Green tensor $\mathbf{C}_S = \mathbf{F}_S^T \cdot \mathbf{F}$, and the right and left stretch tensors $\mathbf{U}_S$ and $\mathbf{V}_S$ of the interface. Moreover, suppose that γ is the function of $\mathbf{C}_S$. Then, as for the usual three-dimensional case, we can also define the Piola–Kirchhoff stresses of the first and second kinds $\mathbf{S}_S$ and $\mathbf{T}_S$, and the Cauchy stress $\boldsymbol{\sigma}_S$ of the interface. These stresses can be expressed as the functions of the interface free energy γ (Huang and Wang, 2006)

$$\mathbf{S}_S = 2\mathbf{F}_S \cdot \frac{\partial (J_2\gamma)}{\partial \mathbf{C}_S}, \tag{1}$$

$$\mathbf{T}_S = 2\frac{\partial (J_2\gamma)}{\partial \mathbf{C}_S}, \tag{2}$$

$$\boldsymbol{\sigma}_S = \frac{1}{J_2}\mathbf{F}_S \cdot \mathbf{T}_S \cdot \mathbf{F}_S^T, \tag{3}$$

where $J_1 = \mathrm{tr}\mathbf{U}_S = \mathrm{tr}\mathbf{V}_S$, $J_2 = \det\mathbf{U}_S = \det\mathbf{V}_S$. It should be emphasized that the above relations are for generally anisotropic interfaces, and they can also degenerate into the isotropic case. It is well known that the material symmetry (e.g. the isotropic interface) requires specification of a reference configuration. Here, the anisotropic (or isotropic) interface is specified to be relative to κ_0, since any discussion on the constitutive relations of the surface/interface based on the fictitious stress-free configuration will be meaningless. For an isotropic interface, formulae (1) and (3) can be written as

$$\mathbf{S}_S = J_2\mathbf{F}_S \cdot \left[\frac{\partial \gamma}{\partial J_1}\mathbf{U}_S^{-1} + \left(J_2\frac{\partial \gamma}{\partial J_2} + \gamma\right)\mathbf{C}_S^{-1}\right], \tag{4}$$

$$\boldsymbol{\sigma}_S = \frac{\partial \gamma}{\partial J_1}\mathbf{V}_S + \left(J_2\frac{\partial \gamma}{\partial J_2} + \gamma\right)\mathbf{1}(\mathbf{x}), \tag{5}$$

where $\mathbf{1}(\mathbf{x})$ is the second-rank identity tensor in two-dimensional space.

It can be proved that the generalized Young–Laplace equations, both in Lagrangian and Eulerian descriptions, can be derived from the stationary condition of a new energy functional proposed by Huang and Wang (2006). They are

$$\left.\begin{array}{l}\mathbf{A}_3 \cdot [\mathbf{S}^0] \cdot \mathbf{A}_3 = -\mathbf{S}_S : \mathbf{b}_0 \\ \qquad\qquad\qquad\qquad \text{(on the interface in } \kappa_0) \\ \mathbf{P}_0 \cdot [\mathbf{S}^0] \cdot \mathbf{A}_3 = -\mathbf{S}_S \cdot \nabla_{0S}\end{array}\right. \tag{6}$$

$$\mathbf{a}_3 \cdot [\boldsymbol{\sigma}] \cdot \mathbf{a}_3 = -\boldsymbol{\sigma}_S : \mathbf{b}$$

$$\text{(on the interface in } \kappa) \qquad (7)$$

$$\mathbf{p} \cdot [\boldsymbol{\sigma}] \cdot \mathbf{a}_3 = -\boldsymbol{\sigma}_S \cdot \nabla_S,$$

where $\mathbf{S}^0$ and $\boldsymbol{\sigma}$ are the Piola–Kirchhoff stress of the first kind and the Cauchy stress in the bulk material, respectively, the square bracket denotes the difference of the quantity across the interface, $\mathbf{A}_3$ and $\mathbf{a}_3$ are the unit normal vectors to the interface in the reference and current configurations, respectively, $\mathbf{P}_0 = \mathbf{I} - \mathbf{A}_3 \otimes \mathbf{A}_3$, $\mathbf{p} = \mathbf{I} - \mathbf{a}_3 \otimes \mathbf{a}_3$, ∇_{0S} and ∇_S denote the surface gradients, and $\mathbf{b}_0$ and $\mathbf{b}$ are the curvature tensors in κ_0 and κ, respectively.

4 Infinitesimal Deformation Analysis

In order to obtain analytical expressions of the size-dependent effective moduli of a nanocomposite, infinitesimal deformation approximations are usually needed. However, as mentioned above, the boundary-value problems involving the surface/interface energy effect should be formulated in the framework of finite deformation in the first place. This is because even in an infinitesimal deformation analysis, the Piola–Kirchhoff stress of the first kind $\mathbf{S}_S$ and the Cauchy stress $\boldsymbol{\sigma}_S$ of the interface are not the same, as will be seen in Equations (9) and (10). This situation is completely different from that in three-dimensional problem without the interface energy effect. For simplicity, let us consider an isotropic interface in the following. To this end, denote

$$\gamma\big|_{\substack{J_1=2\\J_2=1}} = \gamma_0, \qquad \frac{\partial \gamma}{\partial J_1}\bigg|_{\substack{J_1=2\\J_2=1}} = \gamma_1, \qquad \frac{\partial \gamma}{\partial J_2}\bigg|_{\substack{J_1=2\\J_2=1}} = \gamma_2, \qquad \gamma_0^* = \gamma_0 + \gamma_1 + \gamma_2. \tag{8}$$

Then, the quantity

$$J_2\left(\frac{\partial \gamma}{\partial J_1} + J_2\frac{\partial \gamma}{\partial J_2} + \gamma\right)$$

can be approximated by $\gamma_0^* + (\gamma_0^* + \gamma_1^*)\mathrm{tr}\,\mathbf{E}_S$, where $\mathbf{E}_S$ is the infinitesimal strain of the interface, and γ_1^* is another material parameter related to the interface energy γ. The Piola–Kirchhoff stress of the first kind and the Cauchy stress of the interface can be expressed as

$$\mathbf{S}_S = \gamma_0^*\mathbf{1}_0 + (\gamma_0^* + \gamma_1^*)(\mathrm{tr}\,\mathbf{E}_S)\mathbf{1}_0 - \gamma_0^*\nabla_{0S} \otimes \mathbf{u} + \gamma_1\mathbf{E}_S, \tag{9}$$

$$\boldsymbol{\sigma}_S = \gamma_0^*\,\mathbf{1}_0 + \gamma_1^*(\mathrm{tr}\,\mathbf{E}_S)\mathbf{1}_0 + \gamma_1\mathbf{E}_S, \tag{10}$$

where $\mathbf{1}_0$ is the second-rank identity tensor in the tangent plane in κ_0. It is seen that even under infinitesimal deformation, for an isotropic interface, there are three independent material parameters: γ_0^*, γ_1 and γ_1^*, where γ_0^* corresponds to the residual interface stress in the reference configuration κ_0. If the displacement gradient of the interface $\nabla_{0S} \otimes \mathbf{u}$ is a symmetric second-order tensor in two-dimensional space, $\mathbf{S}_S$ can then be formally written as

$$\mathbf{S}_S = \gamma_0^* \mathbf{1}_0 + \lambda_S (\mathrm{tr}\,\mathbf{E}_S)\mathbf{1}_0 + 2\mu_S\,\mathbf{E}_S, \tag{11}$$

where

$$\lambda_S = \gamma_0^* + \gamma_1^*, \; 2\mu_S = -\left(\gamma_0 + \left.\frac{\partial \gamma}{\partial J_2}\right|_{\substack{J_1=2\\ J_2=1}}\right).$$

It is interesting to note that in most cases, μ_S is negative since γ_0 is the surface/interface energy at κ_0 and $\partial\gamma/\partial J_2$ is the change rate of the surface/interface energy due to the change of the surface/interface area, and the negative μ_S has been confirmed by Shenoy (2005) in his atomistic calculations.

The change of the interface energy-induced residual elastic field due to the change of configuration can be solved from the governing equations in the bulk material as well as the following equations at the interface:

$$\begin{aligned}
&\mathbf{A}_3 \cdot [\Delta\boldsymbol{\sigma}] \cdot A_3 = -\Delta\mathbf{S}_S : \mathbf{b}_0 \\
&\mathbf{P}_0 \cdot [\Delta\boldsymbol{\sigma}] \cdot \mathbf{A}_3 = -\Delta\mathbf{S}_S \cdot \nabla_{0S} \quad \text{where} \\
&\Delta\mathbf{S}_S = (\gamma_0^* + \gamma_1^*)(\mathrm{tr}\,\mathbf{E}_S)\mathbf{1}_0 - \gamma_0^* \nabla_{0S} \otimes \mathbf{u} + \gamma_1 \mathbf{E}_S.
\end{aligned} \tag{12}$$

The above theoretical framework can be used to calculate the effective moduli of composites with the interface energy effect. For example, for a composite containing spherical particles, we can calculate the displacement and stress at the interface on the matrix side, so that a single particle together with the interface can be "mapped" onto a "homogeneous equivalent" particle. Then the effective moduli can be calculated using conventional micromechanical approaches, e.g. the Mori–Tanaka method and the generalized self-consistent method. Detailed discussion is given in Huang and Sun (2006), and will not be reproduced here. In particular, if the interface stress only weakly depends upon the interface strain such that it has a liquid-like properties and can be regarded as a constant, namely, $\gamma = \gamma_0$, the effective bulk and shear moduli of a heterogeneous material containing spherical voids of radius a can be expressed as

$$\bar{K} = \frac{1}{3}\left[\frac{12K_0\mu_0(1-f) + 2(3K_0 + 4\mu_0 f)n}{3K_0 f + 4\mu_0 + 2(1-f)n}\right], \tag{13}$$

$$\bar{\mu} = \frac{\mu_0}{2}\left[\frac{4(1-f)m_1\mu_0^2 + 4(2m_2 - fm_1)\mu_0 n - 42m_4\mu_0 n - (m_1 f + 2m_3)n^2}{2(2fm_3 + m_1)\mu_0^2 + 4(fm_3 + m_2)\mu_0 n - 21m_4\mu_0 n - (1-f)m_3 n^2}\right], \tag{14}$$

where

$$m_1 = 7 - 5\nu_0, m_2 = 5 - 4\nu_0, m_3 = 4 - 5\nu_0, m_1 = 1 - \nu_0$$

and $n = \gamma_0/a$. K_0, μ_0 and ν_0 are the bulk modulus, shear modulus and Poisson ratio of the matrix, and f is the volume fraction of the voids. It is seen that even for a constant surface tension, the change of the curvature of the surface will affect the effective moduli of a porous material. This is different from the results in the literature.

5 Concluding Remarks

The effect of surface/interface becomes very important in nanostructures and nanocomposites due to their large ratio of interface to volume. Taking into account this effect renders the mechanical behavior of a composite size-dependent. In this paper, some basic concepts related to the description of the effect of surface/interface energy on the mechanical behavior of heterogeneous materials are discussed. Then the interface constitutive relations are given in terms of the interface energy in both Lagrangian and Eulerian descriptions within the framework of finite deformation. The generalized Young–Laplace equations taking into account the change of configuration are presented. Finally, following the proposed theoretical framework, the effective properties of a heterogeneous material with the interface energy effect can be obtained. It is shown that the liquid-like constant surface/interface tension can also affect the effective moduli of the nanocomposite.

Acknowledgements

The project is supported by the National Natural Science Foundation of China (10032010, 10372004, 10525209) and Shanghai Leading Academic Discipline.

References

Benveniste, Y. and Miloh, T. (2001) Imperfect soft and stiff interfaces in two-dimensional elasticity, *Mech. Mater.* **33**, 309–323.

Cuenot, S., Frétigny, C., Demoustier-Champagne, S. and Nysten, B. (2004) Surface tension effect on the mechanical properties of nanomaterials measured by atomic force microscopy, *Phys. Rev. B* **69**, 165410-1-5.

Dingreville, R., Qu, J. and Cherkaoui, M. (2005) Surface free energy and its effect on the elastic behavior of nano-sized particles, wires and films, *J. Mech. Phys. Solids* **53**, 1827–1854.

Duan, H.L., Wang, J., Huang, Z.P. and Karihaloo, B.L. (2005) Size-dependent effective elastic constants of solids containing nano-inhomogeneities with interface stress, *J. Mech. Phys. Solids* **53**, 1574–1596.

Gumbsh, P. and Daw, M.S. (1991) Interface stresses and their effects on the elastic moduli of metallic multilayers, *Phys. Rev. B* **44**, 3934–3938.

Gurtin, M.E. and Murdoch, A.I. (1975) A continuum theory of elastic material surfaces, *Arch. Rat. Mech. Anal.* **57**, 291–323.

Hashin Z. (2002) The interphase/imperfect interface in elasticity with application to coated fiber composites, *J. Mech. Phys. Solids* **50**, 2509–2537.

Huang, Z.P. and Sun, L. (2006) Size-dependent effective properties of a heterogeneous material with interface energy effect: From finite deformation theory to infinitesimal strain analysis, *Acta Mech.* (in press).

Huang, Z.P. and Wang, J. (2006) A theory of hyperelasticity of multi-phase media with surface/interface energy effect, *Acta Mech.* **182**, 195–210.

Müller, P. and Saúl, A. (2004) Elastic effects on surface physics, *Surf. Sci. Reports* **54**, 157–258.

Nix, W.D. and Gao, H.J. (1998) An atomistic interpretation of interface stress, *Scripta. Mater.* **39**, 1653–1661.

Shenoy, V.B. (2005) Atomistic calculations of elastic properties of metallic fcc crystal surfaces, *Phys. Rev.* B **71**, 094104-1–11.

Shuttleworth, R. (1950) The surface tension of solids, *Proc. Phys. Soc.* A **63**, 444–457.

Sun, L., Wu, Y.M., Huang, Z.P. and Wang, J. (2004) Interface effect on the effective bulk modulus of a particle-reinforced composite, *Acta. Mech. Sinica* (English series) **20**, 676–679.

Wang, J., Duan, H.L., Zhang, Z. and Huang, Z.P. (2005) An anti-interpenetration model and connections between interface and interphase models in particle-reinforced composites, *Int. J. Mech. Sci.* **47**, 701–718.

Measurements and Simulations of Interface Behavior in Metal Thin Film Peeling Along Ceramic Substrate

Yueguang Wei*, Haifeng Zhao and Siqi Shu

LNM, Institute of Mechanics, Chinese Academy of Sciences, Beijing 100080, China;
**E-mail: ywei@lnm.imech.ac.cn*

Abstract. Peeling experiments for aluminum thin film along the Al_2O_3 substrate are carried out, and the variations of external driving force (energy release rate) at the steady-state delamination of the thin film in the metal film/ceramic substrate system are measured. Additionally, theoretical modeling for the thin film delamination is also performed. Based on the bending model, three double-parameter criteria are used. Three double-parameter criteria include: (1) the interfacial fracture toughness and the separation strength, (2) the interfacial fracture toughness and the interfacial crack tip slope angle of thin film, and (3) the interfacial fracture toughness and the critical von Mises effective strain of thin film at crack tip. Based on the three double-parameter criteria, the thin film nonlinear peeling problems are solved analytically for each case. The results show that the solutions of thin film nonlinear peeling based on the bending model are very sensitive to the model parameter selections. Through analyses and comparisons to different solutions, a connection between solutions based on the bending models and based on the two-dimensional elastic-plastic finite element analysis is obtained. The effective regions of each model can be specified through comparing the present experimental result with model solutions.

Key words: metal thin film, delamination, interfacial toughness, peel test, double-parameter criterion.

1 Introduction

Most advanced materials are inseparable with thin films which with its particular characteristic, has been widely applying to the surface and interfacial engineering areas. The material behaviors of the thin film/substrate systems are mainly dominated by the interfacial adhesion property (strength). In order to evaluate the adhesion behaviors, a simple test method, peel test, was designed fifty years ago [1]. Due to the good advantages of the test method, such as simply operating, the test has been widely applying to many research regions [2–4]. Specifically, when both the thin film and substrate are elastic materials, the interfacial adhesion toughness can be obtained directly through measuring peeling force in peel experiment. However,

Yilong Bai et al. (eds), IUTAM Symposium on Mechanical Behavior and Micro-Mechanics of Nanostructured Materials, 61–70.

when thin film or substrate is a ductile material, the measured peeling force is often much larger than the interfacial adhesion toughness. The phenomenon is come from the plastic dissipation due to material plastic loading and unloading deformation. In order to model the peeling force (or energy release rate) increase due to plastic dissipation, Kim and his collaborators [5, 6] presented a bending model to predict the plastic dissipation. Within the following decade after bending model was presented, most analyses related to the ductile thin film peeling adopted the bending model of Kim et al., e.g., [7, 8]. However, Wei and Hutchinson [9] adopted a different method from that of Kim et al. in analyzing the elastic-plastic thin film peeling problems. In Wei and Hutchinson's analysis, the thin film delamination process was simulated by using the two-dimensional elastic-plastic finite element method (FEM), except the detached part of thin film, which was described still by bending model in order to avoid the difficulty in the two-dimensional large deformation analysis. They obtained a kind of different results from that of bending model, qualitatively and quantitatively. Recently, Wei [10] adopted three different double-parameter criteria based on the bending model to obtain the different relationships between the peel force and the thin film thickness. In the present research, in order to further explore the connection of bending model solution with elastic-plastic FEM solution, and in order to assess the effectiveness for every models, a series of the peel experiments for Al thin film delaminated along the ceramic (Al_2O_3) substrate are carried out. The relationship between the peel force and the thin film thickness at the steady-state delamination is measured. By comparing the experimental curves with the modeling solutions, a primary connection of both the bending model and the two-dimensional FEM model is presented.

2 Model Descriptions

Peeling experiments for Al thin film along the ceramic (Al_2O_3) substrate are carried out for a series of thin film thickness, $t = 20, 50, 80, 100, 200$, and 225 microns. There exists an adhesion layer between film and substrate with a thickness of 20 microns. The interface layer material is Epoxy with two different percentages of curing agent (Polyimide), respectively. A brittle adhesive layer is formed for a compound of Epoxy with Polyimide, 1:1, and a ductile adhesive layer is for the proportion 1:1.5. The adhesive layer thickness is 20 micron in the present experiment. Figures 1a and 1b show the peel force varying with the peeling displacements for thin film thicknesses $t = 20$ and 50 microns, respectively. Three curves in Figure 1a or in Figure 1b correspond to three samples. The adhesive layer corresponds to ductile one. From Figures 1a and 1b, the feature of the peel force variations can be described as follows: firstly, peel force varies linearly with peeling displacement, secondly a steady-state peeling process is quickly arrived when peel force attains a critical value. Figure 1c shows the experimental result of the steady-state peel force (or called energy release rate) varying with several thin film thickness. Two curves correspond to ductile and brittle interface adhesives, respectively.

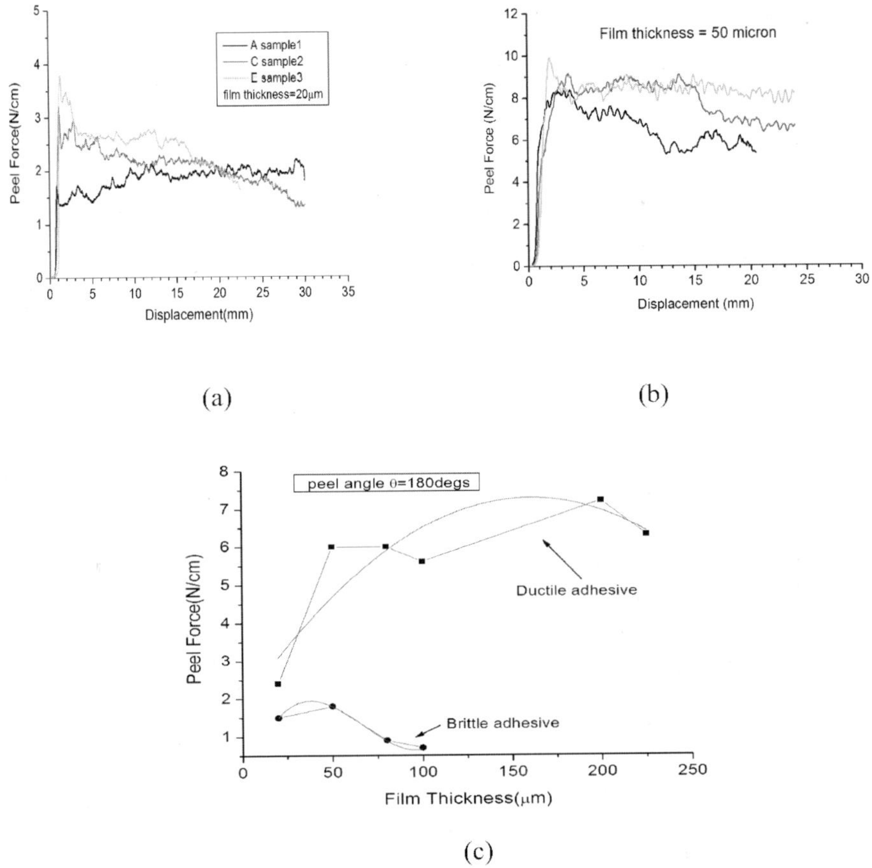

Fig. 1. Peeling experimental results.

3 Bending Models and Delamination Criteria in Peel Test

Delamination process of elastic-plastic thin film in peel test can be described by Figure 2a. The thin film undergoes the delamination and plastically loading and unloading process under the act of the peel force P. The cross-section of the thin film is from a free-stressing state to the loading and unloading processes, as described by OABCDEF, in Figure 2a.

The process of the ductile thin film peeled and delaminated along substrate interface can be characterized by the double-parameter criterion (for elastic delamination case, single-parameter criterion is valid). Two independent parameters are needed to characterize the main characters here, the interfacial adhesion property and the plastic dissipation of the system. In the present research, three double-parameter criteria [10] will be used respectively for describing the elastic-plastic peeling process, which are given in Figures 2b, 2c and 2d, respectively.

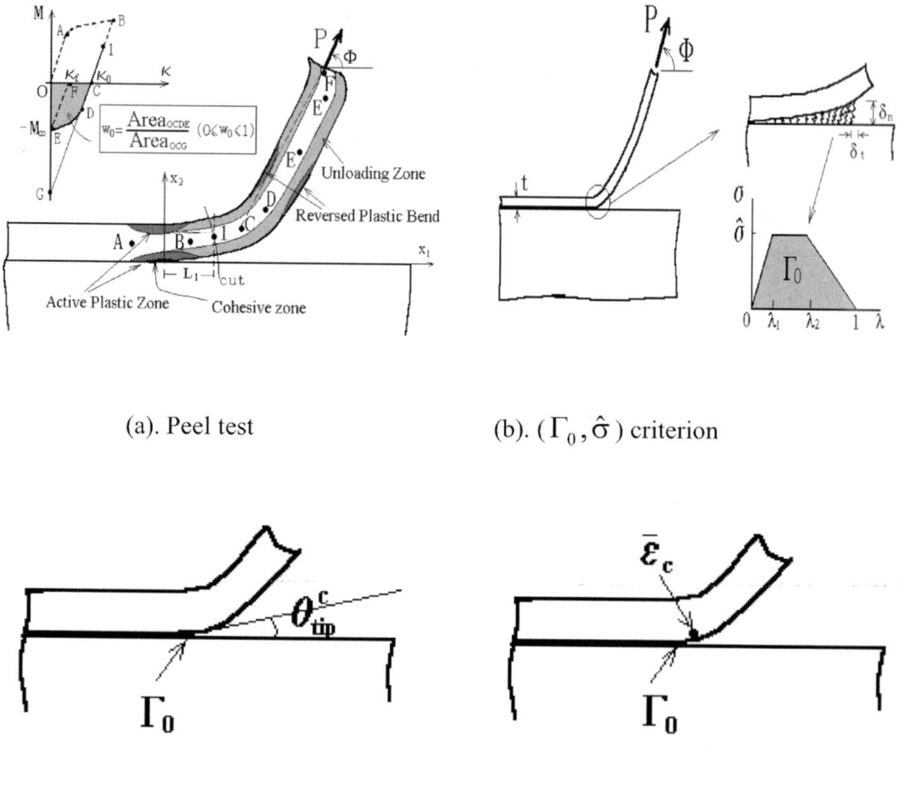

(a). Peel test (b). $(\Gamma_0, \hat{\sigma})$ criterion

(c). $(\Gamma_0, \theta^c_{tip})$ criterion (d). $(\Gamma_0, \bar{\varepsilon}_c)$ criterion

Fig. 2. Peel test sketch (a) and three simplified double-parameter models (b)–(d) [10].

For the thin film peeling process, the relation among the peeling force P per unit width of thin film (or energy release rate of system), the interfacial adhesion toughness Γ_0, as well as the geometrical and physical parameters of thin film and substrate is usually concerned. Under steady-state delamination, the relation can be written as:

$$\begin{aligned} P(1 - \cos\Phi) &= \Gamma_0 && \text{(elastic peeling);} \\ P(1 - \cos\Phi) &= \Gamma_0 + \Gamma^p && \text{(elastic-plastic peeling),} \end{aligned} \tag{1}$$

where Γ^p is the plastic dissipation. Based on the stress-strain analysis for thin film, one can obtain the fundamental relations of the thin film undergoing the nonlinear bending, furthermore, one can also obtain the plastic dissipation relation for Γ^p, as given in next section.

4 Fundamental Relations

Kim and Aravas [5] derived out the fundamental relations based on the bending model for elastic-plastic thin film bending under the incompressible conditions ($\nu = 1/2$). The rigorous derivation based on the general case of the compressible elastic-plastic conditions is given by [10, 11]. The relations of moment and curvature respectively for elastic, plastic and unloading cases can be dictated as follows:

$$\frac{M}{M_0} = \frac{2\kappa}{3\kappa_e};\ \frac{M}{M_0} = \left\{\frac{2}{3} - \frac{2}{N+2}\gamma\right\}\frac{1}{(\kappa/\kappa_e)^2} + \frac{2}{N+2}\gamma\left(\frac{\kappa}{\kappa_e}\right)^N;$$

$$\frac{M}{M_0} = \frac{2}{3}\frac{\kappa - \kappa_0}{\kappa_e}, \tag{2}$$

and curvature relation:

$$\kappa = \sqrt{[1-\cos(\phi-\theta)]\frac{2P}{B} + (1-w_0)\kappa_0^2}, \quad \theta_B \approx \theta_{\text{tip}} \le \theta \le \theta_C. \tag{3}$$

where $M_0 = 3/2M_e$ is the limit bending moment for elastic-perfectly plastic material; M_e and κ_e are the elastic limit moment and elastic limit curvature, respectively,

$$M_e = \frac{2}{3}M_0 = \frac{\sigma_Y t^2}{6\sqrt{1-\nu+\nu^2}}, \quad \kappa_e = \frac{2(1-\nu^2)\sigma_Y}{Et\sqrt{1-\nu+\nu^2}}$$

$$\gamma = 2\sqrt{\frac{1}{3}(1-\nu+\nu^2)}^{\,1-N}(1-\nu)^N. \tag{4}$$

$B = Et^3/12(1-\nu^2)$ is the bending modulus; $w_0 (0 \le w_0 \le 1)$ is defined in Figure 1a which characterizes the inversely plastic behavior (or Bauschinger effect); θ_{tip} is the crack tip slope angle at thin film delamination; N is material strain hardening exponent. For incompressible material $\nu = 0.5$ and $\gamma = 1$ expression (2) comes to the result of Kim and Aravas [5].

Suppose that substrate is rigid or Young's modulus of substrate is much larger than that of thin film, by means of formulae (2) (M-κ relations in sketch of Figure 2a), one can obtain the plastic dissipation relation through calculating the area within the circuit OABCDEO under M-κ curve,

$$\Gamma^P = \frac{1}{2}M_e\kappa_e - \frac{1}{2}M_B(\kappa_B - \kappa_0) + \left(\frac{2}{3} - \frac{2}{N+2}\gamma\right)M_0\left(\kappa_e - \frac{\kappa_e^2}{\kappa_B}\right)$$

$$+ \frac{2\gamma}{(N+1)(N+2)}M_0\left(\frac{\kappa_B^{N+1}}{\kappa_e^N} - \kappa_e\right) + \frac{1}{2}B\kappa_0^2 w_0 \tag{5}$$

5 Bending Model Solutions of Thin Film Nonlinear Peeling

Based on three double-parameter criteria, the peel force can be solved for peeling process. The solving procedures can be dictated as follows. Firstly, from (3) a relation

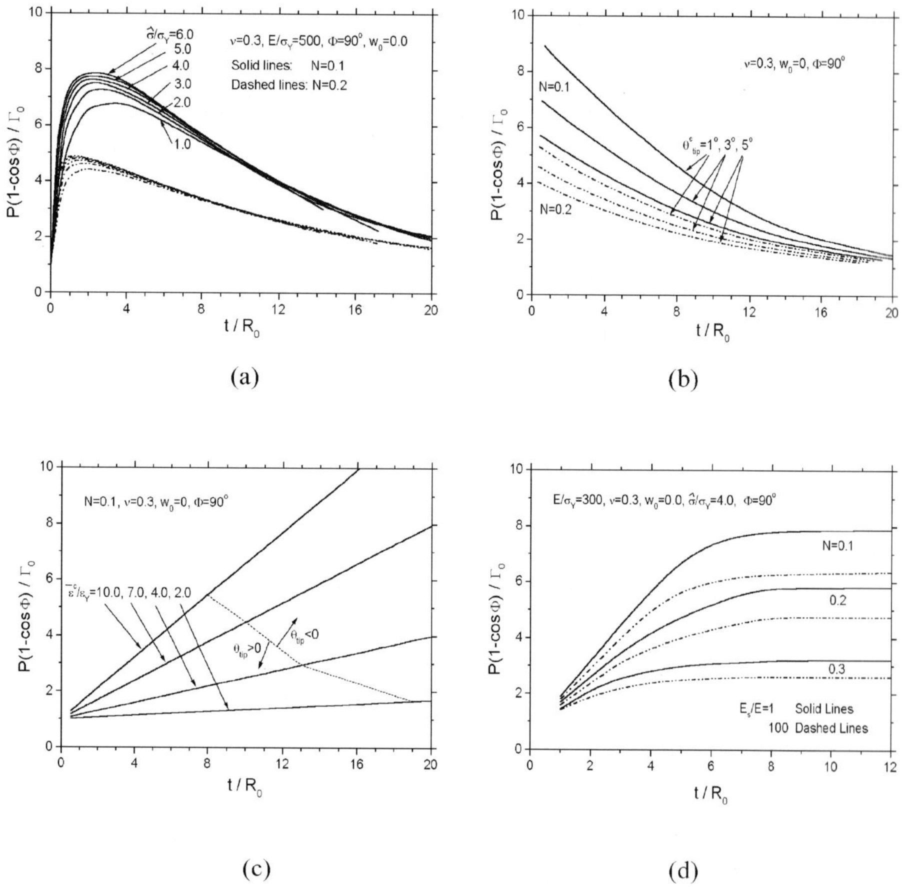

Fig. 3. Results based on three double-parameter criteria (a)–(c) and two-dimensional finite element method (d).

among the parameters ($\kappa_B, P, \theta_{\text{tip}} = \theta_B, \kappa_0$) can be obtained to each criterion. Then the parameters (M_B, κ_B, κ_0) can be solved simultaneously with Equations (2) for $M = M_B$ and $\kappa = \kappa_B$. Finally, the peel force variations with related parameters can be attained from (1) and (5).

The solution forms by adopting three double-parameter criteria can be given through independent parameters by dimensional analysis. They are dictated as

$$\begin{cases} \dfrac{P(1-\cos\Phi)}{\Gamma_0} = f_1(E/\sigma_Y, \hat{\sigma}/\sigma_Y, N, \nu, t/R_0, w_0, \Phi) \\ \theta_{\text{tip}} = r_1(E/\sigma_Y, \hat{\sigma}/\sigma_Y, N, \nu, t/R_0, w_0, \Phi) \end{cases} \quad \text{(for } (\Gamma_0, \hat{\sigma}) \text{ criterion);} \quad (6)$$

$$\frac{P(1-\cos\Phi)}{\Gamma_0} = f_2(N, \nu, t/R_0, w_0, \Phi, \theta^c_{\text{tip}}) \quad \text{(for } (\Gamma_0, \theta^c_{\text{tip}}) \text{ criterion);} \quad (7)$$

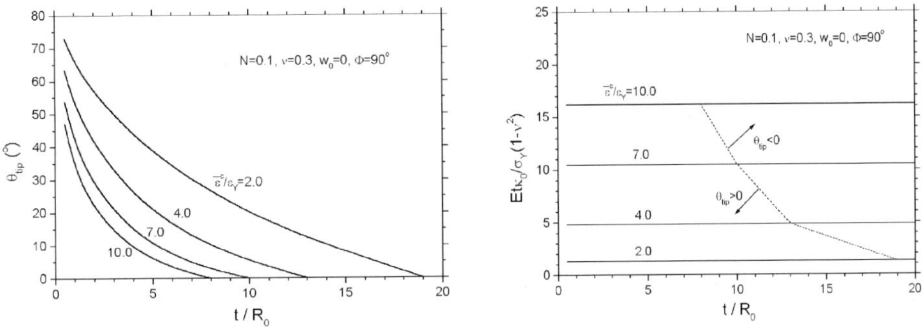

(a) (b)

Fig. 4. Variations of the thin film inclined angle at tip and the residual curvature with thin film thickness based on the criterion $(\Gamma_0, \bar{\varepsilon}_c)$.

$$\begin{cases} \dfrac{P(1-\cos\Phi)}{\Gamma_0} = f_3(N, \nu, t/R_0, w_0, \Phi, \bar{\varepsilon}_c/\varepsilon_Y) \\ \theta_{\text{tip}} = r_3(N, \nu, t/R_0, w_0, \Phi, \bar{\varepsilon}_c/\varepsilon_Y) \end{cases} \quad (\text{for } (\Gamma_0, \bar{\varepsilon}_c) \text{ criterion}), \quad (8)$$

where a length parameter R_0 is introduced and is defined as $R_0 = E\Gamma_0/3\pi(1-\nu^2)\sigma_Y^2$, which characterizes the plastic zone size at crack tip under the small scale yielding case. θ_{tip} is crack tip slope angle. $\varepsilon_Y = \sigma_Y/E$ is yielding strain.

The solutions based on three double-parameter criteria $(\Gamma_0, \hat{\sigma})$, $(\Gamma_0, \theta_{\text{tip}}^c)$ and $(\Gamma_0, \bar{\varepsilon}_c)$ are given in Figures 3a–3c, respectively. Results in Figure 3 show the curves of the normalized energy release rate (or peeling force) versus normalized thin film thickness under elastic-plastic steady-state delamination. If thin film deformation is elastic, $P(1-\cos\Phi)/\Gamma_0 = 1$, see first relation of formula (1). Clearly, the energy release rate is remarkably enlarged by the plastic dissipation. From Figure 3a, the variation of the normalized peeling force (energy release rate) $P(1-\cos\Phi)/\Gamma_0$ with thin film thickness can be described as follows. Its value tends to 1 when thin film thickness is very small, i.e., the effect of plastic dissipation can be neglected. As thin film thickness increases, energy release rate increases sharply and obtains a maximum value at about $t/R_0 = 2$. The plastic dissipation decreases gradually as thin film thickness increases further. The results shown in Figure 3b are based on the $(\Gamma_0, \theta_{\text{tip}}^c)$ criterion. From the results, the variation of $P(1-\cos\Phi)/\Gamma_0$ *versis* t/R_0 is much different from that based on the double-parameter criterion $(\Gamma_0, \hat{\sigma})$. Here, the value of $P(1-\cos\Phi)/\Gamma_0$ increases always as thin film thickness decreases. This trend is consistent with that given by Kim and his collaborators for incompressible material [5, 6]. The results based on the $(\Gamma_0, \bar{\varepsilon}_c)$ criterion are given in Figure 3c. Obviously, the variation of $P(1-\cos\Phi)/\Gamma_0$ *versus* t/R_0 is different from those shown above based on the other double-parameter criteria. In Figure 3c, $P(1-\cos\Phi)/\Gamma_0$ increases linearly with increasing t/R_0. The inclined angles of the straight line increase with increasing the critical von Mises effective strain at tip. In order to investigate the third criterion, Figures 4a and 4b show the variations of the crack tip slope angle and the residual curvature. When thin film thickness is small, the slope angle is large. With increasing thin film thickness, the slope angle decreases sharply and transits to negative value from positive value. Obviously, the negative slope angle of thin film at tip violates the physical requirement. It implies that the criterion is failure and the predicted results, top part of the dashed line in Figure 3c, should be cut off. The reason can be interpreted in referring to Figure 4b that when thin film thickness is small, the large critical Mises effective strain at crack tip is realized by large bending deformation (curvature) in thin film, however when thin film thickness is large, it is difficult or even impossible to realize the large bending curvature of thin film to meet with the high critical Mises effective strain, and in this case the criterion $(\Gamma_0, \bar{\varepsilon}_c)$ is failure.

In order to assess the bending model solutions based on the above three two-parameter critera, the two-dimensional elastic-plastic finite element results [9] are also shown here, see Figure 3d. From Figure 3d, the bending model results based on the third double-parameter criterion have the similar variation trends and quantity with those of the elastic-plastic finite element calculation within the region of about $t/R_0 < 5$ (in Figure 3d, E_s is Young's modulus of elastic substrate). However, as thin film thickness increases continuously, the finite element solution tends to the small scale yielding solution (insensitive to thin film thickness), while the bending model solution based on the double-parameter criterion is failure. Therefore, when thin film thickness is large, i.e., $t/R_0 > 5$, how to set up a governing parameter criterion based on the bending model will be a tough task to need to be explored in the future.

6 Comparison of Experiment Results with Modeling Results

Peeling experimental results of the relationship between the energy release rate and the film thickness at the steady-state delamination are shown in Figure 1c. Modeling results based on the bend model three criteria and the two-dimensional finite element method are shown in Figure 3. Through comparing the experimental results with the modeling results, the following key points can be read:

1. Both parameter criteria, $(\Gamma_0, \hat{\sigma})$ and $(\Gamma_0, \theta^c_{\text{tip}})$, are suitable for a weaker-adhesion interface case, while two-dimensional elastic-plastic analysis (finite element simulation) is suitable for a stronger-adhesion interface case.
2. When film thickness is very thin, such as for $t/R_0 < 5$, the criterion $(\Gamma_0, \theta^c_{\text{tip}})$ is failure, while the other criteria mentioned above are valid.

7 Discussions to Bending Model Solutions

In the present research, the peel experiments for Al thin film delaminated along the ceramic (Al_2O_3) substrate have been carried out. The relationship between the peel force and the thin film thickness at the steady-state film delamination has been measured. By comparing the experimental curves with the modeling solutions, the effective regions for each analysis models has been assessed, and a primary connection of both the bending model and the two-dimensional FEM model has been developed.

The bending model solutions of thin film nonlinear peeling based on the three double-parameter criteria have been obtained in last section. Obviously, the bending model solutions are very sensitive to the selection of the governing parameters. This leads to a question: what is a reasonable selection of the governing parameter group for predicting the thin film nonlinear peeling by using the bending model? From the solutions based on the first and second double-parameter criteria, when normalized thin film thickness is smaller than about 5, $t/R_0 < 5$, the contribution of bending plastic dissipation to energy release rate is very large. This can be confirmed from variation of the residual curvature in [11]. However, from the solutions based on the criterion of the interfacial fracture toughness and the critical Mises effective strain, when thin film thickness is smaller than about 5, $t/R_0 < 5$, the contribution of the bending plastic dissipation to energy release rate is obviously smaller than that based on the first or second criterion.

Through comparing the experimental results with the modeling results, one can conclude that both parameter criteria, $(\Gamma_0, \hat{\sigma})$ and $(\Gamma_0, \theta^c_{\mathrm{tip}})$, are suitable for a weaker-adhesion interface case, while two-dimensional elastic-plastic analysis (finite element simulation) is suitable for a stronger-adhesion interface case, however when film thickness is very thin, such as for $t/R_0 < 5$, the criterion $(\Gamma_0, \theta^c_{\mathrm{tip}})$ is failure, while the other criteria mentioned above are valid.

Acknowledgements

The work is supported by National Science Foundations of China through Grants 10432050 and 10428207.

References

1. Spies, G.J., The peeling test on redux-bonded joints. *J. Aircraft Eng.* **25**, 1953, 64.
2. Feliu-Baez, R., Lockhart, H.E. and Burgess G., Correlation of peel and burst tests for pouches. *Packaging Tech. Sci.* **14**, 2001, 63.
3. Choi, J.W. and Oh, T.S., Peel strength and peel angle measured by the T-peel test on Cr/BPDA-PDA interface. *J. Adhe. Sci. Tech.* **15**, 2001, 139.
4. Asai, H., Iwase, N. and Suga, T., Influence of ceramic surface treatment on peel-off strength between aluminum nitride and epoxy-modified polyaminobismaleimide adhesive. *IEEE Trans. on Adv. Packaging* **24**, 2001, 104.
5. Kim, K.S. and Aravas, N., Elasto-plastic analysis of the peel test. *Int. J. Solids Struct.* **24**, 1988, 417.
6. Kim, J., Kim, K.S. and Kim, Y.H., Mechanical effects of peel adhesion test. *J. Adhe. Sci. Tech.* **3**, 1989, 175.
7. Kinloch, A.J., Lau, C.C. and Williams, J.G., The peeling of flexible laminates. *Int. J. Fracture* **66**, 1994, 45.
8. Moidu, A.K., Sinclair, A.N. and Spelt, J.K., On determination of fracture energy using the peel test. *J. Testing Evalu.* **26**, 1998, 247.

9. Wei, Y. and Hutchinson, J.W., Interface strength, work of adhesion and plasticity in the peel test. *Int. J. Fracture* **93**, 1998, 315.
10. Wei, Y., Modeling nonlinear peeling of ductile thin films – Critical assessment of analytical bending models using FE simulations. *Int. J. Solid & Structures* **41**, 2004, 5087.
11. Wei, Y. and Hutchinson, J.W., Peel test and interfacial toughness, in *Encyclopedia of Comprehensive Structural Integrity*, Vol. 8, W.W. Gerberich and W. Yang (eds.), Chapter 5, Elsevier Science Publisher, 2003, p. 181.

Micro-Cantilevers for Thin Films: Young's Modulus

G.J. McShane[1], M. Boutchich[1], S. Phani[1], D.F. Moore[1] and T.J. Lu[1,2,*]

[1] *Cambridge University Engineering Department, Cambridge CB2 1PZ, U.K.;*
* *E-mail: tjl21@cam.ac.uk*
[2] *School of Aerospace, Xian Jiaotong University, Xian 710049, P.R. China*

Abstract. A simple and effective characterisation technique based on micro-cantilever beams for thin film materials using commonly available equipment – scanning surface profiler – is described. The advantages of this class of techniques are simplicity, speed, cost and a wide applicability. A technique for extracting the Young's modulus from static deflection data is developed and validated in experiments on thin film specimens of silicon nitride deposited on a silicon substrate under different conditions. Finite element analysis is used to assess the influence of factors affecting the bending of thin films, and thus guide the analysis of micro-cantilever deflection data for reliable characterisation of the material.

Key words: MEMS materials, thin films, cantilevers, scanning surface profiler, finite element methods, elastic modulus.

1 Introduction

A large number of applications in the field of micro-electro-mechanical systems (MEMS) rely on the mechanical properties of thin film materials, whereas a wide range of sensors and actuators rely on the properties of elastic flexures for targeted performance specifications. The use of thin film materials in packaging applications, for example passive micro-clips, active bi-morph actuators or bi-stable devices, places a considerable demand on reliable material properties data [1]. Accurate characterisation of these mechanical properties is important for the reliable and optimal design of MEMS devices. Wafer to wafer variations in deposition processes, or alterations to deposition conditions, often require checking the properties of a particular film. A straightforward and reliable technique for extracting materials properties with the least disruption would be extremely valuable.

Osterburg and Senturia [2] suggested to use the electrostatic pull-in phenomenon to measure the Young's modulus, Poisson's ratio and the residual stress in thin films (the M-TEST). The test procedure therefore requires conducting test structures (beams and diaphragms), separated from a conducting substrate by an insulating

Yilong Bai et al. (eds), IUTAM Symposium on Mechanical Behavior and Micro-Mechanics of Nanostructured Materials, 71–83.

layer. This places a degree of restriction on the types of wafer and thin-film material that can be tested.

Noticing the deficiencies of the M-Test, Hopcroft [3] discussed a new test (the MAT-TEST) based on the deflection of micro-cantilevers using scanning profilometry. Cantilever beams can be created using a range of common surface- and bulk-micromachining techniques. They can be fabricated from any material (insulating or conducting) with sufficient stiffness to permit a free-standing structure. The use of statically determinate structures removes the effect of in-plane residual stresses (although through-thickness variations may lead to beam curvature). A technique for extracting the Young's modulus by using a polynomial curve fit to beam deflection data was suggested [3].

Techniques exist for the tensile testing of thin film specimens [4–6]. This allows for the more conventional stress-strain curve analysis of the specimens, and can provide information on both the elastic modulus and the failure stress. These techniques are, however, complex in terms of sample preparation and load application methods. The application of macroscopic testing techniques on the microscopic scale is extremely difficult, especially the accurate measurement of force and displacement with MEMS actuators and sensors [7].

The work presented here aims to develop measurement and analysis techniques applicable to cantilever beam based Young's modulus measurement. The benefits of cantilever beams as test structures in terms of simplicity, reproducibility, reliability of fabrication, time and cost are significant. Beams are tested by scanning from root to tip with a surface profiler, measuring deflection against horizontal position for a prescribed force. The principal material to be investigated is silicon nitride. This is a valuable structural material in MEMS, as it has a high modulus, can endure large strains without permanent deformation, and its resistance to common anisotropic wet etching processes makes it ideal for creating suspended structures without the need for a sacrificial layer.

This paper is organized as follows. The fabrication and testing procedure for thin-film cantilever beams are described in Section 2. Section 3 outlines the method adopted for the extraction of Young's modulus from measurements of force and deflection, subject to the specific constraints of scanning profiler testing of micro-cantilevers. Section 4 uses finite element analysis to quantify beam bending phenomenon which will affect the measured Young's modulus, in particular the effects of large deflections, beam stiffening, loading with a point force and the existence of compliant root conditions. Experimental results for the extraction of Young's modulus are presented and discussed in Section 5.

2 Fabrication and Testing

Cantilever beams are used as the test structures for measuring Young's modulus with scanning profilometry. A cantilever beam can be produced using a range of standard micro-machining techniques. For the test structures described here, the beam outline is cut into the surface film (silicon nitride deposited on a silicon substrate)

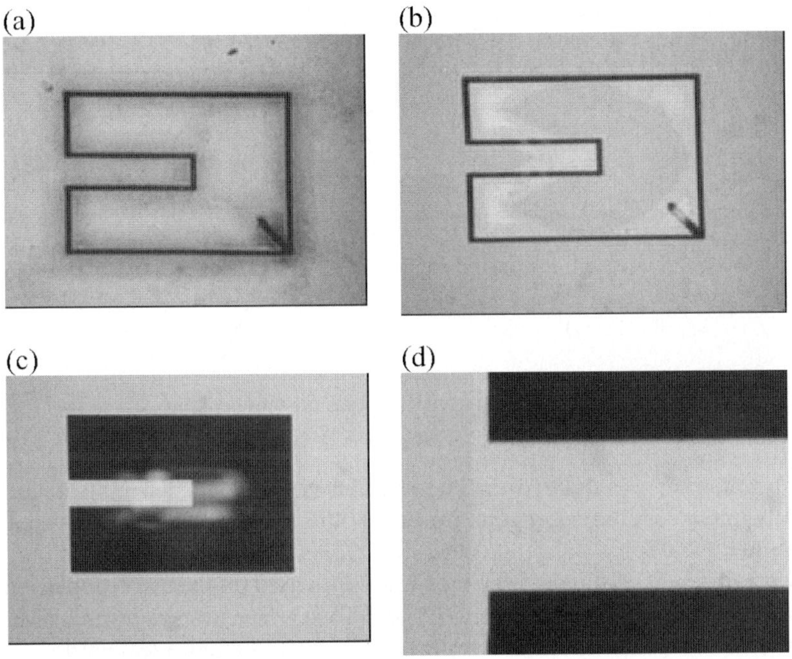

Fig. 1. Fabrication process of a typical cantilever test beam: (a) after laser ablation of beam outline; (b) after one hour of anisotropic wet etching in potassium hydroxide solution; (c) after four hours wet etching; (d) enlarged view showing root undercut.

using focussed laser ablation (Figure 1). The beam structures are released by bulk micromachining of the silicon substrate, exposed by the pattern cut. An anisotropic potassium hydroxide (KOH) wet etch is used (20% solution of KOH and deionised water at 80°C). The silicon nitride structure forms an ideal mask material, as its etch rate is considerably lower than that of the silicon. Upon etching, the beams are put into ethanol before drying, to minimise the possibility of stiction or breaking due to surface tension of the solution.

As the silicon (111) planes provide a natural etch stop, the substrate will etch into a pit bounded by these planes and extremities of the pattern cut into the nitride. Provided the beam pattern is accurately aligned to the (110) planes, the beam will be released with minimum undercut of the cantilever root. Undercut to some extent is unavoidable in practise. If the cantilever root lay in a [110] direction exactly, the etch rate of (111) planes is not zero and hence a small undercut would be experienced. Larger structures require a longer etch time, increasing this effect. More significant is the effect of misalignment with the (110) planes due to imperfect pattern alignment or the tolerance expected in the crystal orientation of the wafer.

When the beam is fully undercut, residual stresses in the film will become apparent. In-plane compressive/tensile residual stresses result in a lengthening/shortening

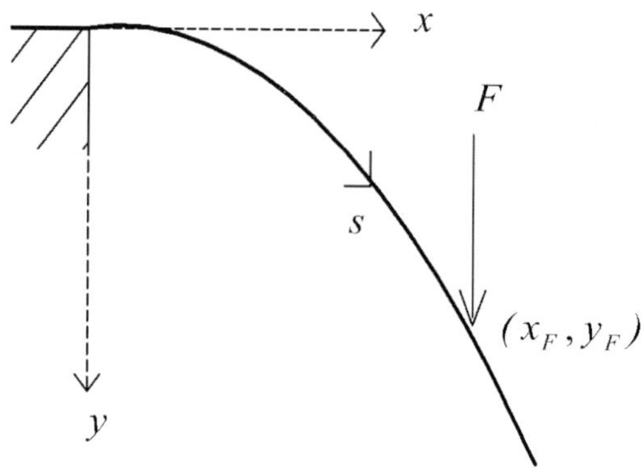

Fig. 2. Notation used to represent beam bending.

of the beam, whereas a through-thickness residual stress gradient leads to beam bending. Any initial curvature of the cantilever needs to be taken into account when looking at subsequent force-deflection data.

Three different specimens of silicon nitride are used for the extraction of Young's modulus, referred to here as TM, STS and TM/O. The silicon nitride TM is silicon rich and deposited at high temperature. It therefore has a low intrinsic stress, but is unsuitable for temperature limited processes. Silicon nitride STS is deposited at lower temperature, increasing its range of applications, but with higher intrinsic stress and reduced mechanical properties. The specimen TM/O is silicon nitride TM deposited on a thermally grown oxide layer. This tests the applicability of the Young's modulus measurement technique to bi-layer materials, which are commonly used in MEMS, for example, thermal actuators.

The film thickness is measured using scanning profilometry. Pieces of silicon nitride released by the etching process (either the redundant boundaries to patterns or squares specifically laser cut into the film for this purpose) can be collected and scanned with a surface profiler to give a thickness measurement. Silicon nitride pieces can be extracted from the etch solution using a pipette and dropped onto a glass slide, or simply caught on the wafer surface if the specimen is dried carefully.

3 Young's Modulus Extraction Method

Scanning a cantilever from a location on the bank across the root to the tip will produce a set of deflection readings for a given stylus force. The conventions used here are to denote horizontal distances measured relative to the start of scan with X and distances measured relative to the beam root with x, as shown in Figure 2. The distance measured along the beam length is given by s. Displacements relative to the horizontal (the bank) are denoted y.

With reference to Figure 2, the curvature of an initially straight cantilever beam with ideal root conditions subjected to a force some distance from the root is given by:

$$\kappa = \frac{\frac{d^2y}{ds^2}}{\sqrt{1-\left(\frac{dy}{ds}\right)^2}} = \frac{\frac{d^2y}{dx^2}}{\left[1+\left(\frac{dy}{dx}\right)^2\right]^{3/2}} = \frac{F(x_F - x)}{EI}, \tag{1}$$

where E is the Young's modulus and I the second moment of area of the cantilever. This can be solved iteratively by assuming a form of $y(x)$ and integrating d^2y/dx^2 twice (taking ideal boundary conditions of zero displacement and rotation at the root) to give a new estimate. A polynomial expression for the deflection at the point of application of the load can be found to the required number of terms:

$$y_F = \left(\frac{F}{EI}\right)\frac{x_F^3}{3} + \left(\frac{F}{EI}\right)^3\frac{x_F^7}{35} + \left(\frac{F}{EI}\right)^5\frac{x_F^{11}}{231} + \left(\frac{F}{EI}\right)^7\frac{x_F^{15}}{1287} + \cdots. \tag{2}$$

For the case where deflections are small, such that higher order terms can be neglected, this expression reduces to the Euler expression (where the horizontal coordinate x_F and the beam length s_F are interchangeable):

$$y_{SD} = \left(\frac{F}{EI}\right)\frac{x_F^3}{3}. \tag{3}$$

Plotting the deflection resulting from a range of applied forces at a given horizontal position results in a straight line with gradient $\gamma = 3EI/x_F^3$, so long as Equation (3) is applicable. However, the horizontal distance to the beam root (x_F) remains an unknown in a practical scanning profiler test, due to the difficulty in locating the start of scan relative to the beam root. This can be solved by considering the gradients (γ_1 and γ_2) of the curves produced at two positions along the beam whose separation is known (x_{F1} and $x_{F2} = x_{F1} + \Delta x$). The required distance can be found by taking the following ratio:

$$1 = \frac{\gamma_1 x_{F1}^3}{\gamma_2 x_{F2}^3} = \frac{\gamma_1}{\gamma_2}\left(1 + \frac{\Delta x}{x_{F1}}\right)^{-3}. \tag{4}$$

This can be repeated for any number of horizontal positions to produce a set of measurements of Young's modulus.

4 Finite Element Analysis

The analysis described in Section 2 depends on a number of key assumptions: (a) beam bending is small and unaffected by higher order deflection terms; (b) bending is entirely two dimensional; (c) root conditions are ideal (i.e., no deflection or rotation permitted at the root). For a given practical beam deflection experiment these need not be the case. Large, non-linear deflections are achievable, depending on the cantilever geometry and the load provided by the surface profiler. Ashwell [8] has shown that anticlastic curvature, a curvature about an axis perpendicular to the main curvature, can cause stiffening in beams and plates (note that when the stylus is applying a load near the root, it is in effect bending a plate). A profiler stylus also has a tip which applies a point load for a practical beam width, which may induce three-dimensional effects. Imperfect root conditions are also a likely consequence of the fabrication issues as discussed in Section 1.

Finite element (FE) analysis, using the commercially available finite element package ABAQUS, is employed to investigate these effects under controllable conditions. Eight-node quadratic shell elements are used in a non-linear analysis to model rectangular cantilever beams. A mesh sensitivity investigation has been carried out to determine the adequate element size to capture the beam bending under the range of loads considered. Beam thickness of 2 μm and widths of 10 μm, 50 μm and 100 μm are used, as realistic values for scanning profiler test specimens. Elastic materials properties are used, with Young's modulus $E = 250$ GPa and Poisson's ratio $\nu = 0.3$, reasonable for a material such as silicon nitride. Loads between 4.91 μN and 147 μN (typical range for a scanning profiler) are applied at 50 μm intervals along the 500 μm beam length. Where required, the horizontal position of the load (x_F) in the deformed configuration is calculated from the known beam length using the reasonable approximation:

$$x_F \approx \sqrt{s_F^2 - y_F^2}. \tag{5}$$

4.1 The Effect of Point Loading

To investigate any three-dimensional effects induced by stylus loading, which is not captured in the previous bending model, the force is applied in the FE model at both a point and a line across the cantilever width. The results, given in Figure 3, show that for small values of s_F/w (ratio of beam length to its width) concentrating the load at a point results in a larger deflection. This is consistent with local 'indentation' occurring when the beam has a high stiffness (due to plate bending). This effect is minimal as long as $s_F/w \approx \sqrt{x_F^2 + y_F^2}/w > 1$.

4.2 Large Deflection and Stiffening Effects

Examination of the ratio of the deflection expression valid for large deflections, Equation (2), and the small deflection approximation, Equation (3), shows that for this approximation to be valid it is important to keep the parameter Fx_F^2/EI small:

$$\frac{y_F}{y_{SD}} = 1 + \frac{3}{35}\left(\frac{Fx_F^2}{EI}\right)^2 + \frac{3}{231}\left(\frac{Fx_F^2}{EI}\right)^4 + \frac{3}{1287}\left(\frac{Fx_F^2}{EI}\right)^6 + \cdots \tag{6}$$

Counteracting the effect of large deflections is the stiffening effect introduced by true three-dimensional plate bending. Aswell [8] found that the effective bending stiffness of a beam, taking into account anticlastic effects, is dependent on a non-dimensional

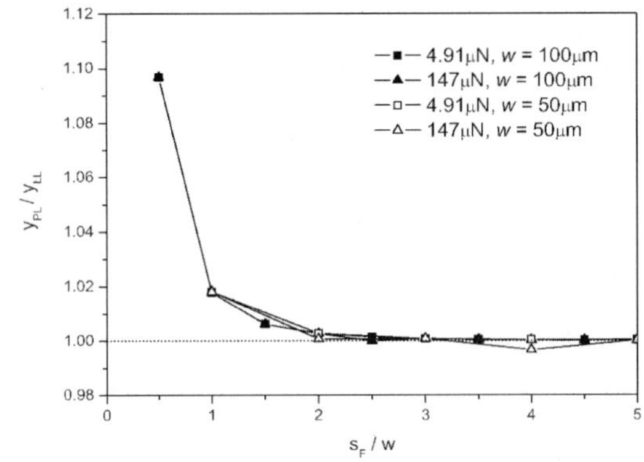

Fig. 3. Finite element calculations for beam deflection utilising a point load (y_{PL}) and a line load (y_{LL}). A practical range of loads and beam widths are shown.

parameter, defined here as ψ, which includes the width (w), thickness (t) and main curvature (κ):

$$\psi = \frac{w^2\kappa}{t} = \frac{12wF(x_F - x)}{Et^4}. \tag{7}$$

For large values of ψ, the effective elastic modulus tends to the plate modulus E_p. For $\psi = 0$, the elastic modulus is simply the Young's modulus E. It should be noted that the analysis carried out by Ashwell assumes that the anticlastic curvature is small (effectively employing small deflection assumptions for the transverse curvature). This assumption can be valid even when the main curvature is large, due to the relationship $\kappa' = \nu\kappa$, where κ' is the anticlastic curvature.

Replacing Young's modulus E with the plate modulus $E(1 - \nu^2)$ in (6) leads to:

$$\frac{y_F}{y_{SD}} = (1 - \nu^2)\left(1 + \frac{3(1-\nu^2)^2}{35}\left(\frac{Fx_F^2}{EI}\right)^2 + \frac{3(1-\nu^2)^4}{231}\left(\frac{Fx_F^2}{EI}\right)^4 + \frac{3(1-\nu^2)^6}{1287}\left(\frac{Fx_F^2}{EI}\right)^6 + \cdots\right). \tag{8}$$

Finite element results along with Equations (6) and (8) are plotted in Figure 4. The first point to note is the tendency toward 'plate-like' behaviour or 'beam-like' behaviour, i.e., whether the deflection is governed by the plate modulus or Young's modulus. At all stylus forces, the results tend towards plate behaviour at lower values of Fx_F^2/EI, corresponding to loading the beam near the root. For higher stylus forces, the higher curvatures induced increase the anticlastic stiffening effect, so the plate-like behaviour extends to higher values of Fx_F^2/EI. The second point to note is the feasibility of the small deflection approximation (3). Given a choice of load such that 'beam-like' behaviour dominates, the selection of Fx_F^2/EI values less than around 0.6 should ensure agreement with small deflection theory to within 5%.

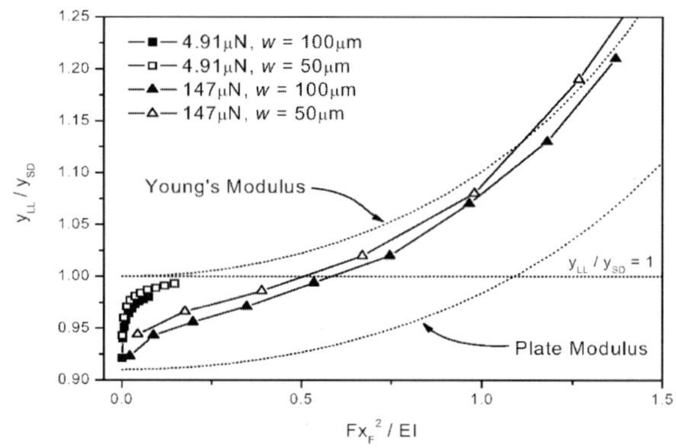

Fig. 4. Finite element calculations for beam deflection (y_{LL}, line load used) normalised with small deflection approximation (y_{SD}). Shown are results for the maximum and minimum typical stylus forces for two beam widths. The analytical curves represent four term expressions using Young's modulus, Equation (6), and plate modulus, Equation (8).

4.3 The Effect of Root Compliance

When deflections are small, the effect of a compliant root can be modelled by assuming the root of the cantilever beam is attached not to a rigid bank but to a second cantilever beam of stiffness $E_R I_R$ and length L_R. This permits the root of the main cantilever to have a finite deflection and rotation. By considering continuity of shear forces and bending moments where the beams join and the principal of superposition, the tip deflection of the main cantilever beam becomes:

$$y_{CR} = \frac{F x_F^3}{3EI} + \frac{F}{E_R I_R}\left(x_F^2 L_R + x_F L_R^2 + \frac{1}{3} L_R^3\right). \tag{9}$$

For a real compliant region, i.e., the undercut at the cantilever root, the material and thickness are the same but the region has a greater width than the main cantilever. If this is modelled by assuming $E_R I_R = E w_R t^3/12$ (where E and t are separately the Young's modulus and thickness of the main beam), the ratio of the deflections for perfect and compliant root conditions is given by:

$$\frac{y_{CR}}{y_F} = 1 + \frac{w}{w_R}\left(3 + \frac{L_R}{x_F} + 3\left(\frac{L_R}{x_F}\right)^2 + \left(\frac{L_R}{x_F}\right)^3\right). \tag{10}$$

This indicates that a compliant root should provide an increase in displacement at a given x_F that is not dependent on the force. Therefore the gradient of a plot of force against deflection for a given horizontal position should be unaffected, the curve being translated parallel by an offset given by (10).

A practical compliant root, caused by the undercut resulting from the wet etch process (Figure 1) will deflect in a more complicated way than a two stage cantilever beam, making it more difficult to isolate the parameters w_R and L_R. One approach

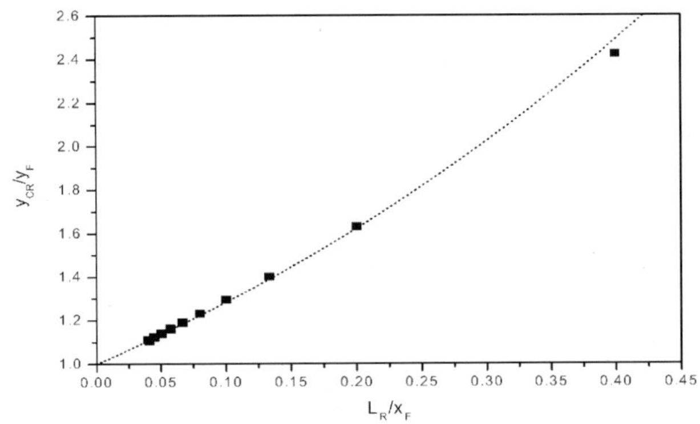

Fig. 5. Finite element calculations for beam deflection for a compliant root model (y_{CR}) and an ideal root model (y_F). The best fit curve for Equation (10) is displayed.

is to take L_R as the undercut distance and w_R as some effective width representative of the undercut bank suffering from bending.

A finite element model of a compliant root has been created using a short (20 μm) wide (500 μm) rectangular plate, clamped on three sides with the cantilever root at the centre of the free edge. The results demonstrate that the model described in (10) proves a good fit to the finite element results when an effective width for the compliant region $w_R = w + 0.85L_R$ is used (Figure 5).

5 Experimental Measurement

5.1 Additional Experimental Considerations

An additional assumption which has not been addressed in Section 3 is that the cantilever beams are initially straight. For real cantilevers formed by thin-film deposition techniques, the resultant bending moment due to any gradient in through-thickness residual stress will cause the cantilever to have an initial curvature once released from the substrate.

5.2 Experimental Procedures

Step 1 – Constant force deflection curves
Each specimen beam is scanned from root to tip centrally along the cantilever beam at constant profilometer stylus force, producing a set of values of x_F and y_F. The stylus is raised before the beam tip is reached to minimise movement of specimen. Repeated scans at different stylus forces are made, using the same data sampling rate.

Step 2 – Force-deflection plots
Fixing the sampling rate permits a value for y_F to be found for a range of forces at

Table 1. Average experimentally measured Young's modulus. The average is taken for each specimen over the range 200 μm $\geq x_F \geq$ 300 μm. Where superposition is applied, the smallest force deflection plot is subtracted from subsequent plots.

Silicon Nitride Specimen	No Superposition Mean (GPa)	No Superposition Standard Dev. (GPa)	Superposition Mean (GPa)	Superposition Standard Dev. (GPa)
TM	155.6	0.6	154.9	0.8
TM/O	140.4	1.0	133.0	1.0
STS	66.5	0.4	62.3	0.4

each value for x_F. This allows a force-deflection plot to be produced for a constant value of x_F, with the gradient providing the Young's modulus [Equation (3)].

Step 3 – Locating beam root
Eight values for the gradient, approximately evenly spaced, are selected. Equation (4) is applied between every pair of the eight data points, giving 28 estimates of the beam root position. An average is then taken, and used as the root position value in subsequent calculations. The standard deviations in the experimentally obtained values of root position calculated by this method are in most cases less than 5% of the average value.

Step 4 – Calculation of Young's modulus
The previous step allows E to be plotted against the horizontal scan position x_F. Results from the finite element analyses can be used to locate a suitable range of x_F over which a representative value for Young's modulus can be expected. The extracted values can be averaged over this range.

5.3 Experimental Results

The approach outlined in the preceding section is now applied to the three specimens of silicon nitride discussed in Section 3. Test cantilevers of width measured at 75 μm for the silicon nitride TM and TM/O and 80 μm for the STS are used. The material thickness is measured as 3.40 μm for the silicon nitride TM, 2.25 μm for the TM/O and 2.42 μm for the STS. In each case, the Young's modulus is calculated with and without the use of superposition to correct for initial curvature. Superposition is applied by subtracting the deflection measured from the lowest force scan.

Figure 6 presents the deflection curves obtained from the scans at different forces on each beam. Five loads are used for each beam, with the magnitude of the force attainable dependent on the bending stiffness of each beam. Figure 7 shows the extracted plots of Young's modulus against horizontal stylus position for the cases where superposition is not applied. Table 1 summarises the averaged values of Young's modulus for each case.

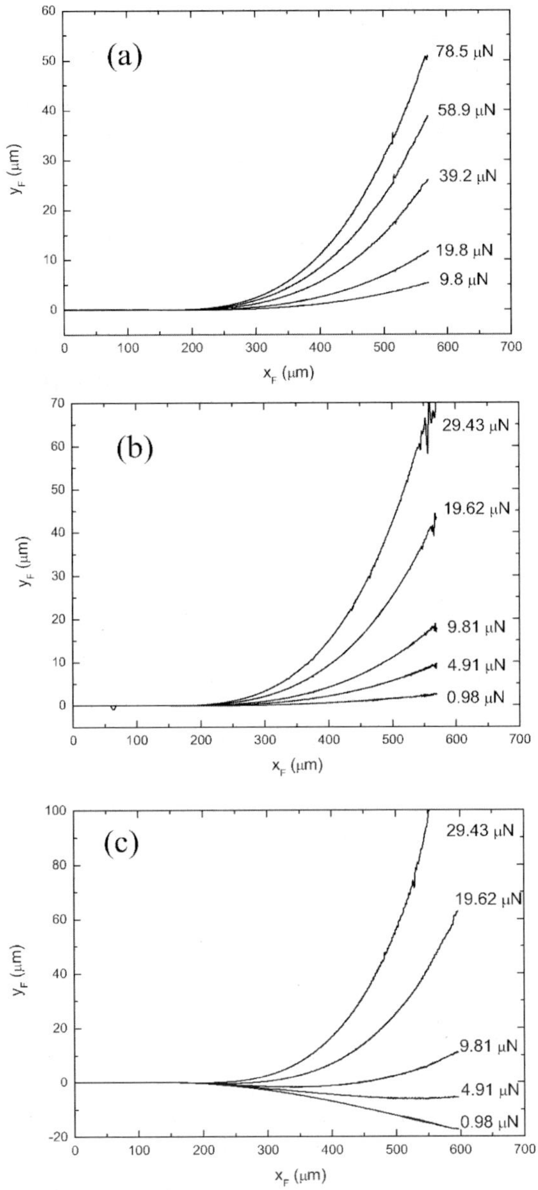

Fig. 6. Experimental deflection curves: (a) SiN TM; (b) SiN TM/Ox; (c) SiN STSMF.

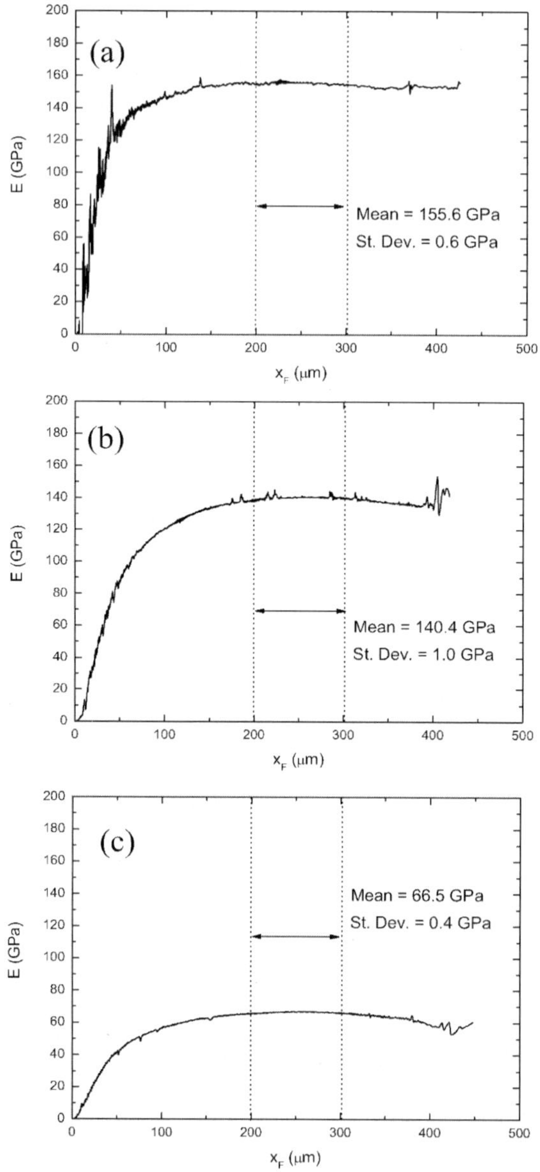

Fig. 7. Calculated Young's modulus (no superposition): (a) SiN TM; (b) SiN TM/Ox; (c) SiN STSMF.

6 Conclusions

The need for a simple and effective technique that can be applied to a wide range of materials using commonly available equipment has prompted the consideration of cantilever beam based materials characterisation. A combined analytical and experimental study has been carried out to demonstrate the applicability of this technique in selected thin film materials. The basis of this approach is the deflection of micro-cantilevers using a scanning surface profiler to extract mechanical properties. The advantages are simplicity, speed, cost and wide applicability.

Acknowledgements

We would like to thank EPSRC for supporting this work through grant 'Active Packaging for Optical and Electronic Microsystems (APOEM)'. TJL wishes to thank the National Basic Research Program (No. 2006CB601202), National 111 Project (No. B06024), and National Natural Science Foundation (Nos. 10328203 and 10572111) of China for partial financial support of this work.

References

1. D.F. Moore, Laser micromachining of thin films for optoelectronic devices and packages, *Proceedings of SPIE* **4941**, 2003.
2. P.M. Osterberg and S.D. Senturia, M-TEST: A test chip for MEMS material property measurement using electrostatically actuated test structures, *JMEMS* **6**, 1997, 107–118.
3. M. Hopcroft, MAT-Test: A new method for thin-film materials characterisation, MPhil Thesis, Cambridge University, 2002.
4. W.N. Sharpe, J. Bagdahn, K. Jackson and G. Coles, Tensile testing of MEMS materials – Recent progress, *Journal of Materials Science* **38**, 2003, 4075–4079.
5. T.E. Buchheit, S.J. Glass, J.R. Sullivan, S.S. Mani, D.A. Lavan, T.A. Friedmann and R. Janek, Micromechanical testing of MEMS materials, *Journal of Materials Science* **38**, 2003, 4081–4086.
6. A. Corigliano, B. De Masi, A. Frangi, C. Comi, A. Villa and M. Marchi, Mechanical characterisation of polysilicon through on-chip tensile tests, *JMEMS* **13**, 2004, 200–219.
7. D.J. Bell, T.J. Lu, N.A. Fleck and S.M. Spearing, MEMS actuators and sensors: Observations on their performance and selection for purpose, *J. Micromech. Microeng.* **15**, 2005, 153–164.
8. D.G. Ashwell, The pure bending of rectangular plates, *Engineering*, 51–52, 21 July 1950, and 76–78, 28 July 1950.

Part 3

Nanomechanics of Biomaterials

Bio-Inspired Mechanics of Bone-Like Hierarchical Materials

Huajian Gao*

Max Planck Institute for Metals Research, Heisenberstrasse 3, 70569 Stuttgart, Germany; E-mail: hjgao@mf.mpg.de

Abstract. Nanotechnology promises to enable mankind to design materials hierarchically via a bottom-up approach, i.e. by tailor-designing materials from atomic scale and up. However, currently we barely have any theoretical basis on how to design a hierarchical material to achieve a particular set of macroscopic properties. To demonstrate the potential of bottom-up design, we consider a model material with self-similar hierarchical structures mimicking the elementary structure of bone. The resulting "fractal bone" exhibits a similar structure at each hierarchical level consisting of staggered hard plates aligned in a soft matrix. Simple analytical models are adopted to evaluate the stiffnesses, strengths and fracture energies of all hierarchical levels. The hierarchical structural sizes are determined based on the principle of flaw tolerance. It is shown that the bottom-up designed fractal bone can tolerate crack-like flaws from nanoscale all the way to macroscopic scales without size limit.

Key words: fracture, flaw tolerance, biomechanics, bio-inspired mechanics, bone, hierarchical materials.

1 Introduction

We have been studying mechanical properties of the elementary structures of hard biological tissues such as bone, tooth, and shells [1, 2]. These materials exhibit hierarchical structures over many length scales. While sea shells exhibit 2 to 3 levels of lamellar structure [3–6], bone has 7 levels of hierarchy [7–10]. Although the higher level structures of biological materials show significant complexity and variations, nature exhibits a convergent evolution towards a generic nanostructure with slender mineral crystals embedded in a soft protein matrix [1–11]. The nanostructure of tooth enamel consists of needle-like (15–20 nm thick and 1000 nm long) crystals embedded in a small volume fraction of protein [12, 13]. The nanostructure of dentin and bone is made of plate-like (2–4 nm thick and up to 100 nm long) crystals embedded

* Present address: Division of Engineering, Brown University, Providence, RI 20912, U.S.A.; E-mail: Huajian_Gao@brown.edu.

Yilong Bai et al. (eds), IUTAM Symposium on Mechanical Behavior and Micro-Mechanics of Nanostructured Materials, 87–94.

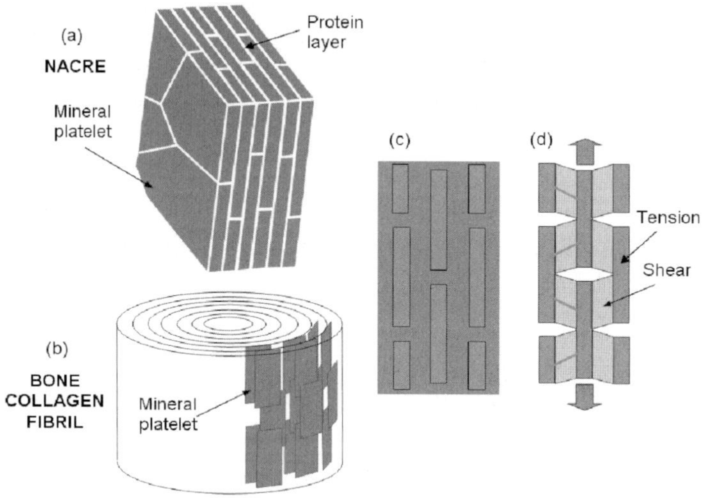

Fig. 1. The elementary structures of nacre and bone. (a) Nacre consists of plate-like mineral crystals 200–500 nm in thickness and a few micrometers in length with a small amount of soft matrix in between. (b) The nanostructure of bone consists of plate-like mineral crystals 2–4 nm in thickness and up to 100 nm in length embedded in a collagen- rich protein matrix. (c) The generic biological nanostructure represents a convergent design of natural evolution. (d) The primary load bearing zones of biological nanostructure show mineral crystals primarily in tension and protein primarily in shear. At the nanostructure level, the load is mainly uniaxial and is transferred via the route of a tension-shear chain as illustrated.

in a collagen-rich matrix [10, 14, 15]. Nacre consists of plate-like crystals (200–500 nm thick and a few micrometers long) with a very small amount of soft matrix [3, 4, 6, 17]. Figure 1 shows the common design of elementary structures of nacre and bone. Similar structures can be found also in the cell walls of wood made of hard cellulose fibrils in a soft hemicellulose-lignin matrix [18, 19]. The toughness of biological materials has been attributed to their multi-level structures [4, 5, 16, 20], the energy dissipation properties of protein [21], the surface asperities of mineral plates [17] and the reduction of stress concentration at a crack tip [22].

2 Flaw Tolerance Criterion

The self-sensing, self-adapting and self-repairing capabilities of bone require constant removal and replacement of old and damaged materials with fresh and healthy materials while an animal is conducting its normal activities, indicating that bone must tolerate crack-like flaws of many size scales. The state of material in which failure occurs not by propagation of pre-existing cracks but by uniform rupture at the limiting strength is referred to as flaw tolerance [1, 2, 23, 24], a concept which has been used to explain the nanostructure designs of bone

[1, 2] and gecko [23, 24]. Flaw tolerance in biological materials can be related to the well known phenomena and concepts of notch insensitivity, fracture size effects and large scale yielding or bridging in fracture mechanics [25–27]. The condition to achieve flaw tolerance can be expressed as [28]

$$\Lambda_{ft} = \Gamma E/(S^2 h) = 1, \tag{1}$$

where Λ_{ft} is called the flaw tolerance number, Γ is the fracture energy, E is the Young's modulus, S is the limiting strength and h is the characteristic size of the material.

3 The Fractal Bone Model

In order to demonstrate the potential of bottom-up design, we consider a hierarchical material with self-similar structures mimicking the nanostructure of bone at all levels of structural hierarchy (Figure 2). The resulting "fractal bone" will be designed to achieve flaw tolerance at all hierarchical levels under uniaxial tension. The fractal bone contains N levels of staggered hard plates in soft matrices that absorb and dissipate fracture energies. The structure design starts at the lowest level. Properties at the next level of hierarchy are determined from the current level, and the flaw tolerance condition is used to determine structural sizes of all hierarchical levels.

At the n-th hierarchical level, the geometrical parameters are the thickness h_n and length ℓ_n of the hard plates with aspect ratio $\rho_n = \ell_n/h_n \gg 1$. The volume fraction of the hard phase is denoted as φ_n. The overall volume fraction of mineral

$$\Phi = \varphi_1 \varphi_2 \ldots \varphi_N = \prod_1^N \varphi_n \tag{2}$$

is assumed to be a material parameter. The mechanical properties of level n are Young's modulus E_n, strength S and fracture energy Γ_n. In the bottom-up approach, the structures are designed in the following sequence (Figure 2)

$$h_1 \to h_2 \to \cdots \to h_N = H \tag{3}$$

by repetitive applications of the flaw tolerance condition in Equation (1).

For the fractal bone, we keep the aspect ratio and volume fraction of the hard phase invariant over all hierarchical levels, $\rho_n = \rho$, $\varphi_n = \varphi$, so that

$$\varphi_n = \varphi = \Phi^{1/N}. \tag{4}$$

In the following, the hierarchical properties of the fractal bone are determined iteratively using a bottom-up approach. More details can be found in a forthcoming paper [29].

The hierarchical stiffnesses of the fractal bone are determined iteratively using the bottom-up equation

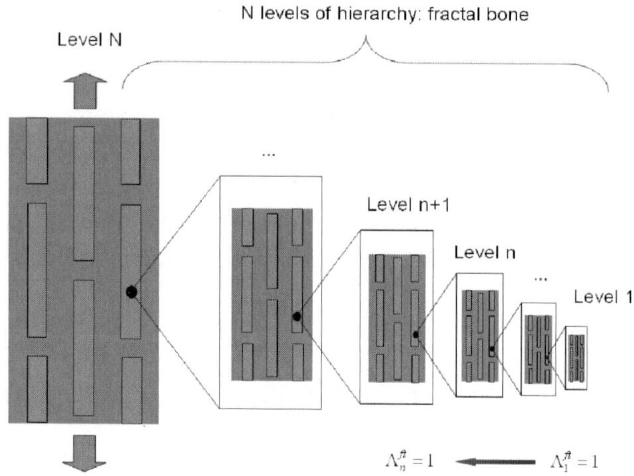

Fig. 2. The N-level hierarchical structures of the fractal bone. Every level of structure is similar to the elementary structure of bone and nacre, with staggered hard plates aligned in a soft matrix. A hard plate of the $(n+1)$-th level is made of staggered hard plates in a soft matrix at the n-th level. The principle of flaw tolerance is used to determine the structural sizes of all levels using a bottom up approach.

$$\frac{1}{E_{n+1}} = \frac{4(1-\Phi^{1/N})}{G_p \Phi^{2/N} \rho^2} + \frac{1}{\Phi^{1/N} E_n}, \qquad E_0 = E_m, \tag{5}$$

where E_m is the Young's modulus of mineral and G_p is the shear modulus of the soft phase assumed to be constant over all structural levels $n = 1, 2, \ldots, N$. For a large number of hierarchical levels, Equation (5) approaches the Voigt upper bound $E_{n+1} \approx \Phi^{1/N} E_n$ with solution $E_n \approx \Phi^{n/N} E_m$.

The hierarchical strengths are determined by the bottom-up equation

$$S_{n+1} = \Phi^{1/N} S_n/2, \qquad S_0 = \sigma_{th}, \tag{6}$$

with solution $S_n = \Phi^{n/N} \sigma_{th}/2^n$. The strength of the soft phase is chosen to be $S_n^p = S_n/\rho$ to ensure that both phases fail at about the same time.

The hierarchical fracture energies of the fractal bone are evaluated assuming that the hard plates are pulled out of the soft phase during fracture (Figure 3),

$$\Gamma_{n+1} = (1-\Phi^{1/N}) h_n S_n \Theta_p, \qquad \Gamma_0 = 2\gamma, \tag{7}$$

where Θ_p denotes the failure strain of the soft phase and γ is the surface energy of mineral.

The hierarchical structural sizes are determined bottom up according to the flaw tolerance condition as

$$\frac{h_{n+1}}{h_n} = \frac{4(1-\Phi^{1/N})\Theta_p E_{n+1}}{S_n \Phi^{2/N}}, \qquad h_0 = \frac{2\gamma E_m}{\sigma_{th}^2}. \tag{8}$$

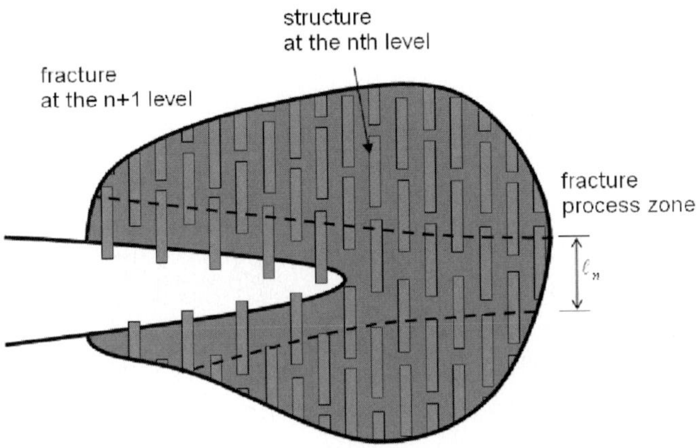

Fig. 3. Fracture at the $(n + 1)$-th level of structural hierarchy is assumed to occur by having the hard plates pulled out of the soft matrix. The hard plates remain intact as the soft phase undergoes large shear deformation until rupture. The width of the fracture process zone is assumed to be on the order of the length of the hard plates.

If we approximate E_{n+1} by the Voigt bound $E_{n+1} \approx \Phi^{1/N} E_n$, Equation (8) can be reduced to a simplified iterative equation

$$\frac{h_{n+1}}{h_n} = \frac{2^{n+2}(1 - \Phi^{1/N})\Theta_p E_m}{\Phi^{1/N}\sigma_{th}}, \qquad h_0 = \frac{2\gamma E_m}{\sigma_{th}^2}, \tag{9}$$

with solution

$$h_n = \left[\frac{2^{(n+3)/2}(1 - \Phi^{1/N})\Theta_p E_m}{\sigma_{th}\Phi^{1/N}}\right]^n h_0. \tag{10}$$

With N levels of structural hierarchy, the flaw tolerance size of the fractal bone will reach the overall dimension

$$H = h_n = \left[\frac{2^{(N+3)/2}(1 - \Phi^{1/N})\Theta_p E_m}{\sigma_{th}}\right]^N \frac{h_0}{\Phi}. \tag{11}$$

Note that $H \to \infty$ as $N \to \infty$ regardless of the values of the material properties E_m, Θ_p, σ_{th}, Θ. Therefore, with increasing hierarchical levels, the fractal bone can tolerate crack-like flaws without size limit.

4 Results and Discussions

Figure 4 shows the calculated properties of the fractal bone as a function of the number of hierarchical levels. In the calculation, we assume typical materials properties of bone $\gamma = 1$ J/m^2, $\Phi = 0.45$, $E_m = 100$ GPa, $\sigma_{th} = E_m/30$ and

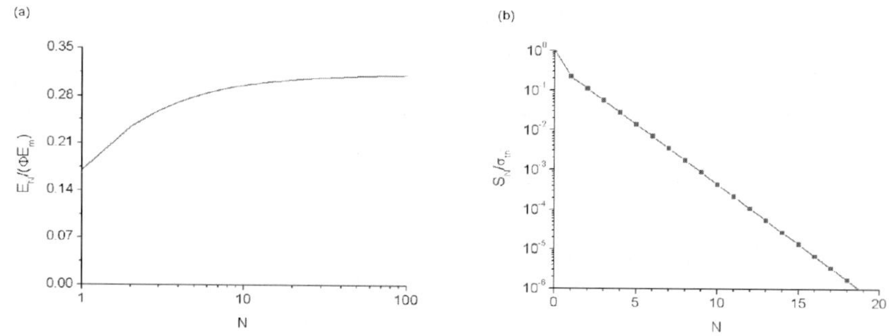

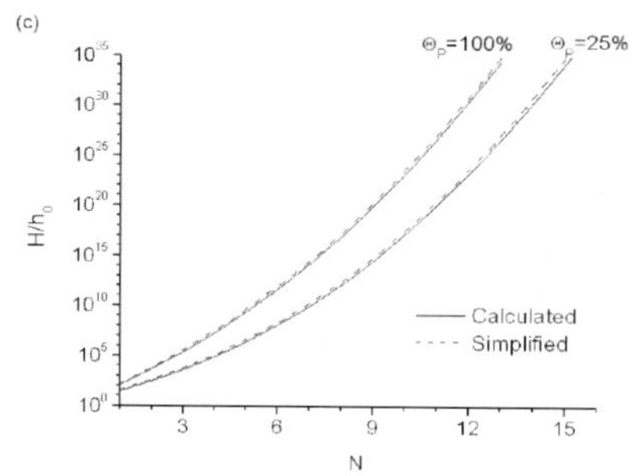

(d)

N	$\Theta_p = 25\%$	$\Theta_p = 100\%$
1	0.27×10^2	0.10×10^3
2	0.30×10^3	0.49×10^4
3	0.40×10^4	0.26×10^6
4	0.76×10^5	0.19×10^8
5	0.22×10^7	0.22×10^{10}
6	0.10×10^9	0.44×10^{12}
8	0.12×10^{13}	0.79×10^{17}
16	0.34×10^{30}	0.15×10^{49}

Fig. 4. Variation of the overall properties of the fractal bone with the number of hierarchical levels. (a) The overall stiffness normalized by the Voigt upper bound shows a moderate increase with increasing hierarchical levels; (b) The overall strength normalized by the theoretical strength of mineral drops by roughly a factor of 2 with each added level of hierarchy. (c) The flaw tolerance size of the fractal bone normalized by the thickness of mineral platelets increases exponentially with structural hierarchy. The solid lines are calculated from Equation (8) and the dashed lines correspond to the simplified solution of Equation (11). (d) Tabulated flaw tolerance sizes of the fractal bone approaching astronomical values as the number of hierarchical levels increases.

$E_m = \mu_p \rho^2 = 1000\mu_p$. We consider two estimates $\Theta_p = 25\%$ and $\Theta_p = 100\%$ for the failure strain of protein.

Figure 4a plots the overall stiffness of the fractal bone normalized by the Voigt upper bound of the composite. The result indicates that hierarchical design only results in a moderate increase in stiffness. After a few levels of hierarchy, the stiffness saturates at about 30% of the Voigt limit. Figure 4b shows that the strength of the fractal bone drops by roughly a factor of 2 with each added level of hierarchy. Assuming the theoretical strength of mineral crystals to be on the order of 3 GPa, the result plotted in Figure 4b indicates a strength reduction by about two orders of magnitude with 6 levels of hierarchy.

On the other hand, the hierarchical structures of the fractal bone exhibit an amazing effect on the toughness of the composite. Figures 4c and 4d plot and tabulate the overall size of the material under flaw tolerance design. The solid lines are calculated from Equation (8) while the dashed lines correspond to the simplified solution in Equation (11). The results show that the flaw tolerance size of the material increases exponentially with the number of hierarchical levels. Under the selected material parameters, the flaw tolerance size of individual mineral platelets is estimated to be $h_0 = 18$ nm. Depending upon the failure strain Θ_p of protein, the flaw tolerance size of the fractal bone increases to about 1 μm with only one level of hierarchy, 10–100 μm with two levels of hierarchy, 100 μm–10 mm with 3 levels of hierarchy, 1 mm–1 m with 4 levels of hierarchy, 100 mm–100 m with 5 levels of hierarchy, 10 m–10 km with 6 levels of hierarchy, and 10^2–10^6 km with 8 levels of hierarchy. With 16 levels of hierarchy, the dimension of the fractal bone reaches astronomical sizes towards the edge of universe! These calculations demonstrate the enormous potential of a bottom-up design methodology on improving the capability of materials against crack-like flaws.

An implicit assumption made in estimating the hierarchical fracture energies is that the soft phase undergoes uniform shear deformation until rupture. This requires the soft phase to be either limited in thickness or sufficiently viscous so that there is no driving force for an interfacial delamination crack to propagate along the interfaces between the hard and soft phases. In other words, the soft phases have been assumed to be nearly perfect energy absorbing materials even at large scales.

The present analysis indicates that bottom-up designed hierarchical materials may be capable of amplifying unique properties of nanostructured materials to macroscopic length scales without size limit. Here we have discussed stiffness, strength and toughness. Similar issues may exist with other mechanical properties such as adhesion, friction, hydrophobicity, corrosion, fatigue, and more generally also with optical, electrical and chemical properties of materials. It can be expected that hierarchical material design will have strong impact on material science in the future.

References

1. Gao, H., Ji, B., Jäger, I.L., Arzt, E. and Fratzl. P., Materials become insensitive to flaws at nanoscale: Lessons from nature, *Proc. Natl. Acad. Sci. USA* **100** (2003), 5597–5600.
2. Ji, B. and Gao, H., Mechanical properties of nanostructure of biological materials, *J. Mech. Phys. Solids* **52** (2004), 1963–1990.
3. Currey, J.D., Mechanical properties of mother of pearl in tension, *Proc. R. Soc. London B* **196** (1977), 443–463.
4. Menig, R., Meyers, M.H., Meyers, M.A. and Vecchio, K.S., Quasi-static and dynamic mechanical response of Haliotis rufescens (abalone) shells, *Acta Mater.* **48** (2000), 2383–2398.
5. Menig, R., Meyers, M.H., Meyers, M.A. and Vecchio, K.S. Quasi-static and dynamic mechanical response of Strombus gigas (conch) shells, *Mater. Sci. Eng. A* **297** (2001), 203–211.
6. Jackson, A.P., Vincent, J.F.V. and Turner, R.M., The mechanical design of nacre, *Proc. R. Soc. London B* **234** (1988), 415–440.
7. Currey, J.D., *The Mechanical Adaptations of Bones*, Princeton University Press, Princeton, 1984, pp. 24–37.
8. Rho, J.Y., Kuhn-Spearing, L. and Zioupos, P., Mechanical properties and the hierarchical structure of bone, *Med. Eng. Phys.* **20** (1998), 92–102.

9. Weiner, S. and Wagner, H.D., The material bone: Structure-mechanical function relations, *Ann. Rev. Mater. Res.* **28** (1998), 271–298.
10. Landis, W.J., The strength of a calcified tissue depends in part on the molecular structure and organization of its constituent mineral crystals in their organic matrix, *Bone* **16** (1995), 533–544.
11. Jäger, I. and Fratzl, P., Mineralized collagen Mbrils: A mechanical model with a staggered arrangement of mineral particles, *Biophys. J.* **79** (2000), 1737–1746.
12. Warshawsky, H., Organization of crystals in enamel, *Anat. Rec.* **224** (1989), 242–262.
13. Tesch, W., Eidelman, N., Roschger, P., Goldenberg, F., Klaushofer, K. and Fratzl, P., Graded microstructure and mechanical properties of human crown dentin, *Calcif. Tissue Int.* **69** (2001), 147–157.
14. Landis, W.J. and Hodgens, K.J., Mineralization of collagen may occur on Mbril surfaces: Evidence from conventional and high voltage electron microscopy and three dimensional imaging, *J. Struct. Biol.* **117** (1996), 24–35.
15. Roschger, P., Grabner, B.M., Rinnerthaler, S., Tesch, W., Kneissel, M., Berzlanovich, A., Klaushofer, K. and Fratzl, P., Structural development of the mineralized tissue in the human L4 vertebral body, *J. Struct. Biol.* **136** (2001), 126–136.
16. Kamat, S., Su, X., Ballarini, R. and Heuer, A.H., Structural basis for the fracture toughness of the shell of the conch Strombus gigas, *Nature* **405** (2000), 1036–1040.
17. Wang, R.Z., Suo, Z., Evans, A.G., Yao, N. and Aksay, I.A., Deformation mechanisms in nacre, *J. Mater. Res.* **16** (2001), 2485–2493.
18. Brett, C. and Waldron, K., *Physiology and Biochemistry of Plant Cell Walls*, Chapman & Hall, London, 1981.
19. Fengel, D. and Wegener, G., *Wood Chemistry, Ultrastructure, Reaction*, Walter de Gruyter, Berlin, 1984.
20. Kessler, H., Ballarini, R., Mullen, R.L., Kuhn, L.T. and Heuer, A.H., A biomimetic example of brittle toughening: (I) steady state multiple cracking, *Comp. Mater. Sci.* **5** (1996), 157–166.
21. Smith, B.L., Schaeffer, T.E., Viani, M., Thompson, J.B., Frederick, N.A., Kindt, J., Belcher, A., Stucky, G.D., Morse, D.E. and Hansma, P.K., Molecular mechanistic origin of the toughness of natural adhesive, fibres and composites, *Nature* **399** (1999), 761–763.
22. Okumura, K. and de Gennes, P.-G., Why is nacre strong? Elastic theory and fracture mechanics for biocomposites with stratified structures, *Eur. Phys. J. E* **4** (2001), 121–127.
23. Gao, H. and Yao, H., Shape insensitive optimal adhesion of nanoscale fibrillar structures, *Proc. Nat. Acad. Sci. USA* **101** (2004), 7851–7856.
24. Gao, H., Wang, X., Yao, H., Gorb, S. and Arzt, E., Mechanics of hierarchical adhesion structure of gecko, *Mech. Mat.* **37** (2005), 275–285.
25. Bazant, Z.P. and Planas, J., *Fracture and Size Effect in Concrete and Other Quasibrittle Materials*, CRC Press, Boca Raton, FL, 1998.
26. Bao, G. and Suo, Z., Remarks on crack-bridging concepts, *Appl. Mech. Rev.* **45** (1992), 355–366.
27. Cox, B.N. and Marshall, D.B., Concepts for bridged cracks in fracture and fatigue, *Acta Metall. Mat.* **42** (1994), 341–363.
28. Gao, H. and Chen, S., Flaw tolerance in a thin strip under tension, *J. Appl. Mech.*, in press.
29. Gao, H., Applications of fracture mechanics concepts to hierarchical biomechanics of bone-like materials, manuscript in preparation.

Force Unfolding Single RNAs: From Equilibrium to Far-From Equilibrium

Fei Liu[1], Huan Tong[1] and Zhong-Can Ou-Yang[1,2,*]

[1] *Center for Advanced Study, Tsinghua University, Beijing 100084, China*
[2] *Institute of Theoretical Physics, The Chinese Academy of Sciences, P.O. Box 2735, Beijing 100080, China; *E-mail: oy@itp.ac.cn*

Abstract. We summarize our recent simulation progress of micromanipulation experiments on RNAs. We mainly consults with two important small RNAs unfolding experiments carried out by Bustamante group. Our results show that, in contrast to protein cases, using the single polymer elastic theory and the well-known RNA secondary structure free energy knowledge, we can successively simulate various behaviors of force unfolding RNAs under different experimental setups from equilibrium to far-from equilibrium.

Key words: continuous time Monte Carlo method, Jarzynski's equality, RNA, single-molecule manipulation.

1 Introduction

Ribonucleic Acid (RNA) is now known to be involved in many biological processes, such as carriers of genetic information (messenger RNAs), simple adapters of amino acids (transfer RNAs), enzymes catalyzing the reactions in protein synthesis, cleavage and synthesis of phosphodiester bonds [1], and regulators of gene expression [2]. These diverse and specific biological functions of RNA are guided by their folding properties. Prediction or measurements of RNA folding and folding dynamics therefore is one of central problems in biological physics.

Single-molecule manipulation technique developed in the past decade provides a fresh and promising way in resolving the RNA folding problem [3–7]. As a concrete example, an optical tweezer setup is sketched in Figure 1. The advent of the single-molecule experiments addresses a challenging issue for theorists: whether or how can we apply the known secondary structural RNA knowledge to explain or predict the phenomena observed in the single-molecule experiments? Many theoretical efforts have been devoted to understanding RNA unfolding behaviors under mechanical external forces [8–11, 12]. However these theories or models are too simple to be applied in experiments; useful free energy data about RNA secondary structure obtained before were often neglected. Moreover, they just studied equilibrium cases, while intriguing nonequilibrium phenomena were beyond their scopes. Simulation

Yilong Bai et al. (eds), IUTAM Symposium on Mechanical Behavior and Micro-Mechanics of Nanostructured Materials, 95–104.

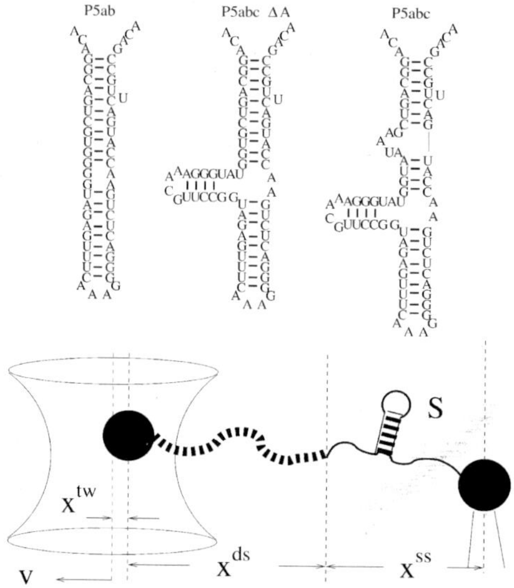

Fig. 1. Sketch of an optical tweezer setup and the native states of RNA molecules, P5ab, p5abcΔA, and P5abc studied in the work. RNA molecules are attached between the two beads (larger black points) with a RNA:DNA hybrid handle (the black dash curves). The center of the light trap is moved with velocity v. Here $z(t) = x^{tw} + x^{ds} + x^{ss}$ is the distance at time t between the centers of the light trap and the bead held by the micropipette. In practice, the RNA is attached between the two beads with two hybrid handles. To simplify the simulation method, we here only consider one handle; it should not drastically change our discussion.

method should be a good choice to overcome these shortcomings. But we noted that the simulations for RNAs are few [3]. To fill this gap, we developed a stochastic kinetic method to investigate the intriguing issue [13, 14]. Here we summarize our previous efforts about it.

2 Model and Method

Figure 1 is the sketch of a mixed ensemble. The position of the center of the light trap is moved according to a time-dependent relationship $z(t) = z_0 + vt$, where z_0 is the offset at time $t = 0$. We denote the system in i-state at time t by three independent quantities, x^{ss}, x^{ds}, and the RNA secondary structure S, i.e., $(S_i, x_i^{ds}, x_i^{ss})_t$. Hence, the unfolding of the single RNA proceeds in a space $\mathrm{S}(S) \times R^{ds} \times R^{ss}$, where $\mathrm{S}(S)$ is the set of all secondary structures of a given RNA sequence S, $R^{ds} = (0, l_{ds})$ and $R^{ss} = (0, l_{ss})$, and l_{ds} and l_{ss} are the contour lengths of the handle and the RNA molecule, respectively. A move set is defined to specify whether two conformations are accessible from each other by an elementary "move" [14],

$$(S_i, x_i^{ds}, x_i^{ss})(t) \rightarrow (S_j, x_i^{ds}, x_i^{ss})(t'), \quad i \neq j,$$

$$(S_i, x_i^{ds}, x_i^{ss})(t) \rightarrow (S_j, x_i^{ds} \mp \delta, x_i^{ss} \pm \delta)(t'),$$
$$(S_i, x_i^{ds}, x_i^{ss})(t) \rightarrow (S_j, x_i^{ds} \pm \delta, x_i^{ss})(t'). \quad (1)$$

The first kind of the moves is the removal or insertion of single base pairs [15] while fixing the extensions x^{ss} and x^{ds}. The other two kinds are to respectively move the positions of the end of the handle and the end of single-stranded RNA with a small displacement δ (1 Å here) while the secondary structure is fixed simultaneously.

Given the system state i at time t, the system energy can be written as

$$E_i(t) = \Delta G_i^0 + u^{tw}(x_i^{tw}) + W^{ds}(x_i^{ds}) + W^{ss}(x_i^{ss}, n_i), \quad (2)$$

where ΔG_i^0 is the free energy obtained from folding the RNA sequence into the secondary structure S_i, and the last three terms are the elastic energies of the optical trap, the handle, and the single-stranded part of the RNA [14], respectively. The light trap here is simply assumed to be a harmonic potential with spring constant k_{tw}. Therefore the loading rate here is $r = k_{tw}v$.

In the real experiments, constant force can be imposed on RNA molecules with feedback-stabilized optical tweezers capable of maintaining a preset force by moving the beads closer or further apart (the constant force ensemble). Therefore the energy of the light trap in Equation (1) is simply replaced by $-f(x_i^{ds} + x_i^{ss})$.

Given the move sets and the unfolding conformational spaces, the RNA unfolding for the two ensembles can be modeled as Markov processes in their respective spaces. We make use of continuous time Monte Carlo algorithm (CTMC) [16, 17] to simulate the processes. To realize simulation, transition probabilities have to be defined. We assume that the transition probabilities satisfy the symmetric rule [18]

$$k_{ij}^t = \tau_0^{-1} \exp[-(E_j(t) - E_i(t))/2\beta], \quad (3)$$

where τ_0 scales the time axis of the unfolding process from the experimental measurements. Apparently, the transition probabilities satisfy the detailed balance condition locally in time [19].

If the moving velocity of the light trap vanishes, an exact partition function method can calculate the molecular average extension and the average force under the given distance z [10]. Different from the experimental measurements of the free energy with slow pulling velocity (quasi-equilibrium process) [6], we obtain the equilibrium information by this exact method.

We simulate and calculate the force unfolding single RNAs at the experimental temperature $T = 298$ K. The elastic parameters used are from single molecule experiment: $p_{ds} = 53$ nm, $l_{ds} = 320$ nm, $b_{ss} = 0.56$ nm, Kuhn length of single-stranded part of RNAs $K_{ss} = 1.5$ nm, and $k_{tw} = 0.2$ pN/nm [5]. The free energy parameters for the RNA secondary structures are from the Vienna package 1.4 [20].

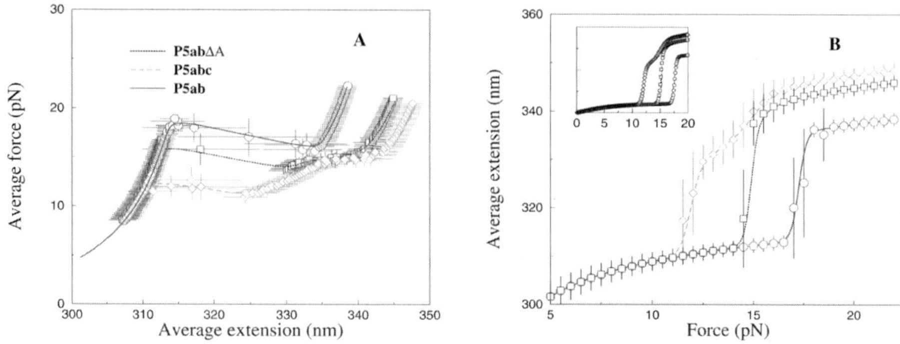

Fig. 2. Comparison of the exact and simulation force-extension curves in equilibrium for P5ab, P5abcΔA and P5abc on the mixed (A) and the constant force (B) ensembles. The different symbols are from the simulation methods, and the different lines are from the exact methods. Inset, force-extension curves for the same ensemble recalculated by another move set.

3 Results and Discussion

3.1 Single RNAs Thermodynamics

A comparison between our simulation in equilibrium and the prediction of the exact partition function method should be helpful in confirming the correctness of our method. Hence we first simulate the average force-extension curves of the three RNA molecules for the two ensembles with standard approach: the average physical quantity A is calculated according to $\langle A \rangle = \tau^{-1} \int_0^\tau A(t)dt$, here $\tau = 10^6$ and we let $\tau_0 = 1$ (see Figure 2). We find that these two independent calculations agree very well.

3.2 Single RNAs Kinetics

3.2.1 The Mixed Ensemble

Force-extension curves. As an example, we stretch P5ab molecule with the velocity $v = 5 \times 10^{-3}$ Å from the offset $z_0 = 350$ nm to 450 nm, and then relax it with the same velocity. One of the time trajectories is showed in Figure 3A.

Record of the force and extension at given times with a slow velocity is a more common method in the experimental equilibrium measurement. Hence we simulate the three curves with the two slow velocities 1×10^{-4} Å and 1×10^{-5} Å. We only show the data per unit times 10^5 and 10^6 (see Figures 3C and 3D). For the faster velocity, we find that, except P5ab case, the unfolding forces for the others do not equal the equilibrium values; whereas for the later, the curves of simulations consist with the exact curves. It means that the unfolding of the three molecules with 1×10^{-5} Å is or near equilibrium. We note that, although the whole extension $z(t)$ monotonically increases with time, the extensions of the molecules may still jump between two values, such as P5ab and P5abc molecules. Indeed similar phenomena were also

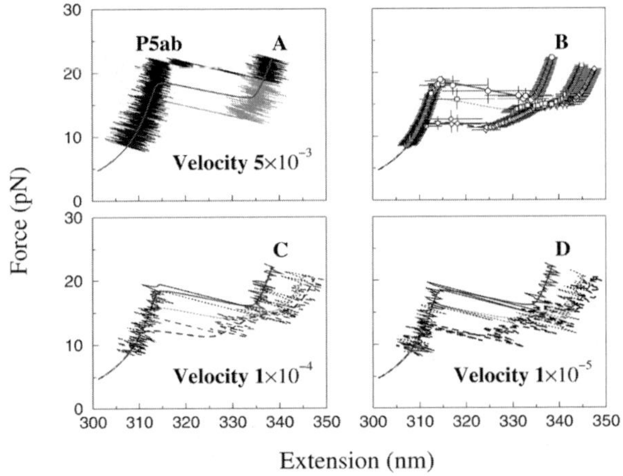

Fig. 3. (A) One of the time trajectories of unfolding and refolding for P5ab with velocity 5×10^{-3}. (B) Figure 2A is showed here again. (C) The unfolding force-extension curves recorded at unit time for the molecules with velocity 1×10^{-4} Å. (D) The force-extension curves recorded at time unit 10^6 with velocity 1×10^{-5} Å. Apparently, the unfolding and refolding trajectories are not coincident, i.e., a force-hysteresis occurs. It also indicates that the molecule is driven from thermodynamic equilibrium [5].

observed in the experiment [5]. They indicate the fluctuations of the extension and RNA structures under the force.

In the experiment [5, 6], the unfolding P5abc are near-equilibrium and far from equilibrium at the loading rates 2–5 pN/s and 34–52 pN/s, respectively (similar values for PabcΔA). And our simulations also show that the unfolding the same molecule are near-equilibrium and far from equilibrium at the velocities 1×10^{-5} Å and 1×10^{-4} Å, respectively. Let them be equal correspondingly we then can estimate the constant $\tau_0 \approx 10^{-7}$ sec. We will scale the time with this parameter below for convenience.

Free energy reconstruction. Hummer and Szabo [21] extended the remarkable Jarzynski equality [22] to extract unperturbed molecular free energy landscape $G_0(x)$ along the molecular extension x by the following expression

$$G_0(x) - G(0) = -\beta^{-1} \ln\langle \delta(x - x_t) e^{-\beta \Delta w_t} \rangle, \tag{4}$$

where $\Delta w_t = w_t - k_{tw}(x(t) - vt)^2/2$, $G(0)$ is the free energy of the whole system in equilibrium at initial time $t = 0$, and

$$w_t = k_{tw} v(vt^2/2 + z_0 t - \int_0^t x(t')dt'). \tag{5}$$

We use their result to reconstruct the free energy landscapes of P5ab and P5abc at two loading rates 20 and 40 pN/s (see Figure 4). The precisions of reconstructions

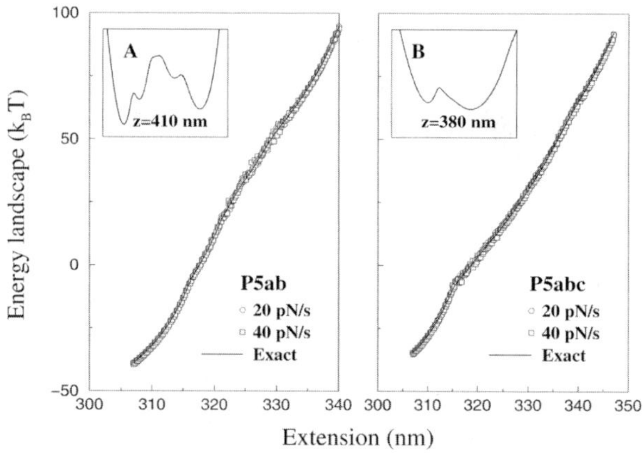

Fig. 4. Comparison of the free energy landscapes of the two molecules P5ab and P5abc reconstructed by Equation (5) and the exact landscapes calculated from the partition function method. The number of trajectories for each case is 1000. The insets are the free energy landscapes of the system composed of the molecules and the light trap potential, which are also from partition function method.

are satisfactory. We note that the landscapes are unexpectedly trivial: neither of them presents energy barrier. Ritort et al. [23] have investigated the Jarzynski's equality by modeling RNA molecules as a two-level system with an intermediate barrier. Our calculations apparently contradict their assumption. In fact, the strong unfolding-refolding cooperativity observed in the experiments [5, 6] arises from the coupling of the RNA molecules and the light trap; the addition of their potentials is a two-level system (see the respective insets in the figure). Therefore, although the two-level system is a good approximation in RNA folding study, should not be simply copied to the force unfolding cases.

3.2.2 Constant Force Ensemble

Liphardt et al. [5] imposed a constant force on P5ab. They found that, when the force was held constant at the transition within about 1 pN, P5ab switched back and forth with time from the folded hairpin to the unfolded single strand. We choose another move set to investigate the phenomenon to enhance simulation efficiency [13]: we keep the first kind of moves in Equation (1) as the new move set. Therefore in the ensemble Equation (2) is simplified as $\Delta G_i^0 - W^{ss}(n_i, f)$, where the second term is the elastic free energy contribution of the single-stranded part of the RNA in the structure S_i under force f. Here we neglect the double-stranded handle since under constant force the handle can be viewed as one part of the feedback mechanism. Note the extension now recorded is $x_{ss}(f, n_i)$.

The inset in Figure 2B shows the force-extension curves for the three RNA molecules with new move set in equilibrium. Except the regions before transitions where the elastic property of the handle dominates, the shapes of the curves and the values

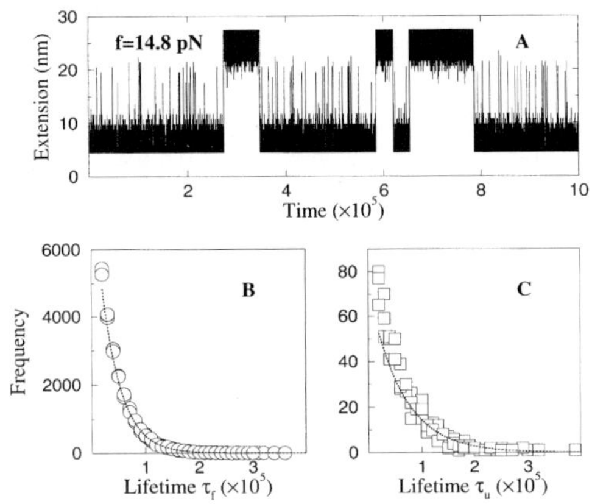

Fig. 5. Simulations for RNA p5ab kinetics: (A) Extension versus time traces of the molecule at a force in equilibrium, here the unit of time is τ_0; (B) and (C) are the frequency distributions of the lifetimes of the single stranded at 14.2 pN, respectively.

of unfolding force obtained by the two different simulation methods are almost the same. We then record the extension-time traces of the RNA molecules at different constant forces in equilibrium, here ionic correction are took into account to compare with the real data [24]. For example, one extension-time traces at force 14.8 pN for P5ab without Mg^{2+} is shown in Figure 5A.

We indeed find that the extension of the molecule jumps between about two values, 5 nm and 22 nm around the unfolding force. Considering that the jumps are extremely rapid with respect to the lifetimes of the molecule in the two states, we simply classify the states whose extensions are larger than 15 nm as the single stranded states, and the others as the hairpin states. Around the unfolding the frequencies of the different lifetimes at the two states can be obtained by a long time simulation (the simulation time is $10^9\tau_0$ for each trajectory after equilibrium). Figures 5B and 5C show the frequency distributions of a typical simulation at force 14.2 pN. These distributions can be well fit to an exponential function $\propto \exp(-t\langle\tau_i\rangle)$ very well, where $\langle\tau_i\rangle$, $i = u, f$ denote the force-dependent average lifetimes at the two states, respectively. We calculate all average lifetimes near the unfolding force of P5ab, and their corresponding values with different forces are shown in Figure 6.

We find that the logarithms of the lifetimes for the two states are strikingly consistent with linear functions of the forces. Because the reaction rate constants are the inverse of the average lifetimes, we fit τ_0 by making $\langle\tau_u\rangle(f^*) = \langle\tau_f\rangle(f^*)$ equal to the experimental value $1/k^*$, here $k^* \equiv k_f = k_u$. We have $\tau_0 = 2.6 \times 10^5\ \text{sec}^{-1}$. A comparison of the simulation results and the experimental data is listed in Table 1. The striking consistence of our results with the experiment assures us that the RNA folding and unfolding model proposed here has grabbed the main physics.

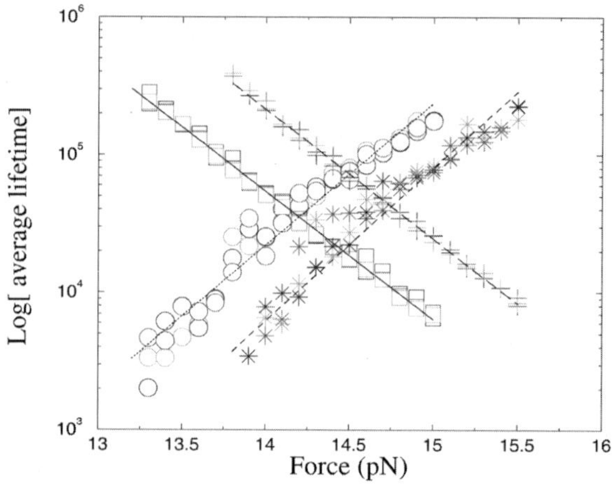

Fig. 6. Logarithm of the average lifetimes of the single stranded and hairpin states for p5ab molecule at difference forces around the unfolding with (cross symbols) and without Mg^{2+} (close symbols). The time is in unit τ_0.

Table 1. Simulation results for P5ab compared to the experimental data from [5] (in bold). The minor difference between the simulation data listed here and the data in [13] is due to different ionic corrections applied.

Molecule	$\langle \Delta x \rangle$ (nm)	f^* (pN)	$\ln k_f(f)$ (s^{-1})	$\ln k_u(f)$ (s^{-1})
P5ab, Mg^{+2}	10	14.5	$41 - 2.8f$	$-39 + 2.9f$
P5ab, by us	20.0	14.7	$39.4 - 2.6f$	$30.1 + 2.2f$
P5ab, EDTA	18	13.3	$37 - 2.7f$	$-32 + 2.6f$
P5ab, by us	20.0	14.2	$35.7 - 2.4f$	$-28.3 + 2.1f$

4 Conclusion

Compared to the enormous kinetic simulations of the force unfolding proteins, the effort contributed to RNA is relatively little. To fit the gap, we developed a kinetic stochastic simulation to the force unfolding single RNAs. The most advantage in study of force unfolding RNAs is that the knowledge accumulated in the past forty years for RNA secondary structure provides a solid fundament for theoretical predictions (including kinetics and thermodynamics) in practice. Therefore, we believe that our model would be useful in RNA biophysical studies in the future.

Acknowledgement

F.L. was supported by the National Natural Science Foundation of China, grant No. 10447117.

References

1. Cech, T.R., The chemistry of self-splicing RNA and RNA enzymes, *Science* **236**, 1987, 1532.
2. Couzin, J., Small RNAs make big splash, *Science* **298**, 2002, 2296.
3. Harlepp, S., Marchal, T.M., Robert, J., Leger, J.F., Xayaphoummine, A., Isambert, H. and Chatenay, D., Probing complex RNA structures by mechanical force, *European Physical Journal E* **12**, 2003, 605.
4. Rief, M., Clausen-Schaumann, H. and Gaub, H.E., Sequence-dependent mechanics of single DNA molecules, *Nature Structural & Molecular Biology* **6**, 1999, 346.
5. Liphardt, J., Onoa, B., Smith, S.B., Tinoco, I. and Bustamante, C., Reversible unfolding of single RNA molecules by mechanical force, *Science* **292**, 2001, 733.
6. Liphardt, J., Dumont, B.S., Smith, S.B., Tinoco, I. and Bustamante, C., Equilibrium information from nonequilibrium measurements in an experimental test of Jarzynski's equality, *Science* **296**, 2002, 1832.
7. Onoa, B., Dumont, S., Liphardt, J., Smith, S.B., Tinoco, I. and Bustamante, C., Identifying kinetic barries to mechanical unfolding of the T. thermophila Ribozyme, *Science* **299**, 2003, 1892.
8. Montanari, A. and Mezard, M., Hairpin formation and elongation of biomolecules, *Physical Review Letter* **86**, 2001, 2178.
9. Zhou, H., Zhang, Y. and Ou-Yang, Z.C., Stretch-induced hairpin-coil transitions in designed polynucleotide chains, *Physical Review Letter* **86**, 2001, 356.
10. Gerland, U., Bundschuh, R. and Hwa, T., Mechanically probing the folding pathway of single RNA molecules, *Biophysical Journal* **84**, 2003, 2831.
11. Lubensky, D.K. and Nelson, D.R., Single molecule statistics and the polynucleotide unzipping transition, *Physical Review E* **65**, 2002, 031917.
12. Liu, F., Dai, L.R. and Ou-Yang, Z.C., Theory for the force-stretched double-stranded chain molecule, *Journal of Chemical Physics* **119**, 2003, 8112.
13. Liu, F. and Ou-Yang, Z.C., Unfolding single RNA molecules by mechanical force: A stochastic kinetic method, *Physical Review E* **70**, 2004, 0409011.
14. Liu, F. and Ou-Yang, Z.C., Monte Carlo simulation for single RNA unfolding by force, *Biophysical Journal* **88**, 2005, 76.
15. Flamm, C., Fontana, W., Hofacker, I. and Schuster, P., RNA folding at elementary step resolution, *RNA* **6**, 2000, 325.
16. Gillespie, D.T., *Markov Process: An Introduction for Physical Scientists*, Academic Press, San Diego, California, 1992.
17. Jansen, A.P.J., Monte Carlo simulations of chemical reactions on surface with time-dependent reaction-rate constants, *Computer Physics Communications* **86**, 1995, 1.
18. Kawasaki, K., Diffusion constants near the critical point for time-dependent Ising models, *Physcial Review* **145**, 1966, 224.
19. Crooks, G.E., Entropy production fluctuation theorem and the nonequilibrium work relation for free energy differences. *Physical Review E* **60**, 1999, 2721.
20. Hofacker, I.L., The Vienna RNA secondary structure server, *Nucleic Acids Research* **31**, 2003, 3429.
21. Hmmer, G. and Szabo, A., Free energy reconstruction from nonequilibrium single-molecule pulling experiments, *Proceedings of National Academy of Sciences, USA* **98**, 2001, 3658.
22. Jarzynski, C., Equilibrium equality for free energy differences, *Physical Review Letter* **78**, 1997, 2690.

23. Ritort, F., Bustamante, C. and Tinoco, I.A., Two-state kinetic model for the unfolding of single molecules by mechanical force, *Proceedings of National Academy of Sciences, USA* **99**, 2002, 3544.
24. Cocco, S., Marko, J.F. and Monasson, R., Slow nucleic acid unzipping kinetics from sequence-defined barriers, *European Physical Journal E* **10**, 2003, 153.

Modelling the Thermal Conductivity of Nanofluids

P. Tillman and J.M. Hill*

*Nanomechanics Group, School of Mathematics and Applied Statistics, University of Wollongong, Wollongong, NSW 2522, Australia; *E-mail: jhill@uow.edu.au*

Abstract. The interaction of nano particulates with conventional materials generally has the effect of dramatically changing all the physical parameters of the material, which normally characterize the bulk material. The nanoparticles themselves constitute highly reactive isolated sites, to the extent that the electronic structure of the nano composite is changed, and accordingly all the physical properties, such as thermal, mechanical, electrical, magnetic and optical become different from those of the bulk materials. In fact generally, the smaller the particles, the greater the quantum effects, which means greater changes to the bulk physical properties of the nano-composite, and this phenomenon is widely accepted as not being properly understood. Nanofluids are simply standard fluids such as water, engine oil, ethylene glycol and toluene, but including a small volume percentage, usually less than 5% of evenly dispersed nanoparticles, which are usually metallic. In this paper, we present a survey of some of the attempts to model the enhanced thermal conductivity of such nanofluids, and address issues such as the nanoparticles themselves, the surrounding layer, cluster structure, the fluid environment, and the different heat transport processes at the micro and nano scales.

Key words: heat transfer, thermal conductivity, mathematical model, nanoparticle, nanolayer, cluster, nanofluid.

1 Introduction

Engineers have been working for decades to develop efficient heat transfer fluids for car motors and industrial equipment. With the ongoing technology development of miniaturization of both microelectro mechanical and microfluidic devices, there is a growing interest in higher thermal conductivities of heat transfer fluids in industry. Traditional heat transfer fluids, such as water, oil, and ethylene glycol mixture are inherently poor heat transfer fluids. Since most solid materials have higher thermal conductivities than those of fluids by many orders of magnitudes, one way of improving the thermal properties of energy transmission fluids is to suspend solid particles in the fluids. Conventionally, the suspended particles are usually micrometers or even millimeters dimensions, but such large particles may cause some severe

Yilong Bai et al. (eds), IUTAM Symposium on Mechanical Behavior and Micro-Mechanics of Nanostructured Materials, 105–118.

abrasion, settling and clogging problems in micro channels. Therefore, conventional working fluids with suspended large particles have limited practical applications in heat transfer enhancement, and are unsuitable for the newly emerging "miniaturized" technologies. With the advent of nanotechnology, it has become possible to manufacture nano size particles and the resulting nanofluids constitute a new class of fluids which were first defined by Choi (1995). Nanofluids are formed by the dispersion of nanoparticles into conventional heat transfer liquids. The most common nanoparticles for nanofluids are generally metallic, metallic oxide and non-metallic. The size of the nanoparticles used are generally below 100 nm and the nanoparticle volume fraction in the nanofluids ranges from the almost vanishing value 0.00026% to 10%. These particles, due to their extreme size, can form a very stable colloidal system. The concept of using nanofluids in heat transfer equipments is attractive and straightforward and there are two important features: (i) high thermal conductivity; (ii) high stability and no sedimentation problem. The primary reason for these features arises from larger surface area of nanophase powders relative to those of conventional powders, which not only markedly improves heat conduction transfer capabilities, but also increases the stability of suspensions.

The first significant heat transfer enhancement with nano-sized particles was reported by Masuda et al. (1993) in Japan. They demonstrated that the thermal conductivity of ultra fine suspensions of alumina with an average diameter 13 nm, silica and other oxides in water increased the thermal conductivity by a maximum amount of 30% for particle volume fraction of 4.3%.

In a series of experiments performed at the Argonne National Laboratory to investigate superior heat transfer capabilities of this new class of engineered fluids, Lee et al. (1999) measured the thermal conductivity behaviour of four oxide nanofluids (Al_2O_3 of diameter 38 nm and CuO of diameter 24 nm in both water and ethylene glycol), and reported more than 20% of enhancement of thermal conductivity at a volume fraction of 4% for the copper oxide/ethylene glycol system. They also noticed that for nanofluids using the same nanoparticles, the conductivity ratio increases for the ethylene glycol nanofluids system are always higher than those of the water nanofluids system.

Wang et al. (1999) reported enhanced thermal conductivities for the Al_2O_3 powder (diameter 28 nm) and CuO powder (diameter 23 nm) with base fluids water, ethylene, engine oil and vacuum pump fluid. The experimental data showed that for a given volume fraction, and the same nanoparticle, the increases in thermal conductivity in ethylene glycol and engine oil are their highest, whereas for the pump fluid it is the lowest. Both Wang et al. (1999) and Lee et al. (1999) observed that the thermal conductivity enhancement increases as the particle size decreases, and that enhancement increases almost linearly with volume fraction in the low volume fraction range up to 10%.

Eastman et al. (2001) showed that 10 nm copper particles in ethylene glycol could enhance the thermal conductivity by 40% with very small particle loading fraction 0.3%. Using CuO (35 nm diameter), the maximum enhancement was 20% for a volume fraction of 4%.

Xie et al. (2002) measured the thermal conductivities of Al_2O_3 suspensions to investigate the effects of the pH value on the aqueous suspension, the specific surface area of the dispersed particles and the crystalline phase of the solid phase. Their results showed that the enhanced thermal conductivity ratio decreased with an increase in pH value. For the Al_2O_3-ethylene glycol suspension, the thermal conductivity can be enhanced by more than 29% at a volume fraction of 5%. For the suspensions using the same base fluid, the enhancements of the thermal conductivities are dependent on the specific surface area, with an optimal specific surface area for the highest thermal conductivity.

Choi (2001) showed that multi walled carbon nanotubes in a synthetic poly oil can increase the effective thermal conductivity of the base fluid by a factor of 2.5, using only a volume percentage of nanotubes of approximately 1%. Patel et al. (2003) reported an 11% enhancement of heat conductivity for almost vanishing volume fraction of 0.008% for gold particles with a thiolate covering in toluene. Hong et al. (2005) found that thermal conductivity of a Fe nanofluid is increased nonlinearly up to 18% as the volume fraction of particles is increased up to 0.55%. Fe nanofluids showed a more effective thermal transport property than Cu nanofluids dispersed with little agglomeration.

In the investigations mentioned above, the thermal conductivities were measured at room temperature. Das et al. (2003) measured Al_2O_3 and CuO suspensions in water for different temperature ranging from 20°C to 50°C for particle loading of 1% to 4%. A two to four fold increase in thermal conductivity enhancement of nanofluids with temperature increase was observed.

The above examples illustrate the possible high increases of thermal conductivity of nanofluids and the practical significance of properly understanding this phenomenon. However, the observed conductivity enhancements for nanoparticle suspensions are greater than those predicted by the existing classical theories such as the Maxwell theory (Maxwell, 1904) and Hamilton–Crosser model (Hamilton and Crosser, 1962). For example, the measured enhancement in thermal conductivity for 1.0 vol% nanotubes in oil is 160%, while the enhancements predicted by the theoretical model are not more than 10% (Choi et al., 2001). It is also found that the thermal conductivity of a nanotube nanofluid varies nonlinearly with the nanotube volume fraction, while all theoretical predictions clearly show a linear relationship between thermal conductivity enhancements and nanotube volume fraction. These results demonstrate that traditional theoretical models are not capable of properly explaining the thermal conductivity enhancement of nanofluids, and the fundamental limitations of the conventional models. Thus, understanding certain fundamental phenomena of heat transfer in nanofluids becomes essential. In this paper, we undertake a systematic survey of modelling endeavours for thermal conductivity of nanofluids and we outline existing models.

2 Theoretical Models of Thermal Transport in Nanofluids

Considerable research has been undertaken on modelling the effective thermal conductivities of liquids with solid particle inclusions dating back to the pioneering of Maxwell work more than a century ago. In recent years, numerous phenomenological models have been developed in an attempt to accurately predict the thermal conductivity of nanoparticle suspensions. Experimental data shows that the thermal conductivity behaviour of a composite fluid can be strongly dependent on the following factors:

(i) thermal conductivity of particles,
(ii) thermal conductivity of liquid,
(iii) particle volume concentration,
(iv) particle size,
(v) particle shape,
(vi) temperature.

Modelling the nanofluid structure has been closely examined by many investigators in recent years. Strictly speaking, a nanofluid is a multi-phase system consisting of the host liquid and percolation patterned cluster inclusions. We assume there are core- nanoparticles, and then there is the molecule-levelled layering of liquid called a nanolayer, formed at the particle/liquid interface. The transport among components of equivalent nanoparticles (nanoparticle-nanolayer) leads to inter-particle collisions, and attachments of colliding nanolayers leads to nanoparticle clusters. The enhancement of thermal conductivity can be considered as heat conduction from these liquid mediated clusters with different shapes, sizes and surface areas.

The main reasons for the enhanced heat transfer characteristics may be summarized as follows:

(a) the suspended nanoparticles increase the surface area and the heat capacity of the fluid,
(b) the suspended nanoparticles themselves have higher thermal conductivity than that of the base fluid,
(c) the organized nanolayer developed on the solid/liquid interface facilitates the transport of energy across the interface,
(d) the interaction and collision among particles, fluid and the flow passage surface are strengthened,
(e) the mixing fluctuation and turbulence of the fluid is intensified,
(f) thermal diffusion becomes ballistic heat conduction due to the characteristic length scales associated with heat carriers, i.e., the mean free path and the wavelength, are comparable to the characteristic lengths of the nanostructure.

The development of existing theoretical models for the thermal conductivity of nanofluids may be divided into two categories:

(1) those using core-layer-medium theory for multiphase systems,
(2) those recent phenomenological models based on the effect of random Brownian motion.

In the following sections, we discuss each of the heat transfer mechanisms and we give details of the some models for nanofluid structure, starting from the nanoparticles themselves, from the nanoparticle body to an adjacent layer, then clusters, and finally thermal conductivity of a fluid-clusters mixture.

3 Thermal Conductivity of Nanoparticles

The first step in formulating a theoretical model of thermal conductivity for nanofluids is to determine the thermal conductivity of nanoparticles k_p correctly. On the macroscale, heat conduction is typically treated as a diffusion process which is governed by Fourier's law. It is well known that on the microscale regime, the thermal conductivity of a thin film material/superlattices is much less than its bulk value k_b, which is due to the scattering of the primary carriers of energy (phonons and/or electrons) at its boundary (Majumdar, 1998). The intrinsic thermal conductivity of the nanoparticles should therefore be reduced in comparison to that of bulk materials. When the mean free path of the heat carriers is comparable with the size of nanoparticles, i.e. 10-100nm, there is not enough scattering events in the medium for the phonons to transport energy. Hence, the applicability of the Fourier's law becomes questionable and we need to turn to the Boltzmann equation to describe the heat transfer process. Chen (1996) showed that the thermal conductivity of non-metallic nanoparticles k_p can be approximated from the bulk value k_b by

$$k_p = \frac{3r_p}{3r_p + 4\ell}, \tag{1}$$

while for metallic nanoparticles Nimtz et al. (1988) showed that

$$k_p = \left(\frac{2r_p}{5 \times 10^{-6}} \right)^3 k_b, \tag{2}$$

where r_p is the radius of the nanoparticle and ℓ is the mean free path of the phonons.

4 Maxwell's Theory

The work of Maxwell (1904) was the first theoretical approach to calculate the effective thermal conductivity for a random spherical particle suspension. Heat conduction is the transfer of heat from one substance to another by direct contact. Just as for the movement of electric current through a material in response to an electric field, the underlying mechanism for the movement of heat depends on the two materials. Maxwell's equation, Equation (3), was first derived from the electric permeability calculation for a compound medium on the basis of potential theory. Maxwell's model is applicable to dilute suspensions for volume fraction $\varphi < 1$ in a homogeneous host medium, where the particles are considered to be isolated in the host medium, so that no interactions exist among them. The prediction of the effective thermal conductivity k_{eff} from Maxwell's equation is accurate to order φ.

$$\frac{k_{\mathrm{eff}}}{k_f} = 1 + \frac{3(k_p - k_f)\varphi}{k_p + 2k_f - (k_p - k_f)\varphi}, \tag{3}$$

where k_f is thermal conductivity of liquid. Much later, Maxwell's equation was extended to the second-order by several authors. Bruggeman (1935) analysed the interaction among the randomly distributed particles and proposed a model using a mean field approach. For a binary mixture of homogeneous spherical inclusions, the Bruggeman model is described by the equations below

$$\varphi\frac{k_p - k_{\mathrm{eff}}}{k_p + 2k_{\mathrm{eff}}} + (1-\varphi)\frac{k_f - k_{\mathrm{eff}}}{k_f + 2k_{\mathrm{eff}}} = 0, \tag{4}$$

where the solution of the above quadratic equation is given by

$$k_{\mathrm{eff}} = [(3\varphi - 1)k_p + (2 - 3\varphi)k_f + \sqrt{\Delta}]/4, \tag{5}$$

and where Δ is defined by

$$\Delta = [(3\varphi - 1)k_p + (2 - 3\varphi)k_f]^2 + 8k_p k_f. \tag{6}$$

We note that the Bruggeman model has no restriction on the suspended particle volume concentration φ, and that Maxwell's model fails to predict experimental results for high particle concentration, while the Bruggeman model fits the experimental data well.

It has been found that the thermal conductivity of suspensions depends not only on the volume concentration of the particles, but also on the shape of the dispersed particles. Hamilton and Crosser (1962) develop an elaborate model to include the shape of the dispersed particles in Maxwell's equation. Their empirical shape factor is given by $n = 3/\Psi$, where Ψ is sphericity, defined as the ratio of surface area of a sphere (with volume equal to that of the particle) to surface area of the particle, and then the conductivity of two-component mixtures can be calculated as follows:

$$\frac{k_{\mathrm{eff}}}{k_f} = 1 + \frac{n(k_p - k_f)\varphi}{k_p + (n-1)k_f - (k_p - k_f)\varphi}, \tag{7}$$

and Equation (3) arises from the value $n = 3$. Comparison between experimental results and those of the Hamilton–Crosser model show that the model can predict the thermal conductivity of nanofluids containing large agglomerated particles (Lee et al., 1999). The model also shows that non-spherical shapes increase the conductivity above that of spheres. However, the model predictions begin to diverge from the experimental data at low volume fractions, which strongly suggests that not only particle shape but also size is dominant in enhancing the thermal conductivity of nanofluids.

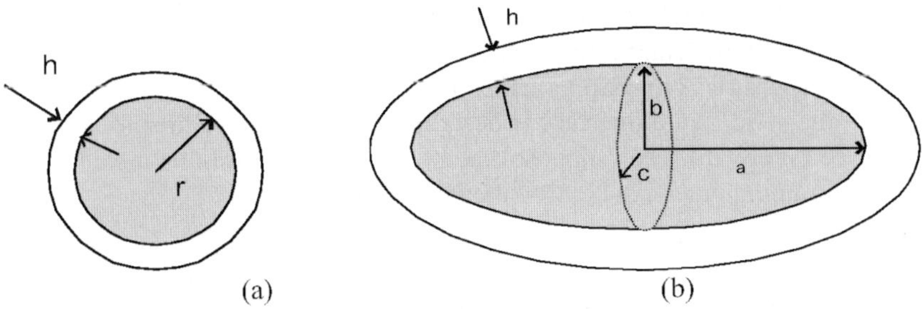

Fig. 1. Nanoparticle with nanolayer at solid/liquid interface, (a) sphere, (b) ellipsoid.

5 Thermal Conductivity of Nanoparticles with Nanolayer

It has long been known that liquid molecules close to a solid surface form layered solid like structure, but little is know about the connection between this nanolayer and the thermal properties of the suspension. Keblinski et al. (2002) explored possible explanations for the anomalous increase of thermal conductivity and their analysis demonstrated that molecular-level layering of the liquid at the solid/liquid interface could play a significant role in interaction of dynamic nanoparticles with base fluid molecules. As the atomic structure of the liquid layering is significantly more ordered than that of bulk liquid, liquid layering at the interface would be expected to lead to a higher thermal conductivity than liquid and becomes a thermal bridge between nanoparticles and the surrounding fluid. In addition, the interfacial layer thickness results in a larger effective volume fraction of the particle-layered-liquid structure which would tend to enhance thermal conductivity.

Yu and Choi (2003) proposed a renovated Maxwell model to include the effect of the nanolayer on the effective thermal conductivity, and they assumed that a solid-like layer of thickness h around the particles of radius r_p which combines with the particle to form an equivalent particle of radius r_p+h (see Figure 1a). Notice however that their assumed layer thickness is a variable.

The increased volume concentration φ_e can be calculated from $\varphi_e = \varphi(1+\beta)^3$, where $\beta = h/r_p$. This nanolayer is more ordered than the bulk liquid and the thermal conductivity of the layer k_{layer} is higher than that of the bulk liquid but lower than that of the nanoparticles. From an effective medium theory, the equivalent thermal conductivity k_{pe} of equivalent particles which include a thermally bridging nanolayer may be calculated as (Schwartz et al., 1995)

$$k_{pe} = \frac{[2(1-\gamma)+(1+\beta)^3(1+2\gamma)]\gamma}{-(1-\gamma)+(1+\beta)^3(1+2\gamma)} k_p, \tag{8}$$

where $\gamma = k_{\text{layer}}/k_p$, and the Maxwell equation can be modified to become

$$\frac{k_{\text{eff}}}{k_f} = 1 + \frac{3(k_{pe}-k_f)(1+\beta)^3\varphi}{k_{pe}+2k_f-(k_{pe}-k_f)(1+\beta)^3\varphi}. \tag{9}$$

This renovated Maxwell model is limited to those suspensions with spherical particles. Recently the Hamilton–Crosser model has been extended by Yu and Choi (2004) for suspensions of nonspherical particles to include the effect of a solid/liquid interface. The solid/liquid interface is described as a confocal ellipsoid with a solid particle (see Figure 1b). They suppose that the outside of the solid ellipsoidal particle and its surrounding layer can be described by the equation

$$\frac{x^2}{a^2+\upsilon}+\frac{y^2}{b^2+\upsilon}+\frac{z^2}{c^2+\upsilon}=1, \tag{10}$$

where $\upsilon = 0$ for the outside surface of the solid ellipsoid, $\upsilon = t$ for the outside surface of its surrounding layer, and a, b and c are the semi-axes of the ellipsoid. The volume ratio is defined by

$$e_0=\frac{\text{volume of particle plus layer}}{\text{volume of particle}}=\frac{\sqrt{(a^2+t)(b^2+t)(c^2+t)}}{abc}, \tag{11}$$

leading to an increased volume fraction $\varphi_e = e_0\varphi$ and we refer the reader to Yu et al. (2004) for further details.

The input data required in the renovated Maxwell model and the renovated Hamilton–Crosser model are the thickness of the nanolayer h and the thermal conductivity of the nanolayer k_{layer}. As $k_f < k_{\text{layer}} < k_p$, by assuming k_{layer} to its lowest and highest values, we may study the lower and upper bounds of the effective thermal conductivity of nanofluids. Yu and Choi (2003) found that the nanolayer thickness is more important to thermal conductivity enhancement than k_{layer}. The dramatic thermal conductivity enhancement with smaller nanoparticles (< 10 nm) is attributed primarily to the increased volume fraction by the nanolayer. This suggests that the existence of a thermally bridging nanolayer is a new mechanism for enhancing thermal conductivity of nanofluids.

The adsorption of liquid molecules on the particle surface can be considered as a monolayer with a hexagonal closed-packed style. The following evaluation method for the thickness of the adsorption layer developed by Yan et al. (1986) has been used in the model of Wang et al.'s model (2003)

$$h=\frac{1}{\sqrt{3}}\left(\frac{4M_f}{\rho_f N_a}\right)^{1/3}, \tag{12}$$

where M_f is the molecular weight of liquid, ρ_f is the density of liquid, and N_A is the Avogadro constant (6.023×10^{23}/mol). Experiments and simulation have shown a typical interfacial width is only on the order of a few atomic distances, i.e., that is approximately 1 nm, but 1 nm thickness of layered-liquid is not enough to double the effective volume of a particle. Thus, although the presence of an interfacial layer may play a role in heat transport, it is not likely to be the sole contributor for the enhancement of thermal conductivity.

6 Thermal Conductivity of Nanoparticle Clusters

As mentioned previously, it is believed that clustering could significantly effect the enhancement of thermal conductivity. Keblinski et al. (2002) were first to discover the nature of heat transport in nanoparticles. They used the Debye theory to demonstrate that the macroscopic theory of diffusive heat transport is invalid in a nanoparticle and they suggested that due to Brownian motion, the particles are in constant motion and move close together even at low packing faction, and thus enhance coherent phonon heat flow among the particles. They demonstrated significant enhancement of the effective thermal conductivity from liquid mediated clusters, which have an increased particle surface area and an effective volume of highly conducting clusters for heat transfer. The effective volume of a cluster can be much larger than the physical volume of the particles. Within such an effective volume of cluster, all particles do not need to be in physical contact with each other, but just within a specific distance, which allows for heat flow between them.

Accordingly, it is necessary to include the effect of clustering in the modelling of the effective thermal conductivity of nanoparticles suspension. The results of the nanoparticle clusters should reflect a picture of a nanoparticle cluster whose conductivity is a monotonically decreasing function of radius (or number of cluster particles). Fractal theory can successfully describe the disorder and stochastic process of clustering and polarization of nanoparticles. The basic idea of the model is that a cluster is self-similar on the average, whose nanoparticle portion is connected at every stage of its evolution, and which therefore remains conducting no matter how small the nanoparticle volume fraction (Hui et al., 1986).

Scaling theory is commonly used for the quantitative description of a fractal system, and the fractal dimension D_f and a scalar with unit ε, are the basic variables for the description of a fractal system. The resultant properties such as volume, area, particle number, etc. of the fractal is a function $F(\varepsilon)$ of a scalar unit ε, and is expressed as $F(\varepsilon) = C\varepsilon^{D_f}$, where C is a shape factor which is independent of ε. Different fractal indexes D_f are needed when describing different phenomena of complex fractals. Some authors use fractal theory to describe the cluster of nanoparticle suspensions to predict the effective thermal conductivity. Xuan et al. (2003) introduce the effect of nanoparticle aggregation structure in their theoretical model of thermal conductivity of nanofluids and the radius of the clusters is included in the equation.

Wang et al. (2003) applied the fractal model in the nanofluid CuO/water. They proposed that the nanoparticles are distributed in fractal clusters in composites. The equivalent nanoparticle volume fraction in fractal clusters is a function of the fractal cluster of radius R. Hence, the volume fraction $\varphi_{fr}(R)$ of nanoparticles in the clusters is given by $\varphi_{fr}(R) = (R/r_p)^{D_f-3}$, where D_f is fractal dimension, R is the equivalent radius of cluster and r_p is the radius of nanoparticle. The fractal dimension D_f for the clusters was determined from electron microscopic photos of clusters in suspension and its value is in the range of 1.73 to 1.81.

The drawback of the fractal models is the need for the transmission electron microscopy photo or computer simulation image of the clusters in suspension to reveal an aggregation structure. From the images, the fractal index of cluster D_f is determined, but this can be quite computationally expensive.

Knowledge of the volume fraction $\varphi_{fr}(R)$ of nanoparticles in the cluster and the thermal conductivity of equivalent particles k_{pe}, which include a thermally bridging nanolayer, the effective thermal conductivity of cluster k_{cl} can be determined from an effective medium theory such as the Maxwell or the Bruggeman model for a binary mixture of homogeneous spherical inclusions. On substituting $\varphi_{fr}(R)$ for φ and k_{pe} for k_p into the Maxwell Equation (3) or the Bruggeman Equation (4), the effective thermal conductivity of the cluster can be expressed as a function of the radius of cluster $k_{cl} = k_{cl}(R)$. The final effective thermal conductivity k_{eff} of the nanofluid can be obtained by including the cluster size distribution function in the Maxwell equation through integration (Wang et al., 2003).

7 Recent Phenomenological Models

The heat transfer is a microscale energy transport phenomenon in nature. The common fundamental assumption in the traditional theoretical approaches to the conductivity of solid-fluid suspensions is that the discrete particles are essentially motionless in a continuous material. This assumption works well for particles embedded in a solid matrix or large particles suspended in a fluid matrix. However, the suspended nanoparticles in a base fluid are generally in random motion under various acting forces, such as Brownian force and the London–Van Der Waals force. The random motion of the suspended nanoparticles strengthens energy transport inside the liquid. Many recent researchers deal with investigations on the possible mechanism of energy transport enhancement and have derived theoretical models based on the phenomena in nature. Wang et al. (1999) proposed that the total increase in thermal conductivity by the Brownian motion of particles Δk_{eff} consists of the increases due to both translational $\Delta k_{\text{eff},t}$ and rotational motions $\Delta k_{\text{eff},r}$ respectively. The values of $\Delta k_{\text{eff},t}$ and $\Delta k_{\text{eff},r}$ are dependent on the Peclet number. Their calculation shows that up to a volume fraction of 10%, the thermal conductivity increased by the Brownian motion is less than 0.5% for the Al_2O_3 fluid mixture. It indicates that heat transfer by advection of nanoparticles is less than that transferred by diffusion. In other words, when the particles move in liquid, the temperature of particles quickly equilibrate with that of the surrounding fluids due to small size of the particles. Therefore, Brownian motion does not contribute significantly to the energy transport in nanofluids.

Due to Brownian motion, some recent phenomenological models of effective thermal conductivities are proposed consisting of two parts: (i) stationary model; (ii) moving particle model. Xuan et al. (2003) simply add the heat transport contribution from the random motion of the suspended nanoparticles on the Maxwell equation as below:

$$\frac{k_{\text{eff}}}{k_f} = 1 + \underbrace{\frac{3(k_p - k_f)\varphi}{k_p + 2k_f - (k_p - k_f)\varphi}}_{\text{Maxwell}} + \underbrace{\frac{\rho_p \varphi c_p}{2k_f}\sqrt{\frac{k_B T}{3\pi r_c \eta}}}_{\text{Brownian motion}}, \tag{13}$$

where T is temperature, ρ_p is density of particle, c_p is the specific heat capacity per particle, J/(kg.K), k_B is the Boltzmann constant, r_c is the apparent radius of clusters, η is the fluid viscosity, kg/(s.m). Xuan's model shows that the effective thermal conductivity of nanofluids is proportional to $\sqrt{T}$ and inversely proportional to $\sqrt{r_c}$.

Kumar et al. (2004) proposed a comprehensive model based on the existing understanding of the heat transport under the continuum level phenomenological formulation. In their model, the diffusive heat transport is assumed to valid in both liquid and solid phases. The standard diffusive heat transport equation comprises heat transfer processes occurring in both liquid and solids as below:

$$q = -k_f A_f \left(\frac{dT}{dx}\right)_f - k_p A_p \left(\frac{dT}{dx}\right)_p. \tag{14}$$

The geometrical effect of an increase in surface with a decrease in particle size, rationalized using a stationary particle model, explains the conductivity enhancement. The moving particle model developed from the Stokes–Einstein formula accounts for the temperature effect. The effective thermal conductivity of nanofluids can therefore be expressed as

$$\frac{k_{\text{eff}}}{k_f} = 1 + c\left(\frac{k_B T}{2\pi \eta r_p^2}\right)\frac{\varphi r_f}{k_f(1-\varphi) r_p}. \tag{15}$$

From the kinetic theory, c can be evaluated by $c = n\ell c_p/3$, where n is the particle concentration, ℓ is the mean free path. From the Debye model for the particle, $c_p = 3Nk_B$, where N is the number of atoms. Kumar et al.'s (2004) theoretical model shows that the thermal conductivity enhancement is inversely proportional to the radius of the particle r_p. The temperature dependence is attributed to the variation of Brownian motion velocity for the particles, which varies in proportion to T/η. k_{eff} is a function of the variables k_f, k_p, T, φ, r_f, r_p. Experimental results for Au in water, Al_2O_3 in water and Au in ethylene glycol, strongly support the prediction of their theoretical model. However, this model shows some inconsistencies with Xuan's model (2003) in terms of the relationship with temperature.

8 Conclusions

The aim of the present work has been to outline some existing techniques for theoretical models, which have been developed over recent years for studying the thermal conductivity of nanofluids. The application of the existing models is still in an ongoing development phase and there are limitations, inconsistencies and shortcomings that need to be overcome, and further applications need to be explored.

The thermal conductivity of nanofluids is an effective medium property of a multiphase system. The models reported in this paper show that the effective thermal conductivity of nanofluids is a function of the thermal conductivities of base fluid and nanoparticles, nanoparticle size and shape, particle clustering and aggregation and fluid temperature. The fundamental limitations of conventional heat conduction models for solid/liquid suspensions, which are rooted in macroscopic transport laws, such as the Fourier law of heat conduction diffusion have been recognized. Although the ballistic heat conduction in solid nanostructure is well known, the ballistic/diffusive conduction has only briefly been introduced in solid/liquid system. The core-shell-medium, multi-component models reflect the nanofluid structure and incorporate the correct thermal conductivity of nanoparticles which have been shown to be less than that of bulk material. On the other hand, some phenomenological models are proposed based on the nature of the heat transfer mechanism among fluid and nanoparticles. The fluid temperature is introduced in this type of models by introducing random Brownian motion. Fractal theory technique is adopted in some models to describe self-similar nanoparticle cluster structure and to characterize their size and other physical properties.

However, there still exists a big gap between the measured and the predicted thermal conductivities and some results from different models are inconsistent. We need to develop a more fundamental understanding of heat conduction mechanisms in nanofluids, and further studies on heat conduction in solid nanostructures and nanofluids are needed in the future. There are difficulties in the accurate determination of the nanolayer thickness, the thermal conductivity of the nanolayer k_{layer}, and the cluster size and distribution in the existing models needs to be addressed.

Acknowledgement

This work is supported through the Discovery Project scheme of the Australian Research Council.

References

Bruggeman, D.A.G. (1935) Berechnung Verschiedener Physikalischer Konstanten von Heterogenen Substanzen, I. Dielektrizitatskonstanten und Leitfahigkeiten der Mis-chkorper aus Isotropen Substanzen, *Annalen der Physik, Leipzig* **24**, 636–679.

Chen, G. (1996) Nonlocal and nonequilibrium heat conduction in the vicinity of nanoparticles, *ASME, Journal of Heat Transfer* **118**, 539–545.

Choi, S.U.S. (1995) Enhancing thermal conductivity of fluids with nanoparticles, in *Developments and Applications of Non-Newtonian Flows*, D.A. Siginer and H.P. Wang (eds), Fluid Eng. Div. Vol. 231, ASME, New York, pp. 99–105.

Choi, S.U.S., Zhang, Z.G., Yu, W., Lockwood, F.E. and Grulke, E.A. (2001) Anomalous thermal conductivity enhancement in nanotube suspensions, *Applied Physics Letters* **79**, 2252–2254.

Das, S.K., Putra, N., Thiesen, P. and Roetzel, W. (2003) Temperature dependence of thermal conductivity enhancement for nanofluids, *ASME, Journal of Heat Transfer* **125**, 567–574.

Eastman, J.A., Choi, S.U.S., Li, S., Yu, W. and Thompson, L.J. (2001) Anomalously increased effective thermal conductivity of ethylene glycol-based nanofluids containing copper nanoparticles, *Applied Physics Letters* **78**, 718–720.

Hamilton, R.L. and Crosser, O.K. (1962) Thermal conductivity of heterogeneous two-component systems, *I and EC Fundamentals* **1**, 187–191.

Hong, T.-K. and Yang, H.-S. (2005) Study of the enhanced thermal conductivity of Fe nanofluids, *Journal of Applied Physics* **97**, 064311-3.

Hui, P.M. and Stroud, D. (1986) Complex dielectric response of metal-particle clusters, *Physical Review B* **33**, 2163–2169.

Keblinski, P., Phillpot, S.R., Choi, S.U.S. and Eastman, J.A. (2002) Mechanisms of heat flow in suspensions of nano-sized particles (nanofluids), *International Journal of Heat and Mass Transfer* **45**, 855–863.

Kumar, D.H., Patel, H.E., Kumar, V.R.R., Sundararajan, T., Pradeep, T. and Das, S.K. (2004) Model for heat conduction in nanofluids, *Physical Review Letters* **93**, 144301-4.

Lee, S., Choi, S.U.S., Li, S. and Eastman, J.A. (1999) Measuring thermal conductivity of fluids containing oxide nanoparticles, *ASME, Journal of Heat Transfer* **121**, 280–289.

Masuda, H., Ebata, A., Teramae, K. and Hishinuma, N. (1993) Alteration of thermal conductivity and viscosity of liquid by dispersing ultra-fine particles (Dispersion of γ-Al_2O_3, SiO_2, and TiO_2 ultra-fine particles), *Netsu Bussei* 4 (Japan), 227–233.

Maxwell, J.C. (1904) *Electricity and Magnetism, Part II*, 3rd edn, Clarendon, Oxford, pp. 435–440.

Majumdar, A. (1998) Microscale energy transport in solids, in *Microscale Energy Transport*, C.L. Tien, A. Majumdar and F. Gerner (eds), Taylor and Francis, Washington, DC.

Nimtz, G., Marquardt, P. and Gleiter, H. (1988) Size-induced metal-insulator transition in metals and semiconductors, *Journal of Crystal Growth* **86**, 66–71.

Patal, H.E., Das, S.K., Sundararajan, T., Nair, A.S., George, B. and Pradeep, T. (2003) Thermal conductivities of naked and monolayer protected metal nanoparticle based nanofluids: Manifestation of anomalous enhancement and chemical effects, *Applied Physics Letters* **83**, 2931–2933.

Schwartz, L.M., Garboczi, E.J. and Bentz, D.P. (1995) Interfacial transport in porous media: Application to DC electrical conductivity of mortars, *Journal of Applied Physics* **78**, 5898–5908.

Wang, X., Xu, X. and Choi, S.U.S. (1999) Thermal conductivity of nanoparticle-fluid mixture, *Journal of Thermophysics and Heat Transfer* **13**, 474–480.

Wang, B., Zhou, L. and Peng, X. (2003) A fractal model for predicting the effective thermal conductivity of liquid with suspension of nanoparticles, *International Journal of Heat and Mass Transfer* **46**, 2665–2672.

Xie. H., Wang, J., Xi, T., Liu, Y. and Ai, F. (2002) Thermal conductivity enhancement of suspensions containing nanosized alumina particles, *Journal of Applied Physics* **91**, 4568–4572.

Xuan, Y., Li, Q. and Hu, W. (2003) Aggregation structure and thermal conductivity of nanofluids, *AIChE, Journal of Thermodynamics* **49**, 1038–1042.

Yan, J.M., Zhang, Q.Y. and Gao, J. Q. (1986) *Adsorption and Agglomeration-Surface and Porosity of Solid*, Science Press, Beijing [in Chinese].

Yu, W. and Choi, S.U.S. (2003) The role of interfacial layers in the enhanced thermal conductivity of nanofluids: A renovated Maxwell model, *Journal of Nanoparticle Research* **5**, 167–171.

Yu, W. and Choi, S.U.S. (2004) The role of interfacial layers in the enhanced thermal conductivities of nanofluids: A renovated Hamilton–Crosser model, *Journal of Nanoparticle Research* **6**, 355–361.

Part 4

Mechanical Behaviors of Carbon Nano-Tube, Nano-Wire, Nano-Layers

A Comparison of Different Interatomic Potentials: Radius Effect of Single Wall Carbon Nanotubes

H. Jiang[1], Y. Huang[2,*] and K.C. Hwang[3]

[1]*Department of Mechanical and Aerospace Engineering, Arizona State University, Tempe, AZ 85287-6106, U.S.A.*
[2]*Department of Mechanical and Industrial Engineering, University of Illinois, 1206 West Green Street, Urbana, IL 61801, U.S.A.; *E-mail: huang9@uiuc.edu*
[3]*Department of Engineering Mechanics, Tsinghua University, Beijing 100084, China*

Abstract. We have compared Brenner's (1990) interatomic potential and Brenner et al.'s (2002) second-generation potential on the radius effect of single wall carbon nanotubes. Our analysis is based on the modified Cauchy–Born rule to incorporate the interatomic potential into a continuum analysis. The results have shown that Brenner et al.'s (2002) second-generation potential gives more accurate equilibrium bond length. These potentials also display rather different radius dependence of the Young's modulus and simple tension and pure torsion behavior of single wall carbon nanotubes.

Key words: interatomic potential, carbon nanotubes, radius effect.

1 Introduction

There exist intensive research on the properties of carbon nanotubes (CNTs) in the past decade. Due to the difficulties of manufacturing and control technologies in experiments, the atomistic simulations are playing an indispensable role in atomic scale studies. Most of these atomistic simulations, such as molecular mechanics and classical molecular dynamics, are based on interatomic potentials that provide analytic potential energy functions and thus the interatomic forces. The continuum theories based on interatomic potentials are also established, e.g., Arroyo and Belytschko (2002); Zhang et al. (2002, 2004); Jiang et al. (2003). The most important advantage of using interatomic potentials is computational efficiency since total energy and interatomic forces can be obtained analytically from interatomic potentials. Among interatomic potentials for carbon, Brenner's (1990) interatomic potential for carbon and hydrocarbon molecules stands out and has become the most adapted interatomic potential for carbon. This potential has been widely used to study a wide range of carbon-based structures and chemical processes. Brenner et al. (2002) later developed a "second-generation" interatomic potential for carbon. In this paper,

Yilong Bai et al. (eds), IUTAM Symposium on Mechanical Behavior and Micro-Mechanics of Nanostructured Materials, 121–134.

we compare Brenner's (1990) and Brenner et al.'s (2002) interatomic potentials by studying the radius effects of single wall carbon nanotubes.

This paper is divided into five sections. We first briefly review Brenner's (1990) interatomic potential and Brenner et al.'s (2002) second-generation potential in Section 2. The continuum theory for single wall carbon nanotubes based on interatomic potentials is established in Section 3. We then use Brenner's (1990) and Brenner et al.'s (2002) interatomic potential to study several radius dependent properties for single wall carbon nanotubes, and compare the difference resulting from the use of different interatomic potentials.

2 Brief Review of Brenner's (1990) and Brenner et al.'s (2002) Interatomic Potentials for Carbon

We first summarize Brenner's (1990) interatomic potential and Brenner et al.'s (2002) second-generation potential for carbon. All parameters mentioned below are determined by fitting with known physical properties of various type of carbon and are given in the original papers.

2.1 Brenner's (1990) Interatomic Potential for Carbon

The interatomic potential established by Brenner (1990) for carbon takes the form

$$V = V_R(r_{ij}) - B_{ij} V_A(r_{ij}), \tag{1}$$

where r_{ij} is the bond length between atoms i and j; V_R and V_A represent the repulsive and attractive pair terms, respectively, and are given by

$$V_R(r) = \frac{D^{(e)}}{S-1} e^{-\sqrt{2S}\,\beta(r-R^{(e)})} f_c(r), \tag{2}$$

$$V_A(r) = \frac{D^{(e)}S}{S-1} e^{-\sqrt{\frac{2}{S}}\,\beta(r-R^{(e)})} f_c(r), \tag{3}$$

f_c is a cut-off function to restrict the range of interactions among only the nearest-neighbor carbon atoms,

$$f_c(r) = \begin{cases} 1 & r < R^{(1)}, \\ \frac{1}{2}\left\{1 + \cos\left[\frac{\pi(r-R^{(1)})}{R^{(2)}-R^{(1)}}\right]\right\} & R^{(1)} < r < R^{(2)}, \\ 0 & r > R^{(2)}, \end{cases} \tag{4}$$

B_{ij} represents the multi-body coupling which depends on neighbor atoms through the bond angle,

$$B_{ij} = \left[1 + \sum_{k(\neq i,j)} G(\theta_{ijk}) f_c(r_{ik})\right]^{-\delta}, \tag{5}$$

where k denotes the carbon atoms other than i and j, r_{ik} is the bond length between atoms i and k, and θ_{ijk} defines the bond angle between carbon bonds i–j and i–k. The angular function G is given by

$$G(\theta) = a_0 \left[1 + \frac{c_0^2}{d_0^2} - \frac{c_0^2}{d_0^2 + (1 + \cos\theta)^2} \right]. \tag{6}$$

2.2 Brenner et al.'s (2002) Second-Generation Interatomic Potential for Carbon

The second-generation interatomic potential established by Brenner et al. (2002) for carbon also takes the form in Equation (1). However, the repulsive and attractive pair terms take the forms

$$V_R(r) = \left(1 + \frac{Q}{r}\right) A e^{-\alpha r} f_c(r), \tag{7}$$

$$V_A(r) = (B_1 e^{-\beta_1 r} + B_2 e^{-\beta_2 r} + B_3 e^{-\beta_3 r}) f_c(r), \tag{8}$$

where f_c is the cut-off function given in Equation (4). One obvious improvement of the second-generation interatomic potential is that the repulsive term goes infinity when distance between atoms vanishes; therefore, it can be used to model the process of atomic collision. On the contrary, the repulsive term in Brenner's (1990) goes to a finite value when distance between atoms vanishes.

The multi-body coupling term B_{ij} also takes the same form as that in Equation (5), except that the angular function G is given by a six-order polynomial spline function. The detailed fitting process and data can be found in Brenner et al. (2002).

3 A Continuum Theory Based on Interatomic Potentials for Single Wall Carbon Nanotubes

3.1 Determine the Atomic Position Prior to Deformation

Jiang et al. (2003) developed a method to take into account the effect of CNT chirality in the continuum analysis. Such a method is briefly summarized here. Figure 1a shows a schematic diagram of a CNT with the diameter d_t. Unlike a planar graphene, a carbon atom and its three nearest-neighbor atoms on the CNT are not on a plane but form a tetrahedron because of the curvature effect. Since a CNT can be considered as a rolled graphene, we map the CNT shown in Figure 1a to a two-dimensional, planar sheet shown in Figure 1b. The distance between each pair of carbon atoms in the "unrolled" plane (Figure 1b) is identical to the corresponding arc length on the CNT (Figure 1a). It is important to point out that Figure 1b is different from a graphene since the bond lengths may not equal and the bond angles deviate from 120°.

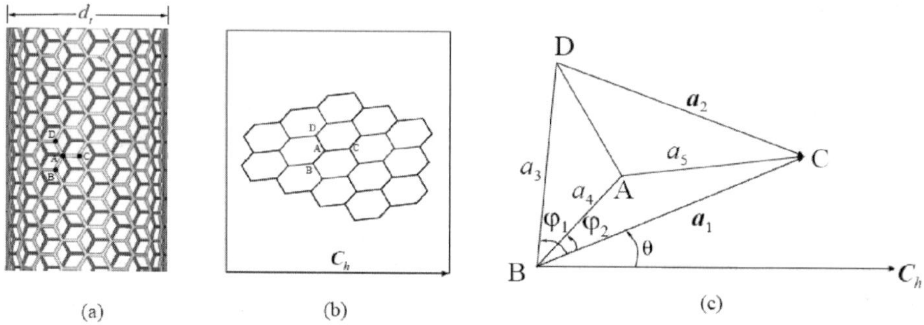

Fig. 1. A carbon nanotube (CNT) prior to deformation; (a) a CNT; (b) a planar, "unrolled" CNT; (c) a representative atom (A) and its three nearest-neighbor atoms (B, C, and D).

Figure 1c shows a representative atom A in the "unrolled" plane along with its three nearest-neighbor atoms B, C, and D. These four atoms A, B, C, and D characterize the positions of all atoms on the planar sheet in Figure 1b since all atoms essentially result from the in-plane translation of these four atoms due to periodicity in the atomic structure of the CNT. Therefore, the lengths and angles between these four atoms completely characterize the planar structure in Figure 1b. Let $\mathbf{a}_1$ and $\mathbf{a}_2$ denote the vectors $\overrightarrow{BC}$ and $\overrightarrow{DC}$ in Figure 1c, respectively, and a_1 and a_2 be the corresponding lengths. The length of BD is denoted by a_3, and the lengths of AB and AC are denoted by a_4 and a_5, respectively (Figure 1c). With these five lengths a_i $(i = 1, 2, \ldots, 5)$, all other lengths and angles in the "unrolled" plane can be completely determined. In order to characterize the cylindrical structure of the CNT shown in Figure 1a, it is necessary to prescribe the CNT diameter d_t and helicity together with these five lengths a_i $(i = 1, 2, \ldots, 5)$. The CNT diameter d_t and angle θ (Figure 1) are related to the chirality (n, m) of the CNT. Following the standard notation for CNTs, the chiral vector $\mathbf{C}_h$, whose length equals the circumference of the CNT, can always be expressed in terms of the base vectors $\mathbf{a}_1$ and $\mathbf{a}_2$ as (Figure 1c) $\mathbf{C}_h = n\mathbf{a}_1 + m\mathbf{a}_2$, where n and m are integers, $n \geq |m| \geq 0$, and the pair (n, m) is called the chirality of the CNT; $(n, 0)$ and (n, n) are called the zigzag and armchair CNTs, respectively. It can be found that the CNT diameter d_t and angle θ as well as the spatial coordinates of atoms A, B, C, and D can all be given in terms of these five lengths a_i $(i = 1, 2, \ldots, 5)$. With these five lengths, the configuration of single-wall CNTs prior to deformation can be completely characterized.

Once all bond lengths and angles are known (in terms of these five lengths a_i $(i = 1, 2, \ldots, 5)$), the energy stored in a bond can be obtained from the interatomic potential for carbon (1). The energy associated with the representative atom A is $1/2(V_{AB} + V_{AC} + V_{AD})$, which also depends on these five lengths a_i $(i = 1, 2, \ldots, 5)$, where the factor $1/2$ results from the equal partition of the bond energy between the pair of atoms in each bond, V_{AB}, V_{AC} and V_{AD} are the energies stored in bonds AB, AC, and AD, respectively. These five lengths a_i $(i = 1, 2, \ldots, 5)$ are determined by minimizing the energy in the representative atom A, i.e.,

$$\partial[V_{AB} + V_{AC} + V_{AD}]/\partial a_1 = 0, \quad i = 1, 2, \ldots, 5.$$

3.2 Continuum Description for Deformed Single Wall Carbon Nanotubes

In this section, we determine atom positions on a CNT via an atomistic-based continuum theory (Zhang et al., 2002, 2004, Jiang et al., 2003). The deformation gradient $\mathbf{F} = \partial \mathbf{x}/\partial \mathbf{X}$ characterizes the deformation of a material point in the continuum analysis, where the material point represents atoms that undergo locally uniform deformation, and $\mathbf{X}$ and $\mathbf{x}$ denote positions of the material point prior to and after deformation, respectively. For a CNT subject to tension, the deformed CNT remains to have a circular cross section such that the deformation gradient $\mathbf{F}$ is intrinsically two-dimensional. Arroyo and Belytschko (2002) and Jiang et al. (2003) accounted for the effect of CNT curvature in the continuum model based on the interatomic potential.

Let $\mathbf{r}_{ij}^{(0)}$ denote the position vector from atom i to atom j prior to deformation. For a material point subject to the deformation gradient $\mathbf{F}$, the position vector $\mathbf{r}_{ij}^{(0)}$ becomes $\mathbf{r}_{ij} = \mathbf{F} \cdot \mathbf{r}_{ij}^{(0)}$ after deformation. Using the Cauchy–Born rule (Born and Huang, 1954; Milstein, 1980) which equates the strain energy at continuum level to energy stored in atomic bonds, we obtain the strain energy density W as a function of deformation gradient $\mathbf{F}$, i.e., $W = W(\mathbf{F})$. Such an approach to obtain W from the interatomic potential, however, is limited to materials with a centrosymmetric atomic structure since the centrosymmetry together with $\mathbf{r}_{ij} = \mathbf{F} \cdot \mathbf{r}_{ij}^{(0)}$ ensure the equilibrium of atoms (Cousins, 1978; Weiner, 1983; Tadmor et al., 1999; Zhang et al. 2002, 2004).

A CNT, however, does not posses centrosymmetry. As shown in Figure 2a, a CNT prior to deformation is composed of two triangular sub-lattices (marked by open and solid circles, respectively), and each sublattice possesses centrosymmetry. Once the deformation is imposed, the Cauchy–Born rule discussed above can be applied to each sub-lattice, but two sub-lattices may undergo a shift vector $\boldsymbol{\zeta}$, as shown in Figure 2b. This shift vector $\boldsymbol{\zeta}$ plays the role of relaxing atoms between two sub-lattices in order to ensure equilibrium of atoms (Cousins, 1978; Weiner, 1983; Tadmor et al., 1999; Zhang et al. 2002, 2004). The position vector between atoms i and j from two different sub-lattices then becomes

$$\mathbf{r}_{ij} = \mathbf{F} \cdot \mathbf{r}_{ij}^{(0)} + \boldsymbol{\zeta}, \tag{9}$$

and their distance is

$$r_{ij} = \|\mathbf{r}_{ij}\| = \sqrt{\boldsymbol{\zeta} \cdot \boldsymbol{\zeta} + 2\boldsymbol{\zeta} \cdot \mathbf{F} \cdot \mathbf{r}_{ij}^{(0)} + \mathbf{r}_{ij}^{(0)} \cdot \mathbf{F}^T \cdot \mathbf{F} \cdot \mathbf{r}_{ij}^{(0)}}. \tag{10}$$

The energy stored in atomic bonds obtained from the interatomic potential now depends on both $\mathbf{F}$ and $\boldsymbol{\zeta}$. The Cauchy–Born rule then gives the strain energy density

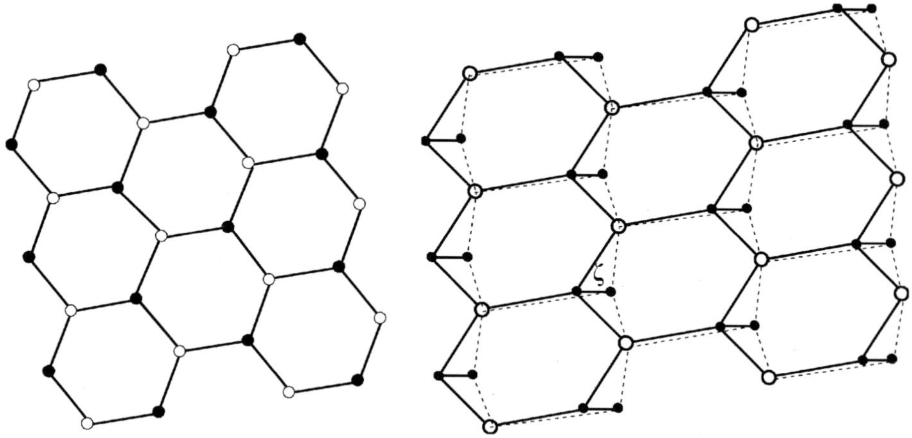

Fig. 2. (a) The decomposition of a hexagonal lattice to two triangular sub-lattices and (b) a shift vector f between two sub-lattices is introduced to ensure the equilibrium of atoms. The solid and dashed lines denote the lattice structures with and without the shift vector $\boldsymbol{\zeta}$, respectively.

W in the continuum analysis in terms of $\mathbf{F}$ and $\boldsymbol{\zeta}$ via the interatomic potential, i.e., $W = W(\mathbf{F}, \boldsymbol{\zeta})$.

The shift vector $\boldsymbol{\zeta}$ is determined by energy minimization, which is equivalent to equilibrium of atoms (Zhang et al., 2002, 2004; Jiang et al., 2003), i.e.,

$$\frac{\partial W}{\partial \boldsymbol{\zeta}} = 0. \tag{11}$$

This is an implicit equation to determine the shift vector $\boldsymbol{\zeta}$ in terms of $\mathbf{F}$, i.e., $\boldsymbol{\zeta} = \boldsymbol{\zeta}(\mathbf{F})$. The strain energy density then becomes

$$W = W[\mathbf{F}, \boldsymbol{\zeta}(\mathbf{F})]. \tag{12}$$

3.3 Stress and Incremental Modulus

The second Piola–Kirchhoff stress $\mathbf{T}$ is the work conjugate of the Green strain $E = 1/2(\mathbf{F}^T \cdot \mathbf{F} - \mathbf{I})$, i.e.,

$$\mathbf{T} = \frac{dW}{d\mathbf{E}} = \frac{\partial W}{\partial \mathbf{E}} + \frac{\partial W}{\partial \boldsymbol{\zeta}} \cdot \frac{\partial \boldsymbol{\zeta}}{\partial \mathbf{E}} = \frac{\partial W}{\partial \mathbf{E}}, \tag{13}$$

where Equation (11) has been used. The incremental modulus tensor $\mathbf{C}$ is obtained by taking the total derivative of the second Piola–Kirchhoff stress $\mathbf{T}$ with respect to $\mathbf{E}$ as

$$\mathbf{C} = \frac{d\mathbf{T}}{d\mathbf{E}} = \frac{\partial^2 W}{\partial \mathbf{E} \partial \mathbf{E}} - \frac{\partial^2 W}{\partial \mathbf{E} \partial \boldsymbol{\zeta}} \cdot \left(\frac{\partial^2 W}{\partial \boldsymbol{\zeta} \partial \boldsymbol{\zeta}} \right)^{-1} \cdot \frac{\partial^2 W}{\partial \boldsymbol{\zeta} \partial \mathbf{E}}. \tag{14}$$

3.4 Equilibrium Equations for Single Wall Carbon Nanotubes

The equilibrium equation for a single-wall CNT has been established by Zhang et al. (2002, 2004), and is given by

$$\frac{1}{R}\frac{\partial}{\partial\Theta}(\mathbf{F}\cdot\mathbf{T})_{R\Theta} - \frac{1}{R}(\mathbf{F}\cdot\mathbf{T})_{\Theta\Theta} + \frac{\partial}{\partial Z}(\mathbf{F}\cdot\mathbf{T})_{RZ} = 0,$$
$$\frac{1}{R}(\mathbf{F}\cdot\mathbf{T})_{R\Theta} + \frac{1}{R}\frac{\partial}{\partial\Theta}(\mathbf{F}\cdot\mathbf{T})_{\Theta\Theta} + \frac{\partial}{\partial Z}(\mathbf{F}\cdot\mathbf{T})_{\Theta Z} = 0,$$
$$\frac{1}{R}\frac{\partial}{\partial\Theta}(\mathbf{F}\cdot\mathbf{T})_{Z\Theta} + \frac{\partial}{\partial Z}(\mathbf{F}\cdot\mathbf{T})_{ZZ} = 0, \tag{15}$$

where (R, Θ, Z) denotes the cylindrical coordinates in the undeformed configuration.

3.5 A Single Wall Carbon Nanotube in Single Tension

The shear stress vanishes in a single wall CNT subject to simple tension along its axial (Z) direction,

$$T_{Z\Theta} = T_{\Theta Z} = 0 \tag{16}$$

Both the deformation gradient $\mathbf{F}$ and second Piola–Kirchhoff stress $\mathbf{T}$ are independent of Θ and Z such that the equilibrium equation (15a) gives

$$T_{\Theta\Theta} = 0, \tag{17}$$

while (15b) and (15c) are satisfied automatically. Equations (16) and (17) can be written in terms of the strain energy density W as

$$\frac{\partial W}{\partial E_{Z\Theta}} = \frac{\partial W}{\partial E_{\Theta Z}} = 0, \quad \frac{\partial W}{\partial E_{\Theta\Theta}} = 0, \tag{18}$$

which gives two implicit equations to determine $E_{\Theta\Theta}$ and $E_{Z\Theta}$ $(= E_{\Theta Z})$ in terms of axial strain E_{ZZ}. The axial force P on the CNT can be obtained by integrating the normal stress traction $\mathbf{e}_z \cdot (\mathbf{F}\cdot\mathbf{T}\cdot\mathbf{e}_z)$ in the cross-section, which gives

$$P = 2\pi R\sqrt{1+2E_{ZZ}}\,T_{ZZ}. \tag{19}$$

3.6 A Single Wall Carbon Nanotube in Pure Torsion

The non-vanishing components of the second Piola–Kirchhoff stress in pure torsion are $T_{\Theta\Theta}$, T_{ZZ}, and $T_{\Theta Z}$ $(= T_{Z\Theta})$, and they are independent of Θ for a CNT in pure torsion. The equilibrium equation (15) projected along the base vector $\mathbf{e}_r$, $\mathbf{e}_\theta$ and $\mathbf{e}_z$ in the cylindrical coordinates becomes

$$F_{\theta\Theta}T_{\Theta\Theta} + F_{\theta Z}T_{\Theta Z} + \kappa R(F_{\theta\Theta}T_{\Theta Z} + F_{\theta Z}T_{ZZ}) = 0,$$
$$\frac{d}{dZ}(F_{\theta\Theta}T_{\Theta Z} + F_{\theta Z}T_{ZZ}) = 0,$$
$$\frac{d}{dZ}(F_{zZ}T_{ZZ}) = 0, \tag{20}$$

where (R, Θ, Z) and (r, θ, z) denote the cylindrical coordinates prior to and after torsion deformation, respectively; κ is the twist for a CNT in pure torsion. The deformation gradient for pure torsion is given by

$$\begin{aligned}\mathbf{F} &= \frac{r}{R}\mathbf{e}_\theta\mathbf{e}_\Theta + \kappa r\mathbf{e}_\theta\mathbf{e}_Z + (1+\varepsilon)\mathbf{e}_z\mathbf{e}_Z \\ &= \frac{r}{R}(-\sin\kappa Z\mathbf{e}_R + \cos\kappa Z\mathbf{e}_\Theta)\mathbf{e}_\Theta \\ &\quad + \kappa r(-\sin\kappa Z\mathbf{e}_R + \cos\kappa Z\mathbf{e}_\Theta)\mathbf{e}_Z + (1+\varepsilon)\mathbf{e}_z\mathbf{e}_Z,\end{aligned} \tag{21}$$

where ε is the axial strain due to finite deformation. Combining with the boundary condition in pure torsion, i.e., $T_{ZZ} = 0$ at the two ends, the substitution of the deformation gradient in Equation (21) into (20) yields

$$T_{\Theta\Theta} = -2\kappa RT_{\Theta Z}, \quad T_{ZZ} = 0. \tag{22}$$

They are two implicit equations to determine the radius r and axial strain ε of the deformed CNT in terms of the twist κ. The torque T on the CNT can be obtained by integrating the shear stress traction $\mathbf{e}_\theta \cdot (\mathbf{F} \cdot \mathbf{T} \cdot \mathbf{e}_Z)$ multiplied by the radius r of the deformed CNT, which gives

$$T = 2\pi r^2 T_{\Theta Z}. \tag{23}$$

4 Comparison between Brenner's (1990) and Brenner et al.'s (2002) Potentials

In this section, we compare Brenner's (1990) interatomic potential and Brenner et al.'s (2002) second-generation potential in several cases, including equilibrium bond length, relation between energy and radius of CNT prior to deformation, Young's modulus, and material response to simple tension and pure torsion of single wall carbon nanotubes.

4.1 Equilibrium Bond Length

For a pair of carbon atoms in a graphene in which the bond angle is 120°, the equilibrium bond length r_0 can be obtained by solving a one-variable equation $dV/dr = 0$, where V is the interatomic potential of one carbon bond and is given in Equation (1). The equilibrium bond length r_0 is 0.145 nm for Brenner's (1990) interatomic potential and 0.142 nm for Brenner et al.'s (2002) second-generation potential. The latter, 0.142 nm for graphene is accurate and consistent with the results reported in other literatures, such as Desch (1934), Brown et al. (1997), and Harrison (1980).

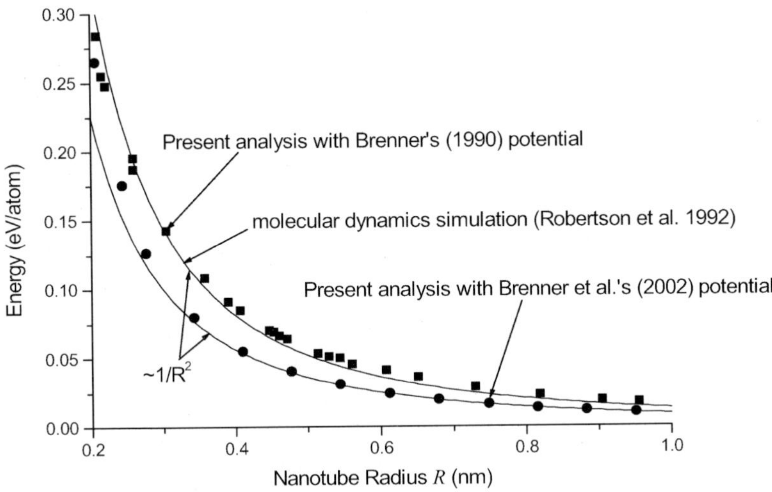

Fig. 3. The energy increase per atom in a carbon nanotube versus the nanotube radius R. A graphene sheet is taken as the ground state.

4.2 Energy versus Radius Prior to Deformation

If we take a graphene as the ground (i.e., zero energy) state, the energy increase per atom in a CNT is given by

$$\frac{1}{2}[V(r_{AB}) + V(r_{AC}) + V(r_{AD})] - \frac{3}{2}V(r_0),$$

where V is the interatomic potential in Equation (1), r_0 is the equilibrium bond length for graphene and obtained in Section 4.1. This energy increase per atom is shown versus the CNT radius in Figure 3 for Brenner's (1990) interatomic potential and Brenner et al.'s (2002) second-generation potential. For comparison, Robertson et al.'s (1992) molecular dynamics simulation results based on Brenner's (1990) interatomic potential is also shown. It is observed that the present analysis with Brenner's (1990) interatomic potential agrees very well with the atomistic studies, because they are based on same interatomic potential. It is also observed that both Brenner's (1990) and Brenner et al.'s (2002) potentials give that the energy scales with $1/R^2$ because both stress and strain are proportional to the curvature $1/R$ at small deformation. It is expected that the present analysis with Brenner et al.'s (2002) second-generation potential will agree with molecular dynamics simulation with the same interatomic potential.

4.3 Young's Modulus of Single Wall Carbon Nanotubes

The linear elastic modulus of a single wall CNT along the axial direction Z can be obtained from incremental modulus tensor $\mathbf{C}$ in Equation (14) for the infinitesimal deformation, i.e., $\mathbf{E} = 0$ and therefore $\zeta = 0$.

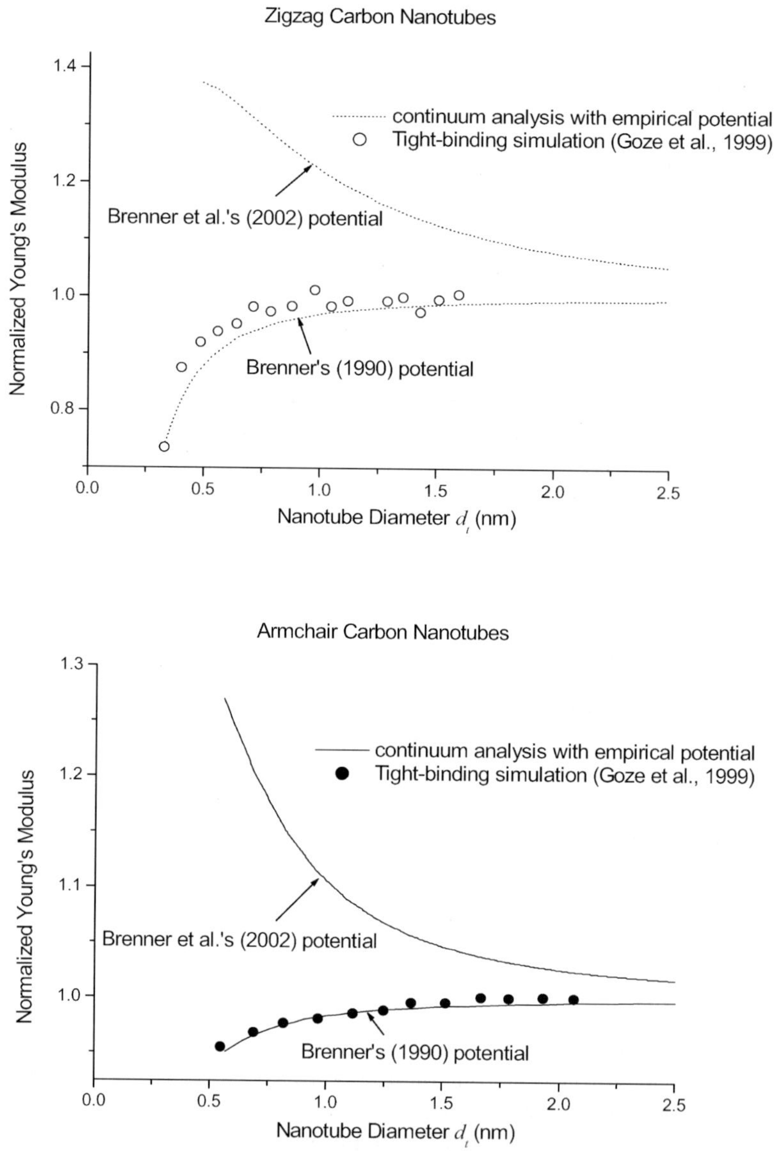

Fig. 4. The Young's modulus of carbon nanotubes (CNTs), normalized by that of the graphene, versus the nanotube diameter d_t for (a) zigzag CNTs and (b) armchair CNTs.

Specifically, the "Young's modulus" along the axial direction of the CNT is

$$\left[C_{ZZZZ} - \frac{C_{ZZ\Theta\Theta}^2}{C_{\Theta\Theta\Theta\Theta}} \right]_{\mathbf{E}=0,\boldsymbol{\zeta}=0}.$$

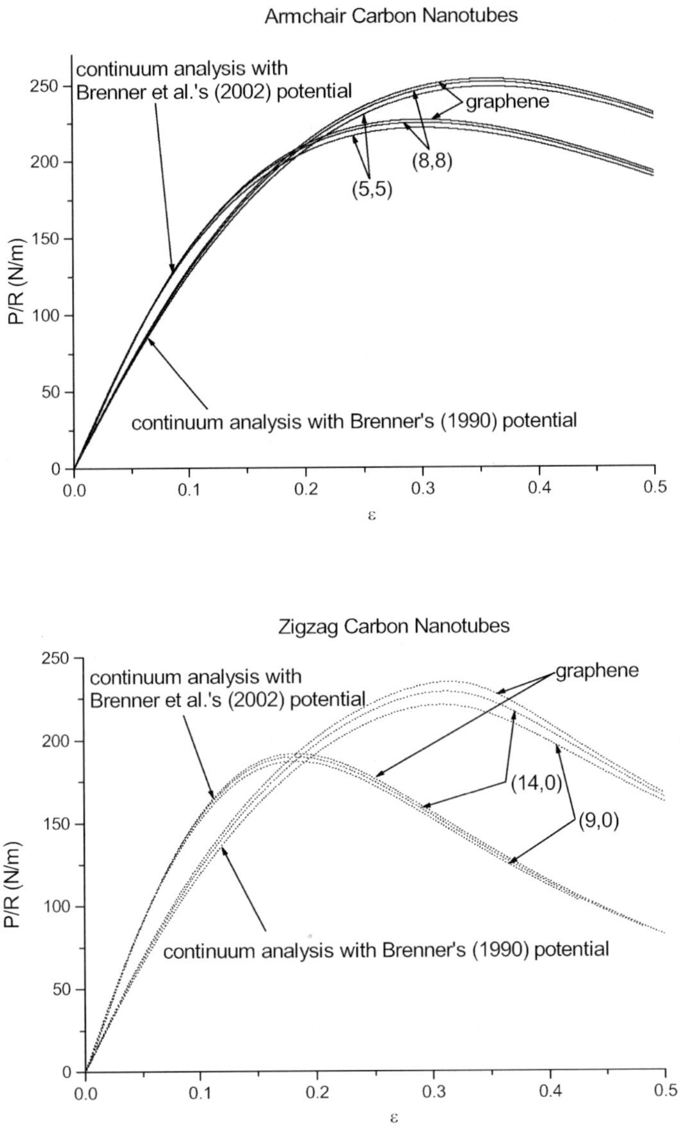

Fig. 5. The axial force P, normalized by the nanotube radius R prior to deformation, versus the engineering strain ε for several zigzag and armchair carbon nanotubes in simple tension.

However, it is important to point out that this "Young's modulus" is in fact the elastic modulus multiplied by the tube thickness. In order to avoid the choice of CNT thickness, we present the Young's modulus of CNT, normalized by that of graphene, versus the CNT diameter in Figures 4a and 4b for zigzag and armchair CNTs, respectively. The tight-binding simulation results (Goze et al., 1999) are also shown. It

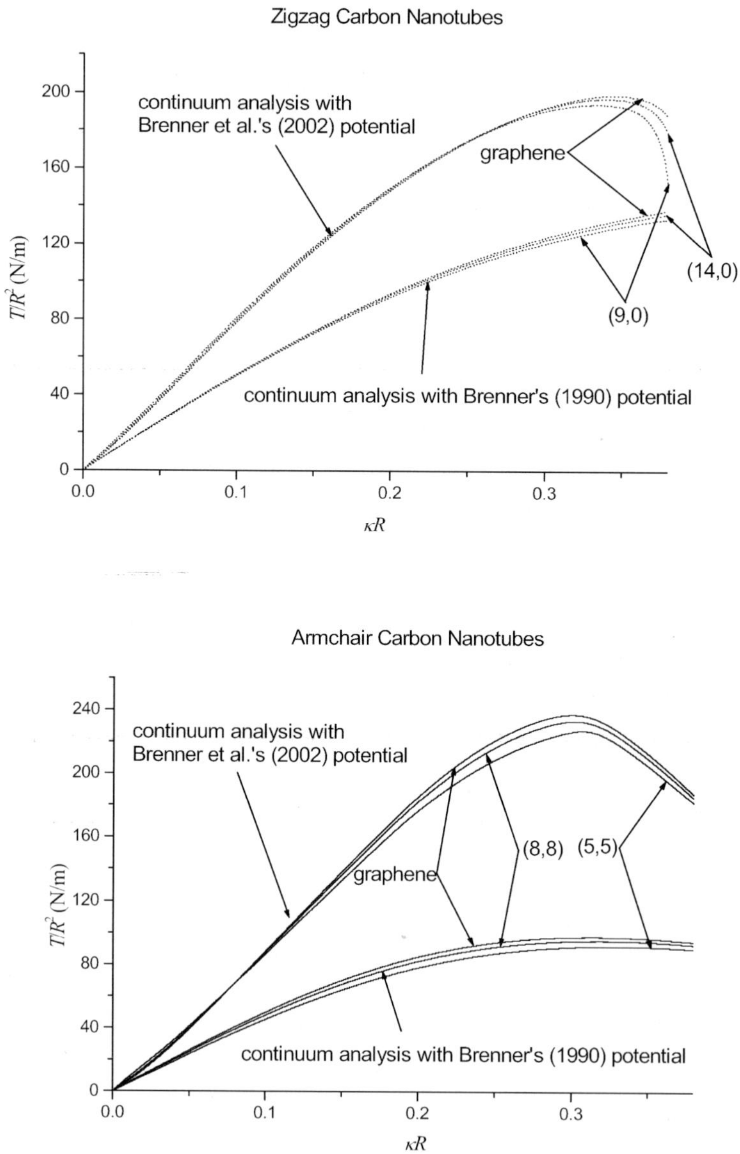

Fig. 6. The torque T, normalized the square of nanotube radius R^2 prior to deformation, versus the normalized twist κR for several zigzag and armchair carbon nanotubes in pure torsion.

is observed that the different potentials give completely different trends of Young's modulus. For Brenner's (1990) interatomic potential, the graphene is the toughest; on the contrary, for Brenner et al.'s (2002) second-generation potential, the graphene is the softest. Moreover, for Brenner's (1990) interatomic potential, Young's modulus of zigzag CNTs has strong diameter dependence and that of armchair CNTs does not. However, for Brenner et al.'s (2002) second-generation potential, Young's modulus of both zigzag and armchair CNTs has strong diameter dependence. It is also observed that the present continuum analysis with Brenner's (1990) interatomic potential agrees well with the tight-binding simulations over a wide range of CNT diameter. It implies that Brenner's (1990) interatomic potential predicts more accurate Young's modulus of CNTs.

4.4 Simple Tension in Single Wall Carbon Nanotubes

Figures 5a and 5b show the axial force P in Equation (19), normalized by the CNT radius R prior to deformation, versus the engineering strain ε for three zigzag CNTs [(9, 0), (14, 0), and $(\infty, 0)$] and three armchair CNTs [(5, 5), (8, 8), and (∞, ∞)]. For Brenner's (1990) interatomic potential, curves for armchair CNTs display little dependence on the CNT radius R and that for zigzag CNTs display strong dependence on CNT radius. However, for Brenner et al.'s (2002) second-generation potential, the curves for both zigzag and armchair CNTs display little dependence on the CNT radius. It is also observed that, when the deformation is small, both zigzag and armchair CNTs for Brenner et al.'s (2002) second-generation potential display stiffer than those for Brenner's (1990) interatomic potential, which is consistent with the results of Young's modulus shown in Figure 4.

4.5 Pure Torsion in Single Wall Carbon Nanotubes

Figures 6a and 6b show the torque T in Equation (23), normalized by the square of CNT radius R^2, versus the normalized twist κR for the same three zigzag and four armchair CNTs as in Figure 5. For Brenner et al.'s (2002) second-generation potential, both zigzag and armchair CNTs give larger torque T that for Brenner's (1990) interatomic potential, but enter the softening stage early than that for Brenner's (1990) interatomic potential.

5 Concluding Remarks

We have compared Brenner's (1990) interatomic potential and Brenner et al.'s (2002) second-generation potential on the radius effect of single wall carbon nanotubes. Our analysis is based on the modified Cauchy–Born rule to incorporate the interatomic potential into a continuum analysis. The modification of the Cauchy–Born rule ensures the equilibrium of atoms. It is observed that, Brenner et al.'s (2002) second-generation potential gives more accurate equilibrium bond length. These potentials also display rather different radius dependence of the Young's modulus and simple tension and pure torsion behavior of single wall carbon nanotubes.

Acknowledgements

The authors acknowledge the financial support from the NSF through Nano-CEMMS (Grant No. DMI 03-28162) at the University of Illinois. YH also acknowledges the support from Office of Naval Research (grant N00014-01-1-0205, Program Manager Dr. Y.D.S. Rajapakse), and NSFC.

References

Arroyo, M. and Belytschko, T. (2002) An atomistic-based finite deformation membrane for single layer crystalline films, *Journal of the Mechanics and Physics of Solids* **50**, 1941–1977.

Brenner, D.W. (1990) Interatomic potential for hydrocarbons for use in simulating the chemical vapor deposition of diamond films, *Physical Review B* **42**, 9458–9471.

Brenner, D.W., Shenderova, O.A., Harrison, J.A., Stuart, S.J., Ni, B. and Sinnott, S.B. (2002) A second-generation reactive interatomic bond order (REBO) potential energy expression for hydrocarbons, *Journal of Physics: Condensed Matter* **14**, 783–802.

Brown, T.L., LeMay, Jr. H.E., Bursten, B.E. and Burdge, J.R. (1997) *Chemistry: The Central Science*, Simon & Schuster, Upper Saddle River, NJ.

Born, M. and Huang, K. (1954) *Dynamical Theory of the Crystal Lattices*, Oxford University Press, Oxford.

Cousins, C. (1978) Inner elasticity, *Journal of Physics C: Solid State Physics* **11**, 4867–4879.

Desch, C.A. (1934) *The Chemistry of Solids*, Cornell University Press, Ithaca, NY.

Goze, C., Vaccarini, L., Henrard, L., Bernier, P., Hernandez, E. and Rubio, A. (1999) Elastic and mechanical properties of carbon nanotubes, *Synthetic Metals* **103**, 2500–2501.

Harrison, W.A. (1980) *Electronic Structure and the Properties of Solids: The Physics of the Chemical Bond*, Freeman, San Francisco, CA.

Jiang, H., Zhang, P., Liu, B., Huang, Y., Geubelle, P.H., Gao, H. and Hwang, K.C. (2003) The effect of nanotube radius on the constitutive model for carbon nanotubes, *Computational Materials Science* **28**, 429–442.

Milstein, F. (1980) Review: theoretical elastic behaviour at large strains, *Journal of Material Science* **15**, 1071–1084.

Robertson, D.H., Brenner, D.W. and Mintmire, J.W. (1992) Energetics of nanoscale graphitic tubules, *Physical Review B* **45**, 12592–12595.

Tadmor, E.B. Smith, G.S. Bernstein, N. and Kaxiras, E. (1999) Mixed 8nite element and atomistic formulation for complex crystals, *Physical Review B* **59**, 235–245.

Weiner, J.H. (1983) *Statistical Mechanics of Elasticity*, Wiley, New York.

Zhang, P., Huang, Y., Geubelle, P.H., Klein, P.A. and Hwang, K.C. (2002) The elastic modulus of single-wall carbon nanotubes: A continuum analysis incorporating interatomic potentials, *International Journal of Solids and Structures* **39**, 3893–3906.

Zhang, P., Jiang, H., Huang, Y., Geubelle, P.H. and Hwang, K.C. (2004) An atomistic-based continuum theory for carbon nanotubes: Analysis of fracture nucleation, *Journal of the Mechanics and Physics of Solids* **52**, 977–998.

Shape Memory Effect and Pseudoelasticity in Cu Nanowires

Wuwei Liang and Min Zhou*

School of Materials Science and Engineering, The George W. Woodruff School of Mechanical Engineering, Georgia Institute of Technology, Atlanta, GA 30332-0405, U.S.A.;
**E-mail: min.zhou@me.gatech.edu*

Abstract. We report the discovery of a novel pseudoelastic behavior in single-crystalline Cu nanowires through atomistic simulations. Under tensile loading and unloading, the nanowires are capable of recovering elongations up to 51%, well beyond the typical recoverable strains of 5–8% for most bulk shape memory alloys (SMAs). This phenomenon is associated with a reversible crystallographic lattice reorientation driven by the high surface-stress-induced internal stresses due to high surface-to-volume ratios at the nanoscale. The temperature-dependence of this behavior leads to a shape memory effect (SME). This behavior is well-defined for wires between 1.76 and 3.39 nm in size over the temperature range of 100–900 K.

Key words: shape memory effect, pseudoelasticity, lattice reorientation, nanowires.

1 Introduction

Shape memory materials have important applications as couplings, sensors, and actuators because of their ability to recover certain configurations under proper thermomechanical conditions. They are sometimes referred to as "smart materials" because they can function simultaneously as sensors and actuators [1, 2]. The SME and pseudoelasticity are considered unique to SMAs, liquid crystal elastomers, and piezoelectric ceramics [2]. Recent research has shown that pseudoelasticity may also be found in other materials at the nanoscale. For example, high resolution transmission electron microscope (HRTEM) observations and atomistic simulations have shown that gold nanowires exhibit reversible changes in electrical conductance, structure, and mechanical properties during cyclic tension and compression [3, 4]. Also, carbon nanotubes are found to completely recover their original shapes after severe deformations with strains up to 15% without inducing residual defects [5, 6]. Here, we report the discovery of a novel pseudoelastic behavior and SME in single-crystalline face-centered-cubic (FCC) Cu nanowires based on results of atomistic simulations. Under tensile loading and unloading, these wires are capable of recovering elongations up to 51%, well beyond the typical recoverable strains of 5–8% for most bulk shape memory alloys (SMAs). The critical temperature associated with

Yilong Bai et al. (eds), IUTAM Symposium on Mechanical Behavior and Micro-Mechanics of Nanostructured Materials, 135–144.

the activation of the SME here is significantly size-dependent, making it possible to design nanoscale components of varying sizes for operation over a wide range of temperature. Such an objective is more difficult to achieve with conventional bulk SMAs since their transition temperatures (martensite start and finish temperatures, austenite start and finish temperatures) only vary with material structure and composition, not size. Moreover, the nanowires also have very short response times which are on the order of nanoseconds due to their extremely small dimensions, making them attractive functional components for biosensors, transducers, actuators, and interconnects in nano-electromechanical systems (NEMS) [7].

2 Configuration

The analysis here focuses on Cu nanowires created experimentally through a "top-down" fabrication approach [8]. These wires have a single-crystalline FCC structure with a $\langle 110 \rangle$ axis and $\{111\}$ transverse surfaces (hereafter denoted as the $\langle 110 \rangle/\{111\}$ wire or configuration). This configuration represents a low energy state for FCC metallic nanowires and has been observed frequently in experiments and atomistic simulations for Au, Cu, and Ag nanowires [8–13]. The specific nanowires analyzed are created computationally in the spirit of the top-down fabrication process by "slicing" square columns of atoms from single-crystalline bulk Cu along the [001], [010], and [100] directions and by allowing them to undergo relaxation. Driven by surface stresses, the nanocolumns spontaneously transform into the $\langle 110 \rangle/\{111\}$ configuration through a lattice reorientation process, exhibiting a contraction in the axial direction and an expansion in the lateral directions. The resulting free-standing wires have the same FCC structure as that of bulk Cu at the same temperature [14], with a rhombic cross-sectional shape as shown in Figure 1a. In this paper, the side length of the rhombic cross-sections (Figure 1a) is used to identify the wire size. All simulations reported here are performed using the embedded-atom method (EAM) potential for Cu [15].

To analyze the wires' mechanical behavior, uniaxial tensile loading and unloading are carried out under quasi-static conditions [16]. Specifically, in each load step, all the atoms are first displaced according to a prescribed uniform strain increment of 0.125% in the length direction. The wires are then relaxed with their ends fixed at constant temperature for 9 picoseconds (ps) to obtain an equilibrium configuration at the prescribed strain. This relaxation process allows structural changes to occur, if the conditions so dictate. This process usually takes less than 6 ps and the averaged stress over the last 3 ps of the relaxation period at each load step is taken as the stress in the wire at the current strain. Unloading is implemented in the same manner, with a negative strain increment of -0.125%.

3 Pseudoelasticity

Upon uniaxial loading and unloading, wires with lateral dimensions between 1.76×1.76 and 3.39×3.39 nm exhibit a pseudoelastic behavior above a critical temperat-

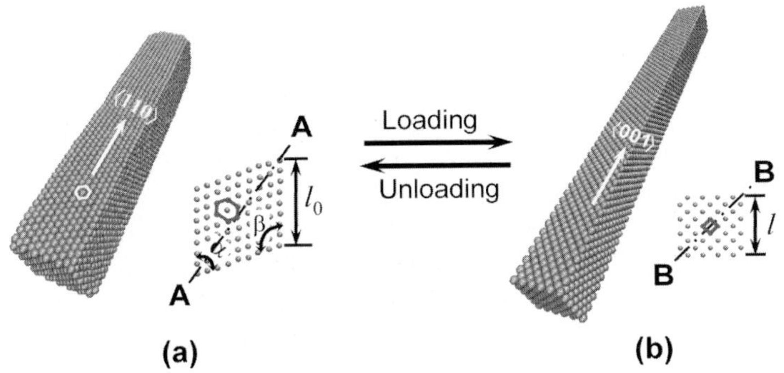

Fig. 1. Reversible lattice reorientations upon loading and unloading in single-crystalline Cu nanowires; (a) original ⟨110⟩/{111} wire with rhombic cross-sections, $\alpha = 70.5°$ and $\beta = 109.5°$, (b) stretched ⟨001⟩/{001} wire with square cross-sections.

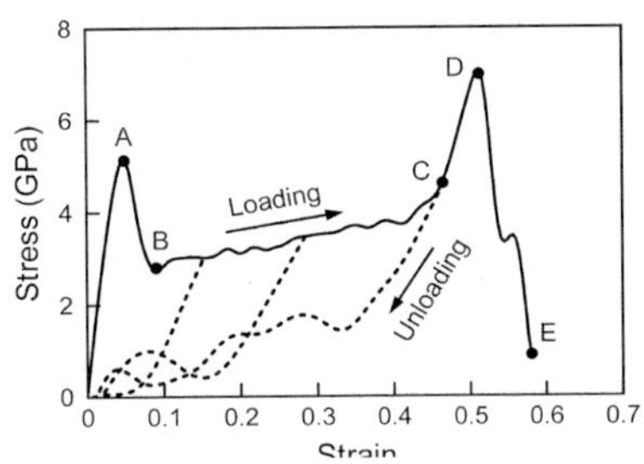

Fig. 2. The stress-strain curve of a 1.76 × 1.76 nm Cu nanowire during loading and unloading at 200 K.

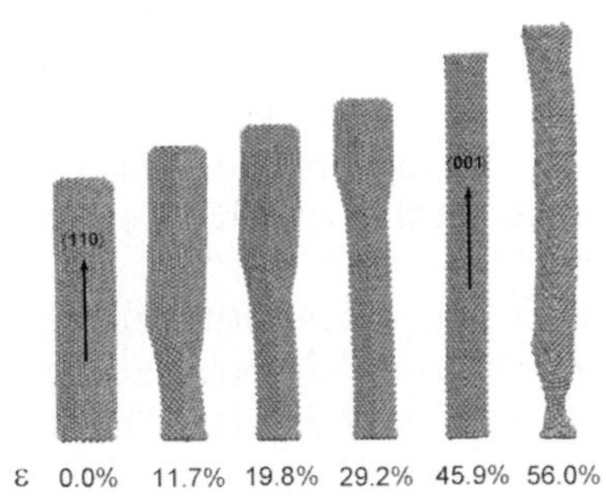

Fig. 3. The progression of the structural transformation during loading.

ure T_{cr} (discussed later) with large recoverable strains of up to 51%. Below T_{cr}, the deformation is not spontaneously recoverable and the wires retain their deformed configurations after unloading.

Figure 2 shows the stress-strain curve of a ⟨110⟩/{111} wire with a lateral dimension of 1.76 × 1.76 nm during loading and unloading at 200 K, and Figure 3 shows the deformation process of the wire. Clearly, the response is drastically different from

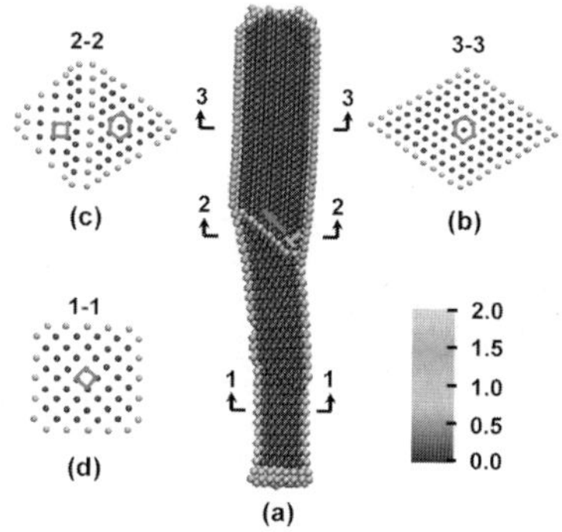

Fig. 4. Lattice orientations on the cross-sections of the wire in Figure 3 at a strain of 0.24; (a) a sectional view along the wire axis and the long diagonal of the cross-section, (b) elongated hexagonal lattice on the cross-section in the unrotated domain with the ⟨110⟩/{111} configuration, (c) a cross-section in the transition region containing both the ⟨001⟩/{001} and the ⟨110⟩/{111} configurations, (d) square lattice on the cross-section in the rotated domain with the ⟨001⟩/{001} configuration. Atoms are colored according to their centrosymmetry values.

that of bulk Cu. Specifically, the wire seems highly ductile with a fracture strain of approximately 58%. The stress-strain curve consists of two linear deformation stages (O→A and C→D) followed by two yield points (A and D, respectively), a stage of slow strain hardening over a wide range of strain (B→C), and a stage of precipitous stress drop (D→E). This behavior arises from a unique underlining deformation process. Between O and A, the ⟨110⟩/{111} wire undergoes elastic stretching. Point A corresponds to the beginning of a lattice reorientation process which leads to a new configuration with a ⟨001⟩ axis and {001} side surfaces (hereafter denoted as the ⟨001⟩/{001} wire or configuration), as shown in Figure 1b. Between C and D, the newly formed ⟨001⟩/{001} wire undergoes elastic stretching. Further loading beyond D causes the wire to yield through the formation and propagation of partial dislocations which ultimately lead to necking and fracture of the nanowire at E [17], as shown in Figure 3.

The unique lattice reorientation process (between point A and C in Figure 2) is completed through the propagation of a twin boundary. Specifically, the twin boundary is formed through the propagation of a $\frac{1}{6}\langle 112\rangle$ Shockley partial dislocation nucleated from an edge at the lower end. This partial dislocation glides across the wire on a {111} plane and leaves behind the twin boundary. Under the tensile loading, the twin boundary sweeps through the wire length and progressively transforms the wire into a new ⟨001⟩ orientation, as shown in Figure 4. Clearly, the twin boundary divides the wire into two domains: one with the initial ⟨110⟩/{111} configuration and the other with the ⟨001⟩/{001} configuration. A cross-section intersecting the twin boundary clearly shows the lattice transition between the two domains (Figure 4c). Upon the arrival of the boundary at the top end of the wire (corresponding to point C in Figure 2), The whole wire is in the ⟨001⟩/{001} state without residual defects.

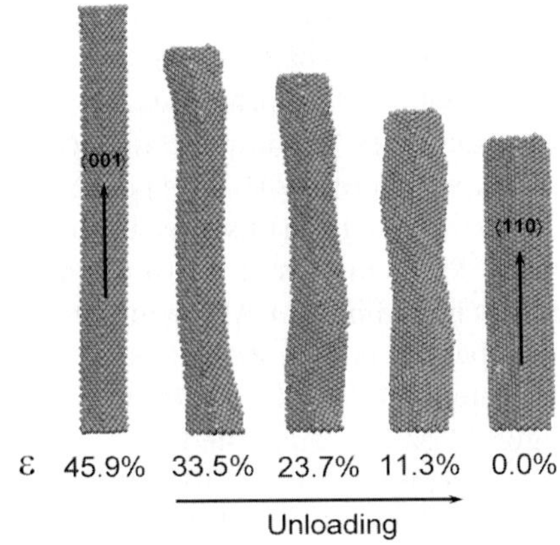

Fig. 5. The reverse lattice reorientation during unloading at 200 K.

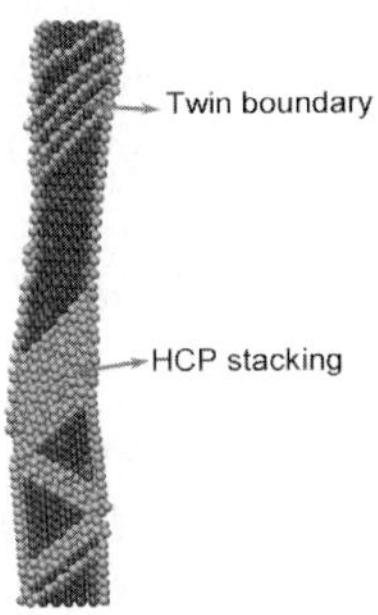

Fig. 6. A sectional view of the wire in Figure 5 showing the defects in the reverse lattice reorientation during unloading ($\varepsilon = 23.7\%$).

The lattice reorientation from ⟨110⟩/{111} to ⟨001⟩/{001} can be reversed upon unloading, allowing the associated deformation to be fully recovered, as shown in Figure 5. This reversibility gives rise to a pseudoelastic behavior of the wire. While the propagation of a single twin boundary is responsible for the lattice reorientation during loading, multiple twin boundaries are formed and propagate during unloading. Sometimes, domains with local HCP stacking are formed when multiple twin boundaries are nucleated next to each other, as shown in Figure 6. These twin boundaries dissociate, propagate, and annihilate during unloading. At the end of the unloading process, all twin boundaries disappear and the wire recover its original ⟨110⟩/{111} configuration without residual defects. The dash lines in Figure 2 represent the unloading paths from three different strains (0.05, 0.30, and 0.464, respectively). Clearly, the deformations are fully recovered when the stress is reduced to zero. The loading and unloading paths together form hysteresis loops typical of shape memory materials [2].

The aforementioned forward (loading) and reverse lattice reorientation (unloading) processes are critical to the SME of the wires because they result in a 41.4% recoverable strain (discussed later). Both processes have been confirmed by experiments. Specifically, FCC nanowires are found to undergo the same forward reorientation during tensile deformation in HRTEM experiments [18]. The spontaneous reverse reorientation process has also been observed in various experiments and computations [8, 9, 14, 19–21]. The same result is also obtained in computations with different atomistic potentials including an EAM, a modified embedded atom method (MEAM), and a surface embedded atom potential (SEAM)) [9].

The large strain associated with the forward and reverse lattice reorientations between A and C in Figure 2 can be quantified by a simple crystallographic analysis. Figures 7a and 7b compare the same $(1\bar{1}0)$ plane in the original undeformed $\langle 110\rangle/\{111\}$ configuration and the deformed $\langle 001\rangle/\{001\}$ configuration. Clearly, the forward and backward lattice reorientations manifest as 90° rotations in opposite directions of the unit cell in the plane. The length and width of the rectangular unit cell in both cases are, respectively, a and $(\sqrt{2}/2)a$; where a is the lattice constant in the stressed states and is assumed to be the same at A and C. The axial strain associated with the lattice reorientation between A and C is given by

$$\varepsilon_{\langle 110\rangle\leftrightarrow\langle 001\rangle} = \left(a - \frac{\sqrt{2}}{2}a\right) \Big/ \frac{\sqrt{2}}{2}a = 0.414. \tag{1}$$

This value of $\varepsilon_{\langle 110\rangle\leftrightarrow\langle 001\rangle}$ is consistent with the value obtained in simulations, as shown in Figure 2. This strain, along with the elastic strain $\varepsilon^e_{\langle 110\rangle}$ associated with the lattice stretching in the $\langle 110\rangle/\{111\}$ configuration between O and A (Figure 2) and the elastic strain $\varepsilon^e_{\langle 001\rangle}$ associated with the lattice stretching in the $\langle 001\rangle/\{001\}$ configuration between C and D, constitutes the total pseudoelastic strain of

$$\varepsilon_r \approx \varepsilon^e_{\langle 110\rangle} + e_{\langle 110\rangle\rightarrow\langle 001\rangle} + \varepsilon^e_{\langle 001\rangle} = 0.05 + 0.414 + 0.05 = 0.514. \tag{2}$$

This recoverable strain is essentially the same for all wires with lateral dimensions between 1.76 × 1.76 and 3.39 × 3.39 nm and endows the nanowires with the ability for pseudoelastic elongations of up to 51.4%, which is many times the typical 5–8% recoverable strains for most bulk SMAs [1].

4 Shape Memory Effect

The pseudoelastic deformations of some shape memory alloys (SMAs) such as Au-Cd, Au-Cu-Zn, Cu-Zn-Al, and Cu-Al-Ni [22–24] proceeds through the reversible movement of twin boundaries. The behavior of these materials is commonly referred to as rubber-like pseudoelasticity due to its resemblance to the behavior of soft rubber [2]. This mechanism is driven by a general tendency for the equilibrium symmetry of the short-range order configuration of lattice imperfections to conform to the symmetry of the lattice [24]. Hence, aging in the martensitic state and the existence of lattice imperfections are necessary conditions. Clearly, this mechanism is

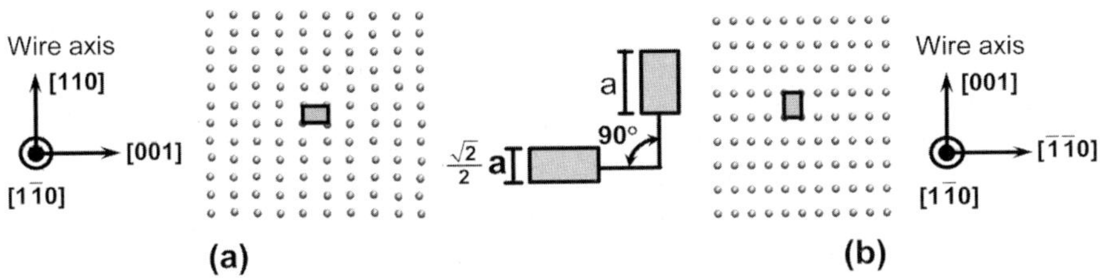

Fig. 7. Lattice rotation associated with the structural transformation during relaxation, (a) a $(1\bar{1}0)$ atomic plane containing the [110] wire axis and the long diagonal AA of the rhombic cross-section in the original wire in Figure 1a, (b) the same $(1\bar{1}0)$ atomic plane as in (a) after lattice reorientation, containing the new wire axis ([001]) and diagonal BB of the new square cross-section in Figure 1b.

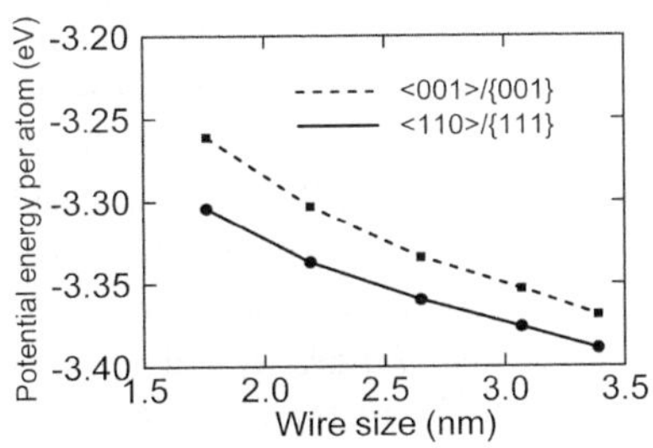

Fig. 8. Comparison of the potential energy per atom of wires with the $\langle 110\rangle/\{111\}$ and $\langle 001\rangle/\{001\}$ configurations at 300 K.

very similar to that of pseudoelastic behavior of the Cu nanowires. However, neither aging nor lattice imperfections are involved in the pseudoelastic behavior of the Cu nanowires. Then what causes the $\langle 001\rangle/\{001\}$ wire to spontaneously revert back to its original $\langle 110\rangle/\{111\}$ configuration upon unloading, since both states have the same FCC crystalline structure and, perhaps, the same "stability"? The answer lies in the surfaces and the extremely high surface-to-volume ratios of nanowires which can significantly affect structural stability. Specifically, the surface energy is 1.280 Jm^{-2} for $\{001\}$ planes and 1.17 Jm^{-2} for $\{111\}$ planes [15], causing the $\langle 110\rangle/\{111\}$ configuration to have a lower energy and to be more stable compared with the $\langle 001\rangle/\{001\}$ configuration. A quantification of the difference in the potential energy as a function of wire size between the two configurations is given in Figure 8. This energy difference primarily results from the energy density difference between $\{111\}$ and $\{001\}$ surfaces. The average potential energy per atom decreases with increasing wire size for each configuration because smaller wires have larger surface-to-volume ratios. On the other hand, regardless of size, $\langle 110\rangle/\{111\}$ wires always have lower energy levels compared with their deformed counterparts with the $\langle 001\rangle/\{001\}$ configuration. Therefore, the $\langle 001\rangle/\{001\}$ wire has a natural tendency for spontaneous reorientation back to the $\langle 110\rangle/\{111\}$ configuration upon unloading. The reorientation

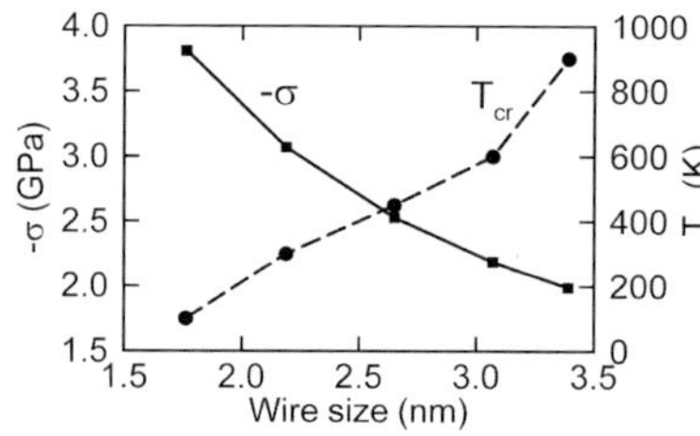

Fig. 9. Variations of surface stress induced compressive stress σ and the critical temperature T_{cr} with wire size.

essentially lowers the surface energy as a result of the increase in atomic density on surfaces when {001} surfaces reorganize into closely-packed {111} surfaces.

The driving force for the spontaneous reorientation can also be viewed as coming from the surface stress which induces a compressive stress in the interior of the wire. This compressive stress is $\sigma = 4fl/A$, where f is the surface stress of the {001} planes in the $\langle 001\rangle/\{001\}$ configuration, l is the side length of the square cross-section (Figure 1b), and A $(= l^2)$ is the corresponding cross-sectional area [16, 20]. Obviously, the magnitude of σ increases as the wire size decreases and can be very high at the nanoscale, as shown in Figure 9. For example, $\sigma = -3.81$ GPa for a $\langle 001\rangle/\{001\}$ wire with $l = 1.45$ nm ($l_0 = 1.76$ nm in the $\langle 110\rangle/\{111\}$ state), sufficient for initiating the reverse reorientation at temperatures above 100 K, even in the absence of externally applied forces. Note, however, that σ is only on the order of Pascals in bulk materials and is negligible, providing an explanation as to why a similar behavior is not seen in bulk Cu.

Like the behavior of normal bulk SMAs, the pseudoelastic behavior reported here is strongly temperature-dependent. Specifically, the reverse lattice reorientation from $\langle 001\rangle$ to $\langle 110\rangle$ occurs only above a size-dependent critical temperature T_{cr} (Figure 9). If unloading takes place at temperatures below T_{cr}, the reverse lattice reorientation does not occur and the wire retains the $\langle 001\rangle/\{001\}$ configuration. When subsequently heated above T_{cr}, the unloaded $\langle 001\rangle/\{001\}$ wire spontaneously returns to its original $\langle 110\rangle/\{111\}$ configuration through the reverse lattice reorientation. This is a novel SME driven by surface stress and the high surface-to-volume ratios of the nanowire. It is a one-way SME that has the $\langle 110\rangle/\{111\}$ configuration as the parent state, as shown in Figure 10.

If the $\langle 110\rangle/\{111\}$ state always has a lower energy than the corresponding $\langle 001\rangle/\{001\}$ state regardless of size, why does the reverse reorientation only occur above T_{cr}? The answer has to do with the energetic barrier and driving force of the process. To initiate the reorientation, partial dislocations nucleate and propagate to accommodate mobile twin boundaries. These defects are of higher energies and thus constitute an energy barrier for the reorientation. Thermal energy can provide the necessary energy for overcoming the barrier [25]. As wire size increases, σ decreases and higher temperatures are needed to initiate the spontaneous reverse reorientation, as shown in Figure 9. For example, T_{cr} is 100 K for a 1.76×1.76 nm wire and 900 K

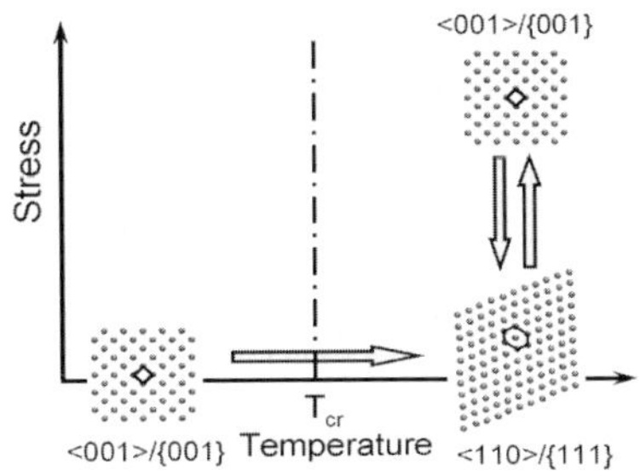

Fig. 10. An illustration of the shape memory effect in Cu nanowires.

for a 3.39 × 3.39 nm wire. For wires thicker than 3.39 × 3.39 nm, T_{cr} approaches a significant fraction of the melting point. Under such conditions, the pseudoelastic behavior and the SME are no longer obvious because the wire behavior becomes disorganized and dominated by random atomic vibrations. Because of this reason, the well-defined rubber-like pseudoelasticity and SME only exist in Cu nanowires with lateral dimensions below 3.39 nm.

In summary, the temperature dependence of the rubber-like pseudoelastic behavior in single-crystalline Cu nanowires leads to an SME which is well-defined for sizes between 1.76 and 3.39 nm over the temperature range of 100–900 K. Tensile strains up to 51% can be recovered. The responsible mechanism is a reversible lattice reorientation driven by the high surface-stress-induced internal stresses at the nanometer scale, explaining why such a behavior is not observed in Cu at higher scales. This unique behavior makes Cu nanowires attractive functional components for biosensors, transducers, actuators, and interconnects in NEMS.

Acknowledgements

This research is supported by NASA Langley Research Center through grant number NAG-1-02054. Computations are carried out at the NAVO, ARL, and ERDC MSRCs through AFOSR MURI number D49620-02-1-0382.

References

1. Otsuka, K. and Ren, X., Recent developments in the research of shape memory alloys. *Intermetallics* **7**, 1999, 511–528.
2. Otsuka, K. and Wayman, C.M., *Shape Memory Materials*. Cambridge University Press, New York, 1998, p. 282.
3. Landman, U., Luedtke, W.D., Salisbury, B.E. and Whetten, R.L., Reversible manipulations of room temperature mechanical and quantum transport properties in nanowire junctions. *Physical Review Letters* **77**(7), 1996, 1362–1365.
4. Landman, U., Barnett, R.N. and Luedtke, W.D., Nanowires: Size evolution, reversibility, and one-atom contacts. *Zeitschrift für Physik D* **40**, 1997, 282–287.

5. Yakobson, B.I., Brabec, C.J. and Bernholc, J., Nanomechanics of carbon nanotubes: Instabilities beyond linear response. *Physical Review Letters* **76**(14), 1996, 2511–2514.
6. Falvo, M.R., Clary, G.J., II, R.M.T., Chi, V., Jr, F.P.B., Washburn, S. and Superfine, R., Bending and buckling of carbon nanotubes under large strain. *Nature* **389**, 1997, 582–584.
7. Büttgenbach, S., Bütefish, S., Leester-Schädel, M. and Wogersien, A., Shape memory microactuators. *Microsystem Technologies* **7**, 2001, 165–170.
8. Kondo, Y. and Takayanagi, K., Gold nanobridge stabilized by surface structure. *Physical Review Letters* **79**(18), 1997, 3455–3458.
9. Diao, J., Gall, K. and Dunn, M.L., Surface stress driven reorientation of gold nanowires. *Physical Review B* **70**, 2004, 075413.
10. Liu, Z., Yang, Y., Liang, J., Hu, Z., Li, S., Peng, S. and Qian, Y., Synthesis of copper nanowires via a complex-surfactant-assisted hydrothermal reduction process. *Journal of Physical Chemistry B* **107**, 2003, 12658–12661.
11. Liu, Z. and Bando, Y., A novel method for preparing copper nanorods and nanowires. *Advanced Materials* **15**(3), 2003, 303–305.
12. Diao, J., Gall, K. and Dunn, M.L., Yield strength asymmetry in metal nanowires. *Nano Letters* **4**(10), 2004, 1863–1867.
13. Rodrigues, V. and Ugarte, D., Structural study of metal nanowires. In *Simulation and Modeling of Mechanical Deformation of Nanowire in Nanowire Materials*, Z.L. Wang (Ed.), Kluwer Academic/Plenum Publishers, 2003, pp. 177–209.
14. Liang, W. and Zhou, M., Pseudoelasticity of single crystalline Cu nanowires through reversible lattice reorientations. *Journal of Engineering Materials and Technology* **127**(4), 2005.
15. Foiles, S.M., Baskes, M.I. and Daw, M.S., Embedded-atom-method functions for the fcc metals Cu, Ag, Ni, Pd, Pt, and their alloys. *Physical Review B* **33**(12), 1986, 7983–7991.
16. Gall, K., Diao, J. and Dunn, M.L., The strength of gold nanowires. *Nano Letters* **4**(12), 2004, 2431–2436.
17. Liang, W. and Zhou, M., Response of copper nanowires in dynamic tensile deformation. *Journal of Mechanical Engineering Science* **218**(6), 2004, 599–606.
18. Rego, L.G.C., Rocha, A.R., Rodrigues, V. and Ugarte, D., Role of structure evolution in the quantum conductance behavior of gold nanowires during stretching. *Physical Review B* **67**, 2003, 045412.
19. Kondo, Y., Ru, Q. and Takayanagi, K., Thickness induced structural phase transition of gold nanofilm. *Physical Review Letters* **82**(4), 1999, 751–754.
20. Diao, J., Gall, K. and Dunn, M., Surface-stress-induced phase transformation in metal nanowires. *Nature Materials* **2**, 2003, 656–660.
21. Hasmy, A. and Medina, E., Thickness induced structural transition in suspended fcc metal nanofilms. *Physical Review Letters* **88**(9), 2002, 096103.
22. Christian, J.W., Deformation by moving interface. *Metall. Trans.* **13A**, 1982, 509–538.
23. Cahn, R.W., Metallic rubber bounces back. *Nature* **374**, 1995, 120–121.
24. Ren, X. and Otsuka, K., Origin of rubber-like behavior in metal alloys. *Nature* **389**, 1997, 579–582.
25. Meyers, C., *Mechanical Metallurgy Principle and Applications*. Prentice-Hall, Englewood Cliffs, NJ, 1984, pp. 320–327.

Instabilities of Carbon Nanotubes Studied Using a Hybrid Atom/Continuum Approach

L.-F. Wang and Q.-S. Zheng*

Department of Engineering Mechanics, Tsinghua University, Beijing 100084, China;
**E-mail: zhengqs@tsinghua.edu.cn*

Abstract. This article describes the methodology of hybrid atom/continuum (HAC) approach, and summarizes some of its applications, particularly on instability behaviors of carbon nanotubes.

Key words: instability, carbon nanotubes, hybrid atom/continuum approach, thin shell model, elastic constants.

1 Introduction

It is known that the carbon-carbon (C-C) bond of graphite is the strongest interaction between atoms in the nature [1, 2]. The atomic structure of a single-walled carbon nanotube (SWNT) can be generated by rolling a graphite monolayer into a cylindrical tubule structures, and then releasing it into an energy optimized, still tubule form [3, 4]. Thus, C-C bonds of SWNTs are found to be similar strong as those of graphite [5]. In contrast, the interaction between constituent SWNTs in multi-walled carbon nanotubes (MWNTs) or in SWNT-bundles is extra-weak, characterized by the van der Waals force. The above extreme properties promise unusual applications of carbon nanotubes (CNTs), such as nanoscale sensors, bearing, gigahertz oscillators and nano-electromechanical systems (NEMs). They also allow us to employ a hybrid atom/continuum (HAC) approach to study various complex mechanical and electromechanical behaviors of these CNT-based devices and NEMS, as well as SWNT-based structures and structural materials, that are beyond the ability scopes of the current first principle calculations, tight-binding analyses, or even semi-empirical potential-based atomics mechanics and molecular dynamics (MD) simulations.

In the HAC approach [6–8], individual SWNTs are modeled as elastic continuous thin shells and the interaction between two SWNTs is simulated using a simple pair-potential (e.g. the 6–12 Lennard–Jones potential) to represent the van der Waals force. In this paper, we explain why and what kind of thin shell model is valid. Using the HAC approach, we have proposed to create gigahertz oscillators based on

Yilong Bai et al. (eds), IUTAM Symposium on Mechanical Behavior and Micro-Mechanics of Nanostructured Materials, 145–152.

MWNTs [7]; we have obtained the full set of elastic parameters of SWNT bundles as bulk materials [8]; and we have also studied the phase transitions of SWNT bundles [8], leading to results that are in excellent agreement with experimental observations. These works are used as examples in the sequel for explaining the power of the HAC approach.

2 Thin Shell Model of SWNTs

Using thin-shell model, Yakobson, Brabec and Bernholc (YBB) [9] have first demonstrated remarkable success in capturing many complex deformation behavior of SWNTs, such as the buckling strains under compression, torsion, and bend. By modeling a SWNT as an isotropic linear elastic cylindrical thin shell rolled from a graphitic monolayer, they have calculated the pre-energies W_0 of SWNTs and the excess energies W_1 of SWNTs under axial tensile strains ε. They further equate the bending rigidity $D_b = Yt^3/12(1-\nu^2)$ and the tensile rigidity $D_a = Yt$ of the modeled shells, that are expressed in terms of the Young's modulus Y, Poisson's ratio ν and shell thickness t, as the second-derivatives of W_0 with respect to the tube curvatures $1/r$ and the second-derivatives of W_1 with respect to ε, respectively, leading to the unexpected super-thin effective thickness $t = 0.066$ nm and super-high Young's modulus $Y = 5.5$ TPa. The value $\nu = 0.19$ is extracted from a reduction of the diameter of one tube stretched in simulations. Since then, the YBB thin-shell model has generated much debate centered at the wall thickness of the continuum models for SWNTs. There have been some proposals to avoid introducing a thickness or set the average interwall spacing of MWNTs as the thickness for SWNT (cf. [10, 11]). Nevertheless, these proposals do not offer the same advantages and self-consistency that the YBB thin-shell model has brought to us [12].

Our recent research [12] has showed that the thin shell model of SWNT could be studied more subtly. For small tubes we have found that the isotropy of thin shell model breaks down and all the model parameters, Young's moduli and thickness for instance, have remarkable size-dependence. Using *ab initio* calculations and atomistic models, we observe that the circumferential strain is one-order larger than the axial strain of a SWNT in reference to their corresponding rolled-up structures. This has therefore motivated us to validate the thin-shell model of SWNTs by considering orthotropic symmetry, instead of isotropic symmetry, with the axial and circumferential directions being privileged. There are five possibly independent elastic constants that relate the in-shell strains to the in-shell stresses. Further we have calculated all the six parameters including the shell thickness, and the results show that SWNTs have the square symmetry, higher than the orthotropy symmetry, and only SWNTs with diameters larger than 1 nm could be modeled as isotropic thin shells with constant thickness and elastic moduli (see Figure 1).

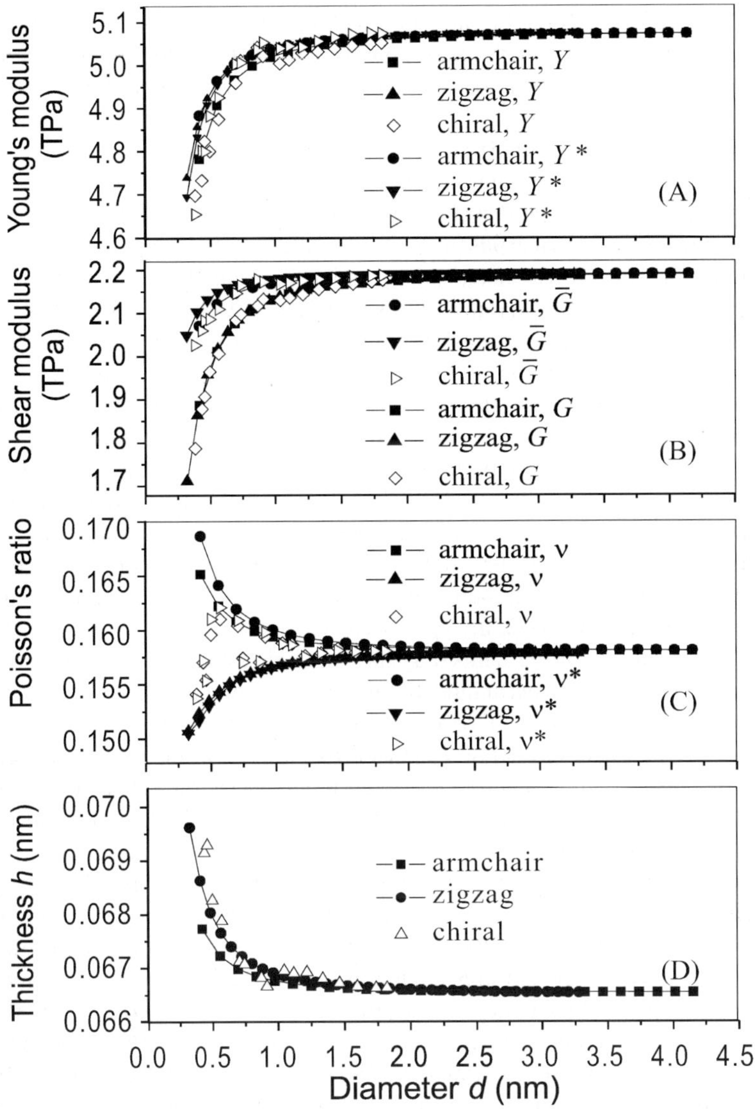

Fig. 1. Diameter-dependence of all the thin-shell model parameters: (A) the axial and circumferential Young's moduli. (B) the shear modulus. (C) Poisson's ratios. (D) the wall thickness. (From [12])

3 Applications of HAC Approach

Using the HAC approach, Zheng and Jiang [7] have proposed to create nanoscale mechanical oscillators of frequencies in the several gigahertz range based on MWNTs. They calculate the restoring force induced by the excess van der Waals energy due to extrusion and the intershell sliding resistance force with assuming the shells of MWNT as rigid ones. The frequencies of such oscillators predicted by this model and the shear strength between two sliding tubes agree well with that of the MD simulations [13–15].

Another application of the HAC approach is to determine the transverse mechanical properties of SWNT bundles [8]. We have calculated the bulk elastic moduli of SWNT bundles in reference to a stable configuration. Owing to their transverse isotropy, the linear elastic properties of SWNT bundles can be completely characterized by five independent elastic constants, i.e., the in-plane bulk modulus or compressibility, in-plane shear modulus, axial Young's modulus, out-of-plane Poisson's ratio and out-of-plane shear modulus. For estimating the in-plane elastic properties, we calculate the intratube and intertube energies using above-mentioned MD-based thin shell model and a 6–12 Lennard–Jones potential in presenting the van der Waals interaction, respectively. Although the van der Waals energy could be estimated by an integral of replacing the discrete distribution of atoms along the tube with a continuum distribution with the same atom density, in calculating the out-of-plane shear modulus c_{44} of SWNT bundles, purely atomic model is needed as illustrated in Figure 2. This allows us to have for the first time predicted a complete set of all the five elastic moduli. For typical SWNT bundles of tube diameter 1.4 nm, the predicted elastic coefficients are $c_{11} = 40.68$ GPa, $c_{12} = 39.32$ GPa, $c_{66} = (c_{11}-c_{12})/2 = 0.68$ GPa, $c_{13} = 12.40$ GPa, $c_{33} = 625.72$ GPa, and (in average) $c_{44} = 1.22$ GPa. We have also examined the elastic coefficients of all more than two hundreds hexagonal crystals listed in the handbook by Every and McCurdy [16]. It is found that the elastic properties of graphite have the highest anisotropy degree, 0.666, and the lowest modulus ratio, 0.39%, compared with those of all other hexagonal crystals. Interestingly, we further find that the predicted elastic properties of SWNT bundles have higher anisotropy degrees and lower modulus ratios than those of graphite. For instance, for SWNT bundles of tube diameter 1.4 nm, the anisotropy degree is 0.836, and the smallest modulus ratio is 0.11%. This reveals new type of extreme materials. All excellent and extreme properties of SWNT bundles, including unique electronic properties, give these new extremely high anisotropic materials more opportunities to become new potential functional materials.

We have also noted that the deformability of tube cross-sections plays the dominant role in characterizing the transverse moduli. Plano-parallel faceting surfaces between adjacent tubes have a lower van der Waals interaction energy than that between two adjacent perfectly circular tubes, a fact that favors polygonizing (see Figure 3). The characteristics of the lattice instability have critical implications on the elastic moduli, resulting from linearization of the elastic behavior of SWNT bundles in reference to a chosen stable state. We have studied the instability characteristics of the lattice of SWNT bundles by analyzing the balance between the intratube atomic interactions and the intertube van der Waals interactions using the same approach described above. We have thus obtained the pressure-strain relations for SWNT bundles with tube diameters ranging from 1 to 2.1 nm shown in Figure 4. It shows that tube cross-sections of a perfect lattice of SWNT bundles turn from perfectly circular to hexagonal with rounded corners, as the tube diameter increases to 2.1 nm, which is

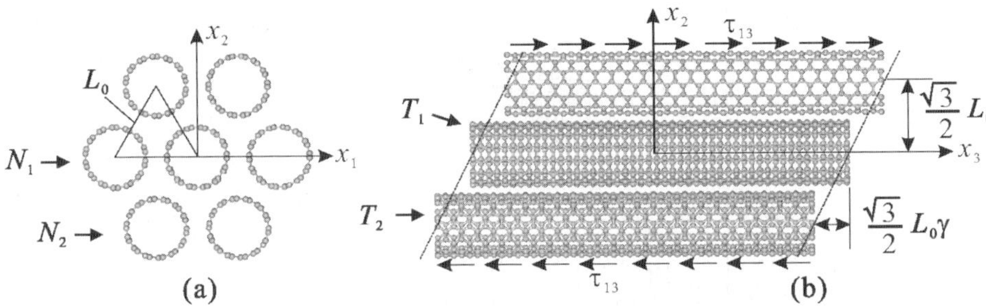

Fig. 2. A schematic illustration for the model used to estimate the out-of-plane shear modulus. (From [8])

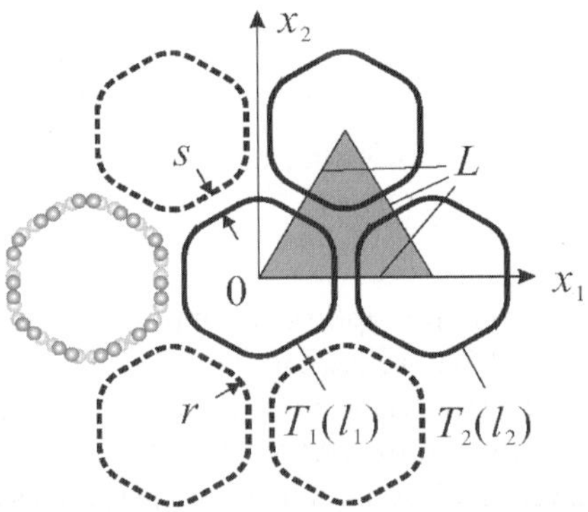

Fig. 3. A schematic illustration of a hexagonal lattice and the triangle lattice unit of SWNT bundles. (From [8])

larger than 1.7 nm as observed by Lopez et al. [17] in their HRTEM study and smaller than 2.5 nm as predicted by Tersoff and Ruoff [18] using an atomistic approach. We have noted that localized structural imperfections often promote instabilities, and using an illustrative example we have showed that a slightly perturbed lattice would result in the onset of cross section polygonization of tubes with a considerably smaller diameter (1.8 nm, for this example). With surprise, we have also observed from Figure 4 that the pressure-strain curves for tubes of different diameters nearly coincide prior to their respective inflection points, and this new finding implies that the in-plane bulk modulus at the origin is nearly independent of tube diameter.

The HAC approach could be also helpful in studying the electromechanical properties of CNT-based nano-devices. For instance, electronic transport properties of the squashed armchair carbon nanotubes have been investigated, using tight-binding molecular dynamics and the Green's function method by Lu et al. [19]. They first process structural optimization without analyzing the electron bands; then using the optimized structure, they employ the TB Green's function method to study the electronic transport properties of the squashed tubes. Instead of their performed tight-

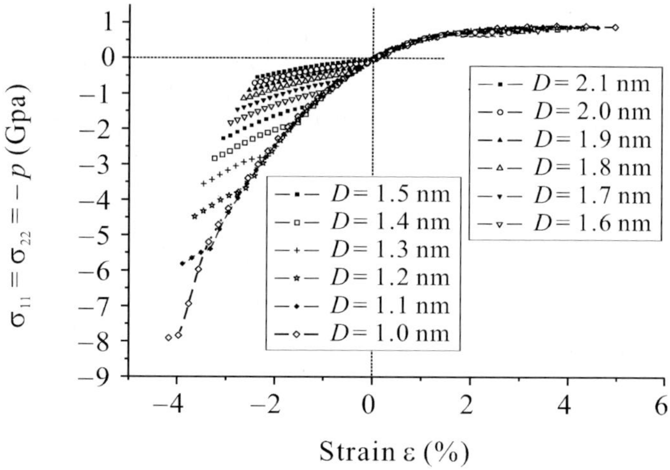

Fig. 4. The pressure-strain relations for SWNT bundles with diameters ranging from 1.0 nm to 2.1 nm. (From [8])

binding calculations, this structural optimization process could alternatively be simulated using the HAC approach to remarkably reduce the computational cost. This is important or even necessary when larger CNTs would be dealt with, such as investigating the electrical properties of bent MWNT.

Finally we note that the HAC approach is not the same as the multiscale method. The latter allows the use of a MD-like method in localized region, where quantities vary quickly on the atomic length scale, and a continuum description in its surrounding material, where small scale variations are insignificant or can be treated in an averaged sense. Two or three methods are seamlessly blended and each simulation is performed on a different region, with a coupling imposed in "handshake" regions where the different simulations overlap. An example of mutiscale approach for MWNTs introduced by Qian et al. [20] show the applicability of this method on CNT systems, in which a meshfree approximation is used for the coarse scale region and meshfree discretization and MD coexist and coupled in the fine scale region. However, the HAC approach does not partition the simulated object into different regions where different methods are applied. It is different from the multiscale method that the HAC approach only uncouples the interactions which affect each other little. There are no handshake regions in HAC model that could avoid the exchanging information about internal force and boundary conditions.

4 Concluding Remarks

The tremendous difference in strength between the intra-wall C-C bond and the inter-wall van der Waals interaction for carbon nanotubes allows to employ a HAC approach to study various complex mechanical and electromechanical behavior of SWNT based structures and structural materials, such as MWNTs and SWNT bundles, that are beyond the ability scopes of the current first principles calculations, tight-binding analyses, or even semi-empirical potential-based atomics mechanics and MD simulations. The characteristic of HAC approach is to model constituent or elementary SWNTs as linear elastic thin shells and, synchronously, take into account of inter-wall interactions atomically. The elastic thin shell model has carefully been assessed, using MD and *ab initio* calculations. It has been found that all the model parameters, Young's moduli and thickness for instance, have remarkable size-dependence, and the isotropy is also broken down for small tubes. Using this approach, we have proposed to creat gigahertz oscillators. We have also studied the phase transitions of SWNT bundles, leading to results that are in excellent agreement with experimental observations. We have obtained the full set of elastic parameters of SWNT bundles as bulk materials. We have also shown that rippling of MWNTs under bending is primarily an effect of multiwalls. MWNTs with two or three shells could form a single or two kinks above a critical curvature when bent; the rippling mode appears when the number of MWNTs shells is larger than four or five. These concepts of the mechanisms of MWNTs ripping are helpful for understanding many unusual experimental observations [21–24]. Further more, we have suggested to study electromechanical properties of CNT-based devices using this approach and pointed out that it is one different approach from the multiscale method.

Acknowledgements

We acknowledge the support of the Chinese National Science Foundation Grants 10172051, 10252001, and 10332020, the 973-Project G2003CB615603, the Ministry of Education of China.

References

1. Coulson, C.A. (1952) *Valence*, Oxford University Press, Oxford.
2. Emri, J. (2003) Use of a novel relationship between bond length and bond order to calculate accurate bond orders for carbon-carbon bonds, *Journal of Molecular Structure: Theochem* **620**, 283–290.
3. Iijima, S. (1991) Helical microtubes of graphitic carbon, *Nature* **354**, 56–58.
4. Robertson, D.H., Brenner, D.W. and Mintmire, J.W. (1992) Energetics of nanoscale graphitic tubules, *Physical Review B* **45**, 12592–12595.
5. Ruoff, R.S. and Lorents, D.C. (1995) Mechanical and thermal properties of carbon nanotubes, *Carbon* **33**, 925–930.
6. Liu, J.Z. (2002) *Studies on Several Mechanical Problems of Carbon Nanotubes*, Thesis, Tsinghua University, Beijing.
7. Zheng, Q.S. and Jiang, Q. (2002) Multiwalled carbon nanotubes as gigahertz oscillators, *Physical Review Letters* **88**, 045503.
8. Liu, J.Z., Zheng, Q.-S., Wang, L.-F. and Jiang, Q. (2005) Mechanical properties of single-walled carbon nanotube bundles as bulk materials, *Journal of Mechanics and Physics of Solids* **53**, 123–132.

9. Yakobson B.I., Brabec C.J. and Bernholc, J. (1996) Nanomechanics of carbon tubes: Instability beyond the linear response, *Physical Review Letters* **76**, 2511.
10. Yakobson, B.I. and Avouris, P. (2001) Mechanical properties of carbon nanotubes, *Topics of Applied Physics* **80**, 287–327.
11. Shenderova, O.A., Zhirnov, V.V. and Brenner, D.W. (2002) Carbon nanostructures, *Critical Reviews in Solid State and Materials Sciences* **27**, 227–356.
12. Wang, L.-F., Zheng, Q.-S. Liu, J.Z. and Jiang, Q. (2005) Size dependence of the thin-shell model for carbon nanotubes, *Physical Review Letters* **95**, 105501.
13. Legoas, S.B., Coluci, V.R., Braga, S.F., Dantas, S.O. and Galvao, D.S. (2003) Molecular-dynamics simulations of carbon nanotubes as gigahertz oscillators, *Physical Review Letters* **90**, 055504.
14. Guo, W.L., Guo, Y.F., Gao, H.J. Zheng, Q.S. and Zhong, W.Y. (2003) Energy dissipation in gigahertz oscillators from multiwalled carbon nanotubes, *Physical Review Letters* **91**, 125501.
15. Tangney, P., Lourie, S.G. and Cohen, M. (2004) Dynamic sliding friction between concentric carbon nanotubes, *Physical Review Letters* **93**, 06653.
16. Every, A.G. and McCurdy, A.K. (1992) Low frequency properties of dielectric crystals. In D.F. Nelson (Ed.), *Numerical Data and Functional Relationships in Science and Technology, Group III: Crystal and Solid State Physics*, Springer-Verlag, Berlin.
17. Lćpez, M.J., Rubio, A., Alonso, J.A., Qin L.-C. and Iijima, S. (2001) Novel polygonized single-wall carbon nanotube bundles, *Physical Review Letters* **86**, 3056–3059.
18. Tersoff, J. and Ruoff, R.S. (1994) Structural properties of a carbon-nanotube crystal, *Physical Review Letters* **73**, 676–679.
19. Lu, J.Q. Wu, J., Duan, W.H., Liu, F., Zhu, B.F. and Gu, B.L. (2003) Metal-to-semiconductor transition in squashed armchair carbon nanotubes, *Physical Review Letters* **90**, 150061.
20. Qian, D., Wagner, G.J. and Liu, W.K. (2004) A multiscale projection method for the analysis of carbon nanotubes, *Computer Methods in Applied Mechanics and Engineering* **193**, 1603–1632.
21. Liu, J.Z., Zheng, Q.-S. and Jiang, Q. (2001) Effect of a rippling mode on resonances of carbon nanotubes, *Physical Review Letters* **86**, 4843.
22. Liu, J.Z., Zheng, Q.-S. and Jiang, Q. (2001) Effect of bending instabilities on the measurements of mechanical properties of multiwalled carbon nanotubes, *Physical Review B* **67**, 075414.
23. Arroyo, M. and Belytschko, T., (2003) Nonlinear mechanical response and rippling of thick multiwalled carbon nanotubes, *Physical Review Letters* **91**, 215505.
24. Pantano, A., Boyce, M.C. and Parks, D.M. (2003) Nonlinear structural mechanics based modeling of carbon nanotubes deformations, *Physical Review Letters* **91**, 145504.

Shallow and Deep Nanoindentation on W/NbN Nanolayers

S.X. Mao[1,*], B.M. Ennis[1] and Y.G. Wei[2]

[1]*Department of Mechanical Engineering, University of Pittsburgh, Pittsburgh, PA 15261, U.S.A.; *E-mail: smao@engr.pitt.edu*
[2]*LNM, Institute of Mechanics, Chinese Academy of Sciences, Beijing 100080, P.R. China*

Abstract. Superlatice W(100)/NbN(100) with bilayer periods ($\Lambda = 5.6$ and 10.4 nm) was non-isostructural superlattice material and fabricated by depositing alternating layers of single crystal tungsten (W), a body-centered cubic metal, and niobium nitride (NbN), a face-centered cubic ceramic, on a MgO single crystal substrate. The lattice constants of the ceramic and metal layers are 0.439 nm and 0.315 nm respectively. The superlattice are nanocomposites that exhibit a hardness at small bilayer repeat periods which exceeds the hardness predicted by the rule of mixtures for normal composites by deep nanoindentation, while shallow nanoindentations does not demonstrate the superhardening. The results indicate that the elastic modulus does not influence the hardness of the superlattice materials. The superhardening results at deeper indentation depths is related to the nature of the interface between the layers in the superlattice materials. Normally, superlattice gains hardness by losing deformability, however, the superlattice demonstrated excellent deformability when reaching the superhardening.

Key words: nanolayers, nanoindentation, superhardening.

1 Introduction

Multilayered materials have been the focus of a significant amount of research. Madan and Barnett [1, 2] have explored several different types of nitride based superlattice thin films and have described the fabrication, the structure and the hardness behavior of various thin films. Chu et al. [3] examined polycrystalline transition metal nitride superlattice films and demonstrated that the hardness for several materials is inversely proportional to the bilayer repeat period and showed that there is an optimum bilayer repeat period that maximizes the hardness for some materials. They also discussed several possible explanations for the hardness behavior such as the supermodulus effect, coherency strains, grain size reduction and Koehler's model. Clemens et al. [4] reviewed the hardness of several metallic and superlattice materials and discussed the strain relaxation, interface morphology and Koehler's model. Anderson et al. [5] examined 50 vol% Cu-50 vol% Ni multilayered samples and discussed the propagation of dislocation loops confined between the layers and

Yilong Bai et al. (eds), IUTAM Symposium on Mechanical Behavior and Micro-Mechanics of Nanostructured Materials, 153–159.

the nucleation of dislocations at the interface. Thus, different classes of multilayered materials have been examined, and several theories have been proposed to explain the mechanisms that may operate to produce this increase in hardness in these materials: the supermodulus effect, coherency strains, the effect of interfacial misfit dislocations on dislocation guide, the reduction of grain sizes as a result of thickness of the alternating layers, and image forces (Koehler's model) at the interface which resist dislocation glide [3, 4]. However, while these articles describe the hardness increases as layer thickness goes down and when layer thickness in the range of a few nanometers, the hardness of the superlattice will research maximum. In this paper, we present the superharening only occurs when indenter penetrates more interfaces. Also we found the superlattice of NbN/W is much hard and deformable than NbN ceramics.

2 Experimental Procedure

The nanocomposite chosen for this investigation was W/NbN which is classified as an immiscible, non-isostructural superlattice material. To evaluate the influence of the layer thickness on the results, two samples of this nanocomposite with different bilayer repeat periods ($\Lambda = 5.6$ nm and $\Lambda = 10.4$ nm) were prepared for experiments. The structure of this superlattice material has been well documented in the literature [1, 2]. The samples were fabricated by depositing alternating layers of single crystal tungsten (W), a body-centered cubic metal, and niobium nitride (NbN), a face-centered cubic ceramic, on a substrate. The lattice constants of the ceramic and metal layers are 0.439 nm and 0.315 nm respectively. A pictorial representation detailing the composition of each sample is shown in Figure 1. Note that the ceramic, NbN, is the surface layer for both nanocomposite samples. Two additional samples were prepared for each individual material comprising the nanocomposites for comparison. All four samples were epitaxially grown as a thin film approximately 1 μm thick on an MgO (001) substrate using reactive DC magnetron sputtering.

The experiments were conducted using Hysitron's Triboscope® Nanoindenter in conjunction with Digital Instruments' Nanoscope IIIa AFM imaging system. Each sample was ultrasonically cleaned with acetone to remove any surface debris. Prior to indentation, each sample was scanned to find suitable areas for indentation. After indentation, the sample surface was scanned *in situ* to record the image of the surface topography. Diamond cubic pyramidal indenter tips with an included angle of 90° were selected for the experiments since a sharp tip radius was necessary to achieve sufficient penetration into the hard samples.

As noted in the literature, the surface roughness of the sample can significantly influence the experimental results when conducting nanoindentation experiments [6, 7]. This is particularly important for shallow nanoindentations where the penetration of the indenter into the material is less than 25 nm. To minimize this influence, the sample surface was scanned *prior to* nanoindentation, the image was recorded, and the actual surface roughness of the scanned area was determined to be less than

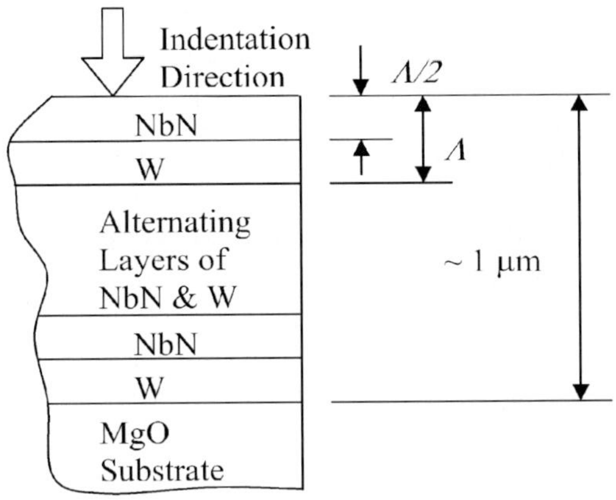

Fig. 1. Structure of the nanocomposite samples used for experimentation. Sample W/NbN ($\Lambda = 5.6$ and 10.4 nm).

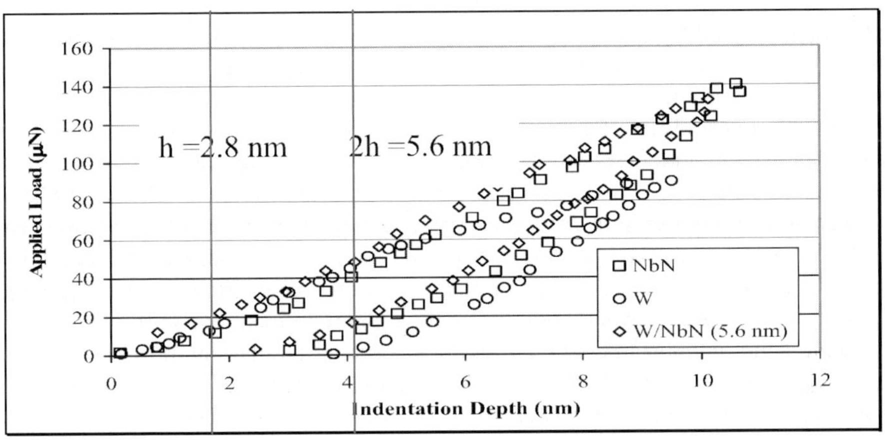

Fig. 2. Load vs. displacement curves during shallow nanoindentation when nanoindentor penetrates the 1st interface at $h = 2.8$ nm and the 2nd interface at $h = 5.6$ for sample W/NbN ($\Lambda = 5.6$ nm). The nanolayer shows very similar behavior to NbN single crystal.

0.5 nm from the recorded image utilizing the Roughness Analysis feature of the NanoScope IIIa software.

3 Experimental Results

A comparison of the load versus displacement curves for NbN and W at shallow indentation depths and the nanocomposites, W (100)/NbN (100) (bilayer thickness $\Lambda = 5.6$ nm), are shown in Figure 2. The comparison of the load versus displacement curves indicate that the superlattice materials, W/NbN ($\Lambda = 5.6$ and 10.4 nm),

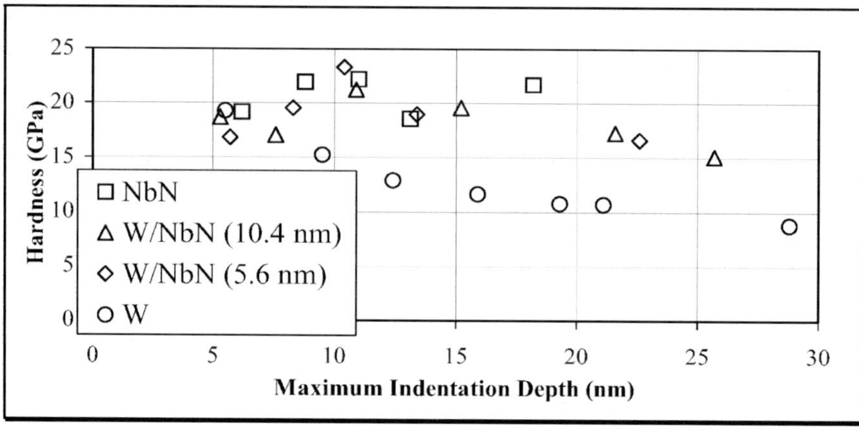

Fig. 3. Hardness as a function of maximum indentation depth (shallow indentations).

exhibit loading and unloading patterns which are very similar to the behavior of NbN. Both comparisons also demonstrate that tungsten initially follows the same loading pattern but starts to deviate from the others at approximately 5 nm. Since the bilayer repeat periods for W/NbN ($\Lambda = 5.6$ nm) and for W/NbN ($\Lambda = 10.4$ nm) are so shallow, the indenter has penetrated only a couple of layers into the superlattice materials.

Hardness is a general measure of the resistance of a material to plastic deformation. As noted earlier, the hardness is normally defined as the ratio of the maximum applied load divided by the corresponding projected contact area. For nanoindentations, the hardness is normally defined as the maximum load divided by the projected area of the indenter in contact with the sample at the maximum load [8]. Thus, $H = P_{\max}/A_C$, where, H, $P_{\max}$ and A_C are the hardness, the maximum applied load, and the projected contact area at the maximum applied load respectively. The experiment was divided into two sets of nanoindentations. The first set of nanoindentations, hereafter referred to as shallow nanoindentations, were conducted to evaluate the influence (if any) of the individual layers on the hardness of the nanocomposites. The second set of nanoindentations, hereafter referred to as deep indentations, was conducted to examine the behavior of the materials as a function of the indentation depth. For the shallow indentations, the applied loads were selected to achieve penetration of the indenter to a depth equal to the theoretical thickness of the individual layers of the nanocomposite materials. Since the focus was only on the first couple of nanolayers, the shallow nanoindentations were designed to achieve a penetration depth of less than 25 nm.

A comparison of the hardness as a function of the maximum indentation depth for all the material samples for the shallow nanoindentations is shown in Figure 3. A comparison of the results shows no significant difference in the hardness for the two superlattice samples (W/NbN, $\Lambda = 5.6$ nm and W/NbN, $\Lambda = 10.4$ nm) and the NbN sample for the shallow indentation depths (less than 25 nm). Also, the in-

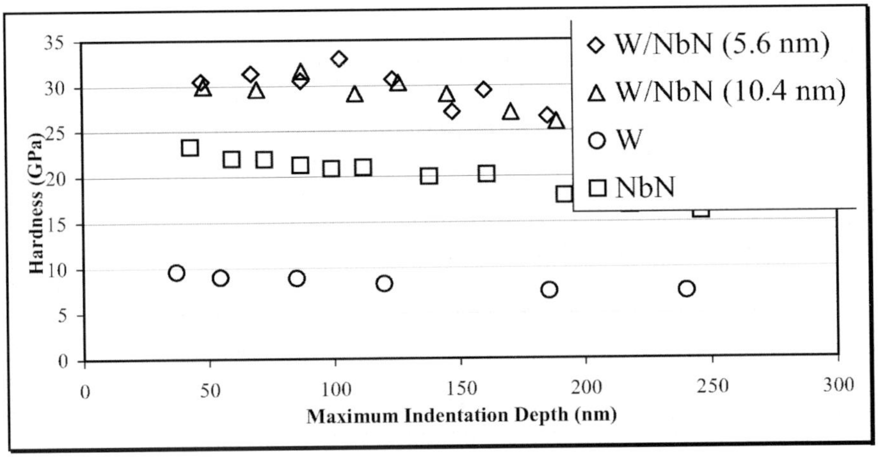

Fig. 4. Hardness as a function of maximum indentation depth (deep indentations).

dentation size effect causes the hardness of W to increase until it is approximately the same as the hardness of NbN (at 5 nm). A comparison of the hardness as a function of the maximum indentation depth is shown in Figure 4 for the deep nanoindentations. An examination of this graph shows several trends. The results for the monolithic material samples, NbN and W, indicate the presence of the Indentation Size Effect (i.e. there is increase in hardness as the indentation depth decreases). The tungsten hardness ranges from approximately 7 GPa at an indentation depth of 240 nm and gradually increases to about 10 GPa at a depth of 40 nm. The niobium nitride hardness ranges from 17 GPa at a penetration depth of 250 nm to about 23 GPa at a depth of 40 nm. In contrast, the behavior of the nanocomposites differs from the behavior for the monolithic materials. The hardness for both nanocomposites is consistent over a range of indentation depths (50 nm to approximately 150 nm) before it begins to taper off slightly. The final observation is that the hardness for each sample at the deeper indentation depths approaches the microhardness reported [1]. An examination of the post-indentation surface topography shows significant pile-up of the material near the indentation site.

4 Discussion

Combining the results of the previous two graphs (Figures 3 and 4), some deductions can be made regarding the factors contributing to the observed hardness increase in the superlattice materials under deep indentation. Both W and NbN exhibit hardness around 20–23 GPa at an indentation depth of 5 nm. This is the approximate thickness of an individual layer in one of the superlattice materials and both superlattice materials exhibit similar hardness at the same indentation depths. The influence of the interfaces is minimal for the shallow indentations because only a few interfaces are in the load zone and the amount of deformation is small. However, at the deeper indentation depths, the superhardeing occurs for the superlattice. The hardness is much higher than the one predicted by the rule of mixtures for conventional composites concept [1, 2]. Since the number of interfaces in the load zone increases at deeper indentation depths, this indicates that the interface between layers is an important factor in the remaining increase in hardness observed in the superlattice materials. At shallow indentation depths, a few interfaces (about 1∼4) are penetrated and the dislocation movement required to accommodate the plastic deformation is minimal. However, as the penetration depth increases, the amount of dislocation activity increases. At some point, the behavior of the interface as a barrier to dislocation motion begins to affect the deformation characteristics of the nanocomposites leading to the observed increase in hardness. Also there are no common slip systems between NbN and W (bcc crystal). In B1 structure NbN, the primary slip system is $\{110\}\langle 1\bar{1}0\rangle$, while the primary slip systems in W are $\{110\}\langle 111\rangle$, with additional slip systems being $\{112\}\langle 111\rangle$ and $\{123\}\langle 111\rangle$. With the 45 degree rotation of the lattices, both the slip planes and slip directions are different. It is extremely difficult for dislocations propagate cross NbN/W interface. Moreover, since the lattice consts. Å for W/NbN are 3.1650/4.39, with 1.36% mismatch, interfacial dislocation or interfacial stress exist along the interface. All these above will make it harder for the indentor to penetrate through the interface. Consequently, hardness will increase with more interface participation.

In general, the elastic modulus should be constant regardless of the indentation depth, although some minor variation is to be expected due to material imperfections. The reduced elastic moduli for the deep and shallow indentations were calculated using the area functions calibrated on a standard quartz sample and compared for all the samples. The measured elastic modulus are between 270∼310 Gpa for both deep and shallow indentations. Since the moduli for all the samples are in the same approximate range, the influence of the elastic modulus on the hardness of the nanocomposites appears to be minimal. Consequently, it is not possible to attribute the increase in hardness observed in the superlattice materials to this particular material property.

Normally, nanolayers gains hardness by losing deformability [9], however, the superlatice demonstrated excellent deformability when reaching the superhardening. The NbN sample exhibits less material pile-up than the other materials. For a ceramic material, the energy of indentation is absorbed by local cracking in the deformation zone under the indenter tip. Hence, little material pile- up occurs during indentation. However, in the nanocomposites, the NbN layers surround the W layers. Due to the ductile properties of the W layers and the fact that the individual layers are only a few nanometers thick, the NbN layers are forced to bend with the W layers since the bending of the NbN layers is easier to achieve than fracture of the material. As a result, the presence of the W layers act to alter the deformation process of the NbN layers from fracture to bending.

5 Conclusions

The shallow indentations show little difference in hardness between the NbN sample and the two superlattice materials. However, an increase in hardness is observed at deeper indentation depths. This indicates that the interface has a strong influence on the increase in hardness.

Acknowledgements

This research was supported from the Oversea Co-operation Project of the National Natural Science Foundation of China through Grant 10428207.

References

1. A. Madan and S.A. Barnett, Fundamentals of nitride-based superlattice thin films, in *Materials Science of Carbides, Nitrides and Borides*, Y.G. Gogotsi and R.A. Andrievski (eds.), pp. 187–204.
2. A. Madan, Yun-yu Wang, S.A. Barnett, C. Engstrom, H. Ljungcrantz, L. Hultman and M. Grimsditch, Enhanced mechanical hardness in epitaxial non-isostructural Mo/NbN and W/NbN superlattices, *J. App. Phys.* **84**(2), 1998, 776–785.
3. X. Chu, M.S. Wong, W.D. Sproul, and S.A. Barnett, Deposition, structure and hardness of polycrystalline transition-metal nitride superlattice films, *J. Mater. Res.* **14**(6), 1999, 2500–2506.
4. B.M. Clemens, H. Kung and S.A. Barnett, Structure and strength of multilayers, *MRS Bulletin*, February 1999, 20–26.
5. P.M. Anserson, T. Foecke and P.M. Hazzledine, Dislocation-based deformation mechanisms in metallic nanolaminates, *MRS Bulletin*, February 1999, 27–33.
6. S.P. Baker, Between nanoindentation and scanning force microscopy: Measuring mechanical properties in the nanometer regime, *Thin Solid Films* **308–309**, 1997, 289–296.
7. W.W. Gerberich, W. Yu, D. Kramer, A, Strojny, D. Bahr, E. Lilleodeen and J. Nelson, Elastic loading and elastoplastic unloading from nanometer level indentations for modulus determinations, *J. Mater. Res.* **13**(2), 1998, 421–439.
8. W.C. Oliver and G.M. Pharr, An improved technique for determining hardness and elastic modulus using load and displacement sensing indentation experiments, *J. Mater. Res.* **7**(6), 1992, 1564–1583.
9. M.Z. Li and S.X. Mao, Length scale (thickness) controlled ductile versus brittle transition in layered materials, *Material Science and Engineering* **A266**, 1999, 73–79.

Part 5

Micro-Mechanics Models and Simulation for the Nanostructured Materials

Cluster Statistical Thermodynamics (CST) – To Efficiently Calculate Quasi-Static Deformation at Finite Temperature Based on Molecular Potential

Ming Hu[1,2,*], Haiying Wang[1], Mengfen Xia[1,3], Fujiu Ke[1,4] and Yilong Bai[1]

[1] *State Key Laboratory of Nonlinear Mechanics, Institute of Mechanics, Chinese Academy of Sciences, Beijing 100080, China; *E-mail: huming@lnm.imech.ac.cn*
[2] *Graduate School of Chinese Academy of Sciences, Beijing 100039, China*
[3] *Department of Physics, Peking University, Beijing 100871, China*
[4] *Department of Applied Physics, Beijing University of Aerospace and Aeronautics, Beijing 100083, China*

Abstract. For quasi-static deformations in engineering practice, molecular dynamics (MD) simulation requires computation resources that are not affordable even with ever-increasing computing power. In order to overcome this weakness, we developed a new method called cluster statistical thermodynamics (CST). By taking the advantage of statistical thermodynamics and adopting finite-element interpolation, the new approach can not only simulate quasi-static deformation but have very high computing efficiency. The new method is based on molecular potentials as MD does, but statistical thermodynamics help us greatly reduce the tedious calculation of thermal fluctuations of molecules. Therefore, the new method appears to be superior to MD in the simulations of quasi-static deformation. Especially CST works much more efficiently than MD with much less storage space and CPU time. In this paper, we illustrate the new methodology by means of some examples of two-dimensional quasi-static tensile process at 300 K. It is found that the results obtained with CST are in good agreement with those obtained by fully atomistic simulations and CST is 600 times faster than MD. Hence, the new method seems to be a very efficient and promising approach to numerical simulations of solid deformations under quasi-static loadings and at finite temperatures, based on molecular potentials.

Key words: cluster statistical thermodynamics, quasi-static deformation, finite temperature, quasi-continuum method, molecular dynamics.

1 Introduction

Macroscopic theories of solid materials are based on statistical average of microscopic interactions and then are insufficient to deal with a system at micro- and nanoscales. In principle, mechanical behaviors of solid materials are determined by their microscopic interactions between molecules and atoms constituting them. So far,

Yilong Bai et al. (eds), IUTAM Symposium on Mechanical Behavior and Micro-Mechanics of Nanostructured Materials, 163–170.

molecular dynamics (MD) simulations, based on the full description of interaction potential of all the molecules or atoms and the Newton equation, seem to be a basic numerical approach to model these behaviors. However, despite ever-increasing computing power, we still encounter two formidable difficulties in these simulations, namely the huge gap in the length and time scales involved [1]. It is well known that even modern computers can, at most, deal with problems up to the spatial and temporal range of micrometers and nanoseconds respectively. Obviously, this can not meet the need to simulate most practical deformation processes, i.e. quasi-static deformations of samples larger than several micrometers at finite temperature. Obviously these two obstacles present a great challenge to current simulations.

During the past decade, in order to solve the length scale problem, many researchers have developed a variety of multiscale methods by imbuing continuum-based models with more-realistic atomistic details. One of the pioneering works in this field is the so-called FEAt method [2], in which the system is partitioned into a "core" (inhomogeneous) described atomistically and the "remainder" (homogeneous) treated as a continuum. Recently, Abraham et al. [3] and Broughton et al. [4] extended the method to include an additional core, handled by the quantum-mechanical tight-binding (TB) approximation. Although this technique has, to some extent, successfully solved the length scale problem, it can only treat dynamical process up to 20 ps so far. Therefore, it is difficult to simulate practical quasi-static processes.

Another important and remarkable approach is the quasi-continuum (QC) method developed by Tadmor et al. [5]. The QC provides a framework whereby degrees of freedom are judiciously eliminated and force/energy calculations are expedited, by means of overlaying the atomic lattice with a finite mesh (represented by repatoms) and recognizing potential energy minimization. It is particularly useful if one is primarily interested in metastable behavior of a defective material. Several researchers have successfully used this method to simulate a variety of problems, such as nanoindentation, crack tip deformation, dislocation interactions and so on. In spite of these successes, at present, the primary disadvantage of the QC method is that it is restricted to processes at absolute zero temperature ($T = 0$ K). Although Shenoy and Phillips made an attempt to extend the QC method to finite temperature [6, 7], its implementation is not efficient and simulations of large problems, other than simple edge dislocations, require a much larger computational effort [7]. They performed one-dimensional Monte Carlo simulations only and noted that the obtained results raise more questions than they answer.

In this paper, we report a different and very efficient approach named cluster statistical thermodynamics (CST), to calculate quasi-static deformation of a system at finite temperature (T) based on molecular potential. In the CST method, a system consisting of atoms is partitioned into a number of clusters determined by their nodes like finite element method. Each cluster is treated by statistical thermodynamics as an equilibrium sub-system consisting of finite atoms or molecules. In particular, we use statistical thermodynamics to greatly reduce the treatment of thermal fluctuations of molecules, which are very time-consuming in MD simulations but not closely relevant to quasi-static deformation. Based on the positions of the nodes, the Helmholtz

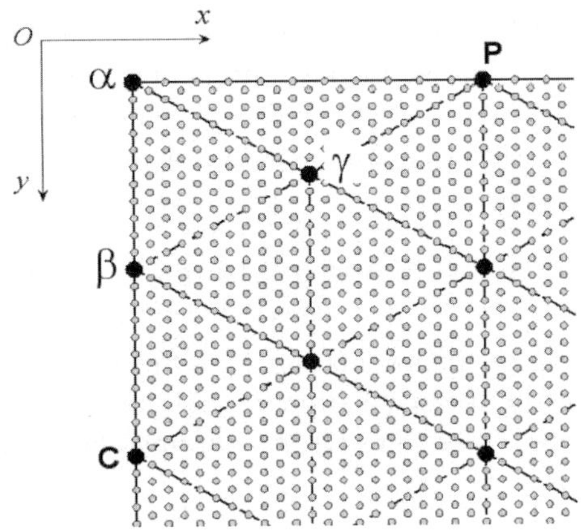

Fig. 1. Schematic of cluster construction. The small gray circles are atoms of the system and the big black circles, like α, β, and γ, are the nodes which construct the clusters.

free energy of each cluster is calculated and the total free energy of the system is obtained by the summation of them. Then, the equilibrium configuration of the system can be obtained by minimizing the total Helmholtz free energy of the system with respect to the positions of nodes under the condition of constant volume or area (in three or two dimensions respectively). We assess the validity of CST by comparing a quasi-static uniaxial tension at room temperature with two fully atomistic models, i.e. molecular dynamics (MD) and 'molecular statistical thermodynamics" (MST), the latter is the limit of CST with single molecule clusters.

2 Method Formulation

Now, we use a two-dimensional case to illustrate the principle and application of the new method. Consider a system of N atoms, whose positions are denoted by $\{\mathbf{x}\}$. As shown in Figure 1, the system is partitioned into a number of clusters constructed by their corresponding nodes (like α, β, γ, etc., see the large black circles in Figure 1).

Thus, the entire system can be characterized by all the nodes whose positions are denoted by $\{\mathbf{X}\}$. If we denote N_{node} as the total of nodes, then, clearly, the freedom degree of the system can be significantly reduced since $N_{\text{node}} \ll N$. The requirement of the cluster construction is that the nodal system should behave as closely as possible to the fully atomistic system. A criterion to this requirement is that the statistical average of an observable F (denoted by $\bar{F}$) calculated according to node positions $\{\mathbf{X}\}$ be equal to the time-average of the observable in the corresponding fully atomistic canonical system. Therefore, the aim of CST can be stated as

$$\bar{F}(\{\mathbf{X}\})_{\text{CST}} = \bar{F}(\{\mathbf{x}\}) = \langle F(\{\mathbf{x}\}) \rangle, \tag{1}$$

where $\bar{F}$ and $\langle F \rangle$ are the time and ensemble average of F, respectively.

When analyzing the mechanical response of a system under *quasi-static* loadings, we are required to calculate the Helmholtz free energy of the system. The free energy serves as a generating function for other mechanical quantities. Firstly, suppose that the total Helmholtz free energy A of the whole system be the summation of the free energy of each cluster, i.e.

$$A = \sum_{\alpha=1}^{N_c} A_\alpha, \tag{2}$$

where A_α is the free energy of cluster α and N_c is the total of clusters.

Secondly, according to the local harmonic (LH) approximation [8], the Einstein model of phonon density of states in statistical thermodynamics gives the Helmholtz free energy of each cluster as

$$A_\alpha = \Phi_\alpha + N_\alpha kT \sum_{\beta=1}^{2} \ln \frac{\hbar\omega_{h\beta}}{kT} = \Phi_\alpha + 2N_\alpha kT \ln \frac{\hbar |D_{\alpha h}|^{1/4}}{kT}, \tag{3}$$

where Φ_α is the potential energy of cluster α, N_α is the number of atoms within cluster α, $\hbar$ is Planck's constant, k is Boltzmann's constant, T is temperature (K), $\omega_{h\beta}$ and $D_{\alpha h}$ are the vibrational frequency and the local dynamical matrix of the atom h adjacent to the center of cluster α, respectively, and β, is either 2 or 3, in accordance with dimensions.

Thirdly, we adopt local mean field approximation, i.e., assuming identical potential energy for atoms in each cluster. Then, the potential energy of cluster α can be expressed as

$$\Phi_\alpha = N_\alpha \Phi_{\alpha h}, \tag{4}$$

where $\Phi_{\alpha h}$ is the potential energy of the atom h adjacent to the center of cluster α and can be calculated in accord with deformed neighbors from the current interpolated displacements in the cluster. At the same time, the determinant of the local dynamical matrix of atom h can be calculated as [9]

$$|D_{\alpha h}| = \frac{\partial^2 \Phi_\alpha}{\partial x_{h1}^2} \frac{\partial^2 \Phi_\alpha}{\partial x_{h2}^2} - \left(\frac{\partial^2 \Phi_\alpha}{\partial x_{h1} x_{h2}} \right)^2, \tag{5}$$

where $x_{h\eta}$, $\eta = 1$ and 2, denote the positions of atom h in x and y directions respectively.

Finally, using Equations (2–5), we can write the total Helmholtz free energy A of the whole system as a function of all the node positions $\{\mathbf{X}\}$, i.e.

$$A = A(\{(X_i, Y_i), i = 1, \ldots, N_{\text{node}}\}). \tag{6}$$

We are now in a position to obtain the equilibrium configuration $\{\mathbf{X}_{\text{eq}}\}$ by minimizing A with respect to the nodes positions $\{\mathbf{X}\}$, when keeping the temperature T and the total area of clusters, $S = \sum_{\alpha=1}^{N_c} S_\alpha$, fixed during the minimizing search. It is important to notice that the CST method described above can be straightforwardly applied to the deformation at absolute zero temperature. In this case, the equilibrium configuration can be easily obtained by minimizing the total potential energy of the system, $\Phi = \sum_{\alpha=1}^{N_c} \Phi_\alpha$, with respect to the nodes positions {}.

Here, it should be pointed out that the CST method is different from the quasi-continuum method developed by Tadmor et al. [5]. The essence of the QC method is repatom (representing a number of atoms) statics/dynamics while the CST is based on the statistical thermodynamical equilibrium of cluster nodes. Hence CST can be readily used to deformations at finite temperature.

3 Some Examples

We validated the CST method by simulating two-dimensional quasi-static tensile processes at 0 K and 300 K. In the present work, a single hexagonally packed lattice of Cu with initial dimensions 68.3 × 16.9 nm was uniaxially tensioned. As shown in Figure 1, the tensile direction is horizontal (x), and the free lateral boundary conditions were adopted in the vertical (y) direction, for both CST and MD simulations. The potential employed in our calculations is the Lennard–Jones potential,

$$e(r) = 4\varepsilon_p \left[\left(\frac{r_0}{r}\right)^{12} - \left(\frac{r_0}{r}\right)^{6} \right]$$

with parameters $r_0 = 2.3276$ Å, $\varepsilon_p = 0.4912$ eV [9], and the truncated distance $r_c = 7.8379$ Å. The total number of nodes and clusters used in CST simulations are 189 and 336 respectively, corresponding to 20549 atoms in full MD simulations. For CST, at each loading state, all the atoms reach their equilibrium positions by minimizing the total Helmholtz free energy of the whole system with respect to the node positions using the conjugate gradient method [10]. For corresponding MD simulations, it was conducted over a period of about 30ps with a time step of 2 fs using the standard Verlet algorithm [11]. For MD simulations, the 30ps period contains two parts: the first 20 ps are used to look for thermo-equilibrium and the remainder for statistical calculation of time-average positions of atoms $\{\mathbf{x}_{\text{eq}}\}$. In order to facilitate the comparison of CST and MD results, we adopt the same stress definition, i.e. stress being the derivative of Helmholtz free energy. According to the LH approximation [8], the Helmholtz free energy of each loading state for MD is then computed by the determinant method (DM) [12]

$$A_{\text{MD}} = \Phi(\{\mathbf{x}_{\text{eq}}\}) + 2kT \sum_{i=1}^{N} \ln \left(\frac{\hbar |D_i|^{1/4}}{kT} \right), \tag{7}$$

where $|D_i| = [\omega_{i1}\omega_{i2}]^2$ is the determinant of the local dynamical matrix of atom i and can be obtained by

$$|D_i| = \left(\frac{\partial^2 \Phi}{\partial x_{i1}^2} \right)_{\text{eq}} \left(\frac{\partial^2 \Phi}{\partial x_{i2}^2} \right)_{\text{eq}} - \left(\frac{\partial^2 \Phi}{\partial x_{i1} \partial x_{i2}} \right)_{\text{eq}}^2 .$$

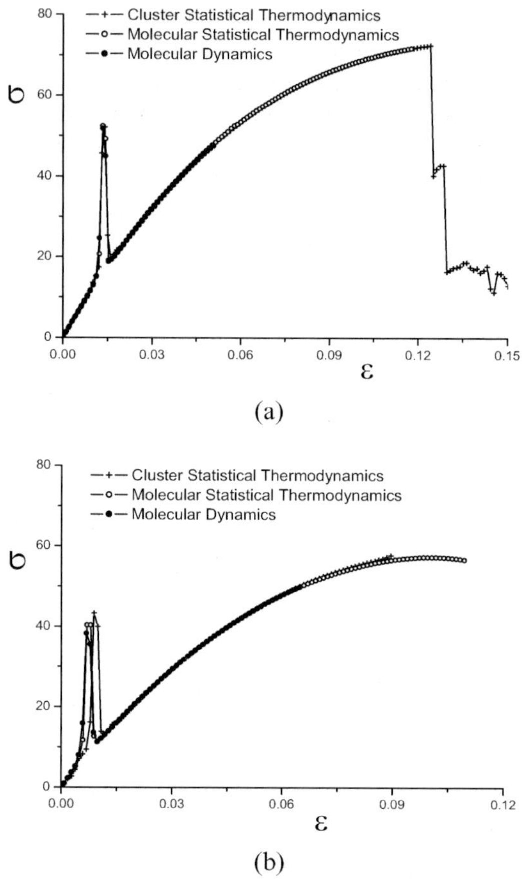

Fig. 2. (a) Stress vs strain curves at absolute zero temperature obtained by CST (+), MST (∘), and MD (•) respectively. The stress σ is calculated by $\sigma = (1/L_{y0})(\partial\Phi/\partial L_x)$, where L_x and L_{y0} are the deformed length in the X direction and the initial length in the Y direction respectively. The unit of stress is $k\Theta/r_0^2$, where $\Theta = 343$ K is the Debye temperature for Cu. (b) Stress vs strain curves at 300K. The stress σ is calculated by $\sigma = (1/L_{y0})(\partial A/\partial L_x)$.

Now, let us start with the comparison of the deformations at absolute zero temperature simulated by CST, MST, and MD respectively. Three stress-strain curves were calculated by differentiating the total potential energy of the system with respect to the deformed length in the x direction, shown in Figure 2a. As one may expect the agreement must be excellent, since the principles of CST, MST and MD become identical at absolute zero temperature. More importantly, Figure 2b shows the comparison of the three stress-strain curves at finite temperature (300K) obtained by CST, MST, and MD respectively.

Figures 2a and 2b clearly demonstrate that, before breaking, the mechanical behavior predicted by the proposed CST method almost coincides with those obtained by two fully atomistic simulations (MST and MD) for both absolute zero temperature and finite temperature. This indicates that the proposed CST method can accurately simulate deformation at finite temperature. In particular, the deformation simulated by CST is a quasi-static process, which corresponds to most practical tests in laboratories. The other advantage of CST simulations is that they are very efficient. For example, for all simulations performed on a PC with CPU Pentium 4 2.6 GHz, the calculations for each loading state took less than 10 s for CST and about 20 minutes for MST but 100 minutes for MD, thus CST is 600 times faster than MD simulations. Therefore, the new method seems to be a very efficient and promising approach to numerical simulation of solid deformations under quasi-static loadings and at finite temperature, based on molecular potentials.

4 Summary

In summary, we report in this paper a novel approach, i.e. cluster statistical thermodynamics (CST) method, to simulate deformations of crystalline solids under quasi-static loadings at finite temperature. The good agreements of uniaxial tensions obtained by CST with fully atomistic simulations validate the new method. Moreover, CST demonstrates much higher efficiency in computing time (about 3 orders less than MD). Therefore, CST appears to be a more realistic (for quasi-static loading at finite temperature) and more efficient method in simulations of solid deformation based on molecular potential.

Acknowledgements

This work is supported by the National Natural Science Foundation of China (10432050), the Chinese Academy of Sciences (KJCX2-SW-L2) and the Major State Research Project of China (G2000077305).

References

1. Kadau, K., Germann, T.C., Hadjiconstantinou, N.G., Lomdahl, P.S., Dimonte, G., Holian, B.L. and Alder, B.J. (2004) Nanohydrodynamics simulations: An atomistic view of the Rayleigh–Taylor instability, *Proc. Natl. Acad. Sci. USA* **101**, 5851–5855.
2. Kohlhoff, S., Gumbsch, P. and Fischmeister, H.F. (1991) Crack-propagation in bcc crystals studied with a combined finite-element and atomistic model, *Philos. Mag. A* **64**, 851–878.
3. Abraham, F.F., Broughton, J.Q., Bernstein, N. and Kaxiras, E. (1998) Spanning the length scales in dynamic simulation, *Comput. Phys.* **12**, 538–546.
4. Broughton, J.Q., Abraham, F.F., Bernstein, N. and Kaxiras, E. (1999) Concurrent coupling of length scales: Methodology and application, *Phys. Rev. B* **60**, 2391–2403.
5. Tadmor, E.B., Ortiz, M. and Phillips, R. (1996) Quasicontinuum analysis of defects in solids, *Philos. Mag. A* **73**, 1529–1563.

6. Shenoy, V., Shenoy, V. and Phillips, R. (1999) Finite temperature quasicontinuum methods, *Mat. Res. Soc. Symp. Proc.* **538**, 465–471.
7. Shenoy, V. (1998) Quasicontinuum models of atomic-scale mechanics, PhD thesis, Brown University.
8. LeSar, R., Najafabadi, R. and Srolovitz, D.J. (1989) Finite-temperature defect properties from free-energy minimization, *Phys. Rev. Lett.* **63**, 624–627.
9. Najafabadi, R. and Srolovitz, D.J. (1995) Evaluation of the accuracy of the free-energy-minimization method, *Phys. Rev. B* **52**, 9229–9241.
10. Press, W.H., Teukolsky, S.A., Vetterling, W.T. and Flannery, B.P. (1992) *Numerical Recipes in Fortran 77: The Art of Scientific Computing*, Cambridge University Press, Cambridge.
11. Allen, M.P. and Tildesley D.J. (1987) *Computer Simulation of Liquids*, Oxford University Press, Oxford.
12. Hoover, W.G., Hindmarsh, A.C. and Holian, B.L. (1972) Number dependence of small-crystal thermodynamic properties, *J. Chem. Phys.* **57**, 1980–1985.

On the Size of the Representative Volume Element for Isotropic Elastic Polycrystalline Copper

F. El Houdaigui[1], S. Forest[1,*], A.-F. Gourgues[1] and D. Jeulin[2]

[1] *Centre des Matériaux – UMR 7633, Ecole des Mines de Paris – CNRS, B.P. 87, 91003 Evry, France; *E-mail: samual.forest@ensmp.fr*
[2] *Centre de Morphologie Mathématique, Ecole des Mines de Paris, 35 rue St Honoré, 77300 Fontainebleau, France*

Abstract. A computational homogenization strategy is developed to determine the number of grains necessary to estimate the effective elastic properties of isotropic polycrystalline copper with a given precision. Finite element simulations of polycrystalline aggregates are presented for both homogeneous and periodic boundary conditions. For different volumes, several realizations are considered. The mean apparent shear modulus and the associated dispersion are estimated as a function of the number of grains. Periodic conditions lead to rapid convergence of the result towards the wanted effective shear modulus. The Representative Volume Element (RVE) size is then related to the evolution of the standard deviation of the apparent shear modulus, via an extension of the notion of integral range A_3. For a precision of 1% and 10 realizations, a minimal RVE size of 445 grains is found. The found value $A_3 = 1.43$ can be compared to the integral range for other microstructures and physical properties.

Key words: representative volume element, homogenization, polycrystal, copper, finite element, integral range.

1 Introduction

Computational homogenization methods are nowadays efficient tools to estimate effective properties of heterogeneous materials. They can take realistic distribution of phases and sophiscated constitutive equations of the constituents into account (Cailletaud et al., 2003). A key-point in such models is the determination of the appropriate size of volume elements of heterogeneous materials to be computed in order to get a precise enough estimation of effective properties. This is related to the long-standing problem of the determination of the size of the Representative Volume Element (RVE) in homogenization theory (Drugan, 1996). It is known that RVE is morphology and property dependent but a well-suited parameter is necessary for quantitative comparisons. Such a parameter was proposed by Kanit et al. (2003).

In the present work, a method is proposed to estimate the size of such a RVE in isotropic linear elastic copper polycrystals. It has three main steps: the choice

Yilong Bai et al. (eds), IUTAM Symposium on Mechanical Behavior and Micro-Mechanics of Nanostructured Materials, 171–180.

of a random model for polycrystalline microstructures containing a finite number of grains, the resolution of boundary value problems on polycrystalline aggregates of increasing sizes and the analysis of the convergence of the calculated apparent properties towards an asymptotic value as a function of the number of grains and of the boundary conditions. The asymptotic value is regarded as the effective property (Sab, 1992). In other words, the objective is to find the minimum number of grains required in a volume element to estimate the effective elastic property with a given accuracy. The size of the RVE for several cubic elastic polycrystals was investigated in 2D by Ren and Zheng (2002) and by Nygårds (2003) using three-dimensional FE simulations and periodic boundary conditions. A relationship between the RVE size and the anisotropy coefficient of each material was identified. Most interestingly, the last author links the notion of representativity of considered material volumes with the decay of the dispersion of calculated apparent properties for increasing grain numbers, as done by Kanit et al. (2003). The present contribution focuses on the dependence of the result on the choice of boundary conditions and on the determination of a statistical parameter quantifying the decrease in scatter with increasing grain numbers and allowing comparisons of RVE sizes for other microstructures and properties.

In the following, vectors are underlined and boldface quantities are second-rank or fourth-rank tensors. The symbol $:=$ defines the quantity on the left of the symbol.

2 Computational Homogenization Method

2.1 Generation of Microstructures

Voronoi mosaics are used here as a random model to represent the polycrystalline morphology, as explained in Barbe et al. (2001) and Kanit et al. (2003). For each realization, one given cubic volume V that contains a given number N_g of Voronoi cells is simulated. In the following, n realizations of volume V are considered. The number of cells for each realization of the microstructure obeys a Poisson distribution with given mean value $\bar{N}_g = N$. The mean volume of one Voronoi cell is equal to 1. No unit length is introduced because the models involved in this work cannot account for absolute size effects. As a result, one has $N = V$. This convention is used throughout the work.

A crystal orientation is attributed to each Voronoi cell which is then regarded as an individual grain of the polycrystal. The crystallographic texture is assumed to be random. It is possible to impose a geometrical periodicity constraint at the boundary of the polycrystalline cube, as shown in Figure 1 (see also Kanit et al., 2003). This condition is enforced in the subsequent FE simulations involving periodicity conditions.

2.2 FE Meshing of Microstructures

The so-called multi-phase element technique is used in order to superimpose a regular 3D FE mesh on the Voronoi tessellation point of each element of the mesh.

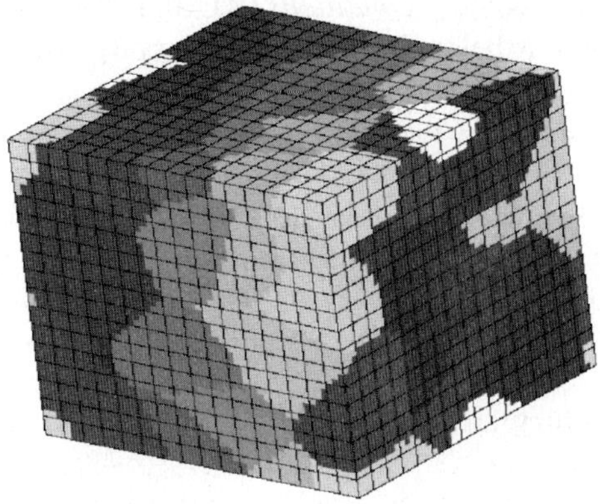

Fig. 1. Regular FE mesh superimposed on a Voronoi mosaic containing 50 grains, using the multiphase element technique.

The crystal orientation of the closest voxel is attributed to every integration point of each element of the mesh. The elements are 20-node quadratic bricks with 27 Gauss points. Figure 1 shows such a mesh made of $16 \times 16 \times 16$ elements. The main drawback of the technique is that one element may contain integration points that belong to several grains. The bias introduced by this meshing technique was investigated in Schmauder (1997), Barbe et al. (2001) and Kanit el al. (2003).

The effect of mesh density, i.e. of the number of elements per grain, was investigated here for elastic polycrystalline copper. FE meshes of a given aggregate made of 50 grains were considered. The convergence of the computed apparent shear modulus μ^{app} was analyzed when the number of mesh elements increases. For each simulation, the geometry of the microstructure is unchanged but the number of degrees of freedom is increased from 5568 to 56355. From these results, a resolution of 16 elements per grain was chosen for the following calculations. The use of finer meshes does not improve the result of more than 1%.

The largest volume computed in this work is a cube with 423 = 74088 elements, i.e. 937443 degrees of freedom. Such computations are made possible in a reasonable time by using parallel computing. The FE program used in this work implements the subdomain decomposition method FETI (Zset, 1996; Feyel, 1999). The mesh is split into 32 subdomains and the tasks are distributed on a platform of 32 processors (768 MB RAM, 800 MHz). Compatibility and equilibrium at interfaces between subdomains are restored by an iterative procedure. The whole resolution requires 21GB of memory.

2.3 Boundary Conditions and Definition of Apparent Moduli

Three types of boundary conditions to be prescribed on an individual volume element V are considered (Zaoui, 1987):

- *Kinematic uniform boundary conditions* (KUBC): The displacement vector $\underline{u}$ is imposed at all points x belonging to the boundary ∂V according to:

$$\underline{u} = \mathbf{E} \cdot \underline{x} \quad \forall \underline{x} \in \partial V \quad \langle \boldsymbol{\varepsilon} \rangle := \frac{1}{V} \int_V \boldsymbol{\varepsilon} dV = \mathbf{E}, \tag{1}$$

where E is a given constant symmetrical second-rank tensor. The macroscopic stress tensor $\mathbf{\Sigma}$ is then defined as the spatial average of the local stress tensor $\boldsymbol{\sigma}$.
- *Static uniform boundary conditions* (SUBC): The traction vector is prescribed at the boundary ∂V according to:

$$\boldsymbol{\sigma} \cdot \underline{n} = \mathbf{\Sigma} \cdot \underline{n} \quad \forall \underline{x} \in \partial V \quad \langle \boldsymbol{\sigma} \rangle := \frac{1}{V} \int_V \boldsymbol{\sigma} dV = \mathbf{\Sigma}, \tag{2}$$

where $\mathbf{\Sigma}$ is a given constant symmetrical second-rank tensor. The outer normal to ∂V at x is denoted by n. The macroscopic strain tensor E is then defined as the spatial average of the local strain ε.
- *Periodicity conditions* (PERIODIC): The displacement field over the entire volume V takes the form

$$\underline{u} = \mathbf{E} \cdot \underline{x} + \underline{v} \quad \forall \underline{x} \in \partial V, \tag{3}$$

where the fluctuation $\underline{v}$ is periodic. $\underline{v}$ (resp. $\sigma \cdot \underline{n}$) takes the same value (resp. opposite value) at two homologous points on opposite sides of V.

The local behaviour at every integration point inside each grain in the simulation is described by the fourth-rank linear elasticity tensor $\mathbf{c}$:

$$\boldsymbol{\sigma}(\underline{x}) = \mathbf{c}(x) : \boldsymbol{\varepsilon}(\underline{x}). \tag{4}$$

No specific behavior is attributed to grain boundaries (Cailetaud et al., 2003). The partial differential equations to be solved using the FE method are the classical stress balance equations without body forces. For a given volume V, and owing to the linearity of the considered boundary value problems, fourth-rank tensors of apparent moduli C_E^{app} and apparent compliance S_Σ^{app} can be defined by the following macroscopic relations:

$$\mathbf{\Sigma} = \langle \boldsymbol{\sigma} \rangle = \frac{1}{V} \int_V \boldsymbol{\sigma} dV = \mathbf{C}_E^{\mathrm{app}} : \mathbf{E}, \quad \mathbf{E} = \langle \boldsymbol{\varepsilon} \rangle = \frac{1}{V} \int_V \boldsymbol{\varepsilon} dV = \mathbf{S}_\Sigma^{\mathrm{app}} : \mathbf{\Sigma}. \tag{5}$$

The first relation is used for KUBC and PERIODIC problems, the second one for SUBC problems. Note that in general, the tensor S_Σ^{app} cannot be expected to coincide with the inverse of C_E^{app}. However, for sufficiently large volumes V, the apparent moduli do not depend on the type of boundary conditions any longer and coincide with the effective properties of the medium (Sab, 1992):

$$\mathbf{S}_\Sigma^{\mathrm{app}-1} = \mathbf{S}^{\mathrm{eff}-1} = \mathbf{C}^{\mathrm{eff}} = \mathbf{C}_E^{\mathrm{app}}. \tag{6}$$

For intermediate volumes V, the following inequalities, written in the sense of quadratic forms, hold (Huet, 1990):

$$\mathbf{S}_{\Sigma}^{\text{app}-1} \leq \mathbf{C}^{\text{eff}} \leq \mathbf{C}_{E}^{\text{app}}. \tag{7}$$

In the next section, both C_E^{app} and the periodic estimations are checked to remain between the bounds defined by (7).

The following two shear loading conditions E_μ and Σ_μ are used in this work:

$$E_\mu = \begin{bmatrix} 0 & \frac{1}{2} & 0 \\ \frac{1}{2} & 0 & 0 \\ 0 & 0 & 0 \end{bmatrix}, \quad \Sigma_\mu = \begin{bmatrix} 0 & \alpha & 0 \\ \alpha & 0 & 0 \\ 0 & 0 & 0 \end{bmatrix} \quad \text{with } a = 1 \text{ MPa} \tag{8}$$

in the particular Cartesian coordinate frame attached to the cubic volume element. In the case of KUBC and PERIODIC conditions prescribed to a given volume V, one defines the apparent modulus μ_E^{app} by the work of internal forces in the volume V subjected to the loading E_μ:

$$\mu_E^{\text{app}}(V) := \langle \boldsymbol{\sigma} : \boldsymbol{\varepsilon} \rangle = \langle \boldsymbol{\sigma} \rangle : \mathbf{E}_\mu = \frac{1}{V} \int_V \sigma_{12} dV. \tag{9}$$

In the case of SUBC boundary conditions, an apparent shear modulus μ_Σ^{app} is defined as the work of internal forces generated in V by the application of the loading Σ_μ:

$$\frac{a^2}{\mu_E^{\text{app}}}(V) := \langle \boldsymbol{\sigma} : \boldsymbol{\varepsilon} \rangle = \Sigma_\mu : \langle \boldsymbol{\varepsilon} \rangle = \frac{2a}{V} \int_V \varepsilon_{12} dV. \tag{10}$$

These definitions remain formal insofar as the apparent elasticity properties of a given material volume element V are not necessarily isotropic.

3 Determination of Apparent Shear Moduli for Polycrystalline Copper

We now consider the special case of linear elastic copper polycrystals with a uniform distribution of crystal orientations. The cubic elasticity constants of pure copper are taken from Gairola (1981):

$$C_{11} = 168400 \text{ MPa}, \quad C_{12} = 121400 \text{ MPa}, \quad C_{44} = 75390 \text{ MPa}.$$

The corresponding value of the anisotropy coefficient $\alpha = 2C_{44}/(C_{11} - C_{12})$ is 3.2.

Due to the uniform distribution of crystal orientations, the effective medium exhibits an isotropic linear elastic behaviour, described by effective bulk and shear moduli k^{eff} and μ^{eff}. For cubic symmetry, the apparent bulk modulus is not a random variable (Gairola, 1981). It is uniquely determined from the single crystal elasticity constants according to the formula $k^{\text{app}} = k^{\text{eff}} = (C_{11} + 2C_{12})/3 = 137067$ MPa. As a result, the homogenization problem reduces to the estimation of apparent shear properties μ^{app} and *in fine* of the effective shear modulus μ^{eff}.

Table 1. Mean apparent shear modulus, associated dispersion and relative error on the mean as a function of the domain size and of the number of realizations for three different boundary conditions.

	V	n	$\bar{\mu}^{\text{app}}$ (MPa)	$D_\mu(V)$ (MPa)	ε_{rel}
KUBC	25	100	52543	3186	1.2%
KUBC	400	50	50088	836	0.4%
KUBC	1000	25	49787	533	0.4%
KUBC	5000	10	49336	222	0.2%
PERIODIC	25	100	49669	3162	1.2%
PERIODIC	123	50	48886	1400	0.8%
PERIODIC	400	50	48784	811	0.4%
PERIODIC	500	50	48764	778	0.4%
SUBC	25	100	43397	3185	1.4%
SUBC	400	50	47308	823	0.4%
SUBC	1000	25	47566	538	0.4%
SUBC	5000	10	48390	178	0.2%

It is shown in Kanit et al. (2003) that the fourth-rank tensor of apparent moduli $C_E^{\text{app}}(V)$ obtained for a finite domain V containing N_g grains is generally not isotropic. However, its ensemble average $\bar{C}_E^{\text{app}}(V)$, i.e. its mean value over a sufficiently large number of realizations, turns out to be isotropic. This has been checked here for polycrystalline copper aggregates. The shear modulus associated with the isotropic elasticity tensor $\bar{C}_E^{\text{app}}(V)$ coincides with $\bar{\mu}_E^{\text{app}}(V)$, the ensemble average of the apparent shear moduli $\mu_E^{\text{app}}(V)$ defined by Equation (9) and computed for a domain V of given size (or equivalently containing $N = V$ grains in average). Accordingly, the estimation of $\bar{\mu}_E^{\text{app}}(V)$ only requires the determination of $\mu_E^{\text{app}}(V)$ for each realization. This is the computation strategy adopted in this work. Similarly, using SUBC conditions, it is sufficient to compute $\mu_\Sigma^{\text{app}}(V)$ for each realization according to Equation (10).

The apparent shear moduli and compliances $\mu^{\text{app}}(V)$ were computed using volume elements V of increasing size, ranging from $V = 25$ to $V = 5000$ grains, with $n(V)$ realizations for every volume. Number n is chosen such that the estimation of the mean $\bar{\mu}^{\text{app}}(V)$ is obtained with a precision better than 1%. This precision is estimated according to the simple sampling rule (13) involving the standard deviation $D_\mu(V)$. All simulation results are shown in Table 1.

Mean values and confidence intervals for the apparent shear modulus $[\bar{\mu}^{\text{app}}(V) - 2D_\mu(V), \bar{\mu}^{\text{app}}(V) + 2D_\mu(V)]$, are plotted in Figure 2, as a function of volume size V. The mean apparent shear moduli strongly depend on the domain size and on the boundary conditions. However, the values converge towards an asymptotic constant as the volume size increase, as expected (Sab, 1992). A striking feature of these results is the very fast convergence of the periodic solution and, in contrast, the very slow convergence associated with homogeneous boundary conditions. The periodic estimate is bounded by the KUBC and SUBC estimates:

$$\mu^{\text{Reuss}} \leq \bar{\mu}_\Sigma^{\text{app}} \leq \bar{\mu}_{\text{periodic}}^{\text{app}} \leq \bar{\mu}_E^{\text{app}} \leq \mu^{\text{Voigt}}, \tag{11}$$

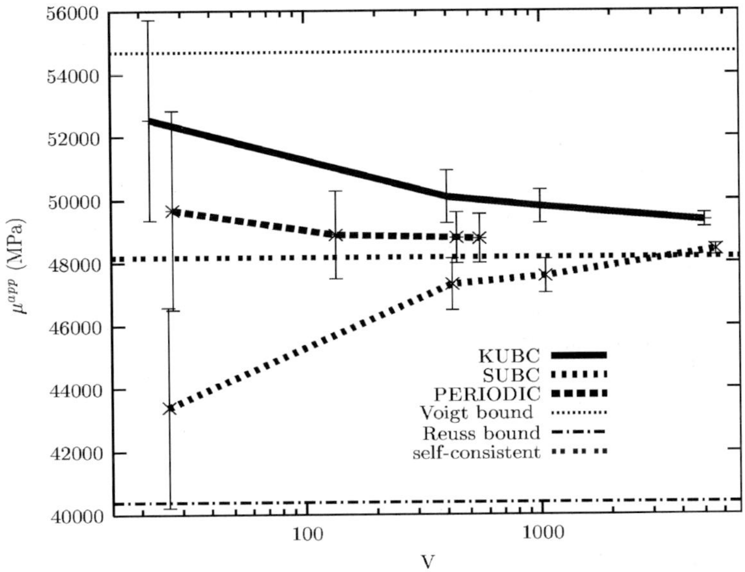

Fig. 2. Mean values and confidence intervals for the shear modulus μ^{app} as a function of domain size, for three different boundary conditions.

where μ^{Reuss} and μ^{Voigt} denote the first order lower and upper bounds for the effective shear modulus of the polycrystal (Zaoui, 1987). For decreasing values of V, the apparent moduli $\mu_E^{\text{Reuss}}(V)(\mu_\Sigma^{\text{Reuss}}(V))$ get closer to the upper (lower) limit $\mu^{\text{Voigt}}(\mu^{\text{Reuss}})$.

4 Determination of the Size of the RVE

The notion of RVE is necessarily related to the choice of a statistical precision in the estimation of the investigated effective property. First, we set a tolerance error α on the bias and find a corresponding volume V_0 such that:

$$|\bar{\mu}^{\text{app}}(V_0) - \mu^{\text{eff}}| \leq \alpha. \tag{12}$$

This condition sets a lower bound for the size of the RVE. Then, the relative precision of the estimation of the mean $\bar{\mu}^{\text{app}}(V)$ of apparent shear moduli for a given volume $V \geq V_0$ and a number of realizations n, can be defined according to the sampling theory by:

$$\varepsilon_{\text{rel}} = \frac{2D_\mu(V)}{\bar{\mu}^{\text{app}}(V)\sqrt{n}}. \tag{13}$$

In turn, the number of realizations required to correctly estimate $\bar{\mu}^{\text{app}}(V)$ is deduced from Equation (13) provided that the variance $D_\mu^2(V)$ is known.

According to homogenization conditions (8), (9) and (10), the apparent shear modulus is obtained by averaging an additive scalar over the volume V. As a result, it is shown by Matheron (1997) that, for asymptotically large volumes, the variance $D^2_\mu(V)$ of $\mu^{app}(V)$ is given by:

$$D^2_\mu(V) = D^2_\mu \frac{A_3}{V}, \tag{14}$$

where A_3 is the integral range, a well-established quantity for additive geometrical properties such as volume fraction. D^2_μ is the point variance of $C_{1212}(\underline{x})$, which depends on crystal orientation at x. For uniform orientation distribution, it can be expressed in terms of the single crystal cubic elasticity constants as follows:

$$D^2_\mu = \langle (c:c)_{1212}\rangle - \langle c_{1212}\rangle^2 \quad \text{with} \quad \langle c_{1212}\rangle = \frac{1}{5}(C_{11} - C_{12} + 3C_{44}), \tag{15}$$

$$\langle (c:c)_{1212}\rangle = \frac{1}{35}(-6C_{44}C_{12} - 4C_{12}C_{11} + 2C^2_{12} + 2C^2_{11} + 6C_{11}C_{44} + 15C^2_{44}), \tag{16}$$

where $\langle\cdot\rangle$ refers to averaging over uniformly distributed orientations. For pure copper, one gets $D_\mu = 13588$ MPa. We choose to identify the integral range A_3 from the results obtained with periodicity conditions because they introduce the smallest bias in the estimated effective shear modulus. We find $A_3 = 1.43$, to be compared with the mean grain size set to 1, and the integral range for the volume fraction of a given orientation $A_3 = 1.17$ given by Kanit et al. (2003).

Equations (12), (13) and (14) can now be used quantitatively to determine a minimal size of RVE for a given precision ε_{rel} and a number of realizations n:

$$V = \frac{4}{n} D^2_\mu \frac{A_3}{\varepsilon^2_{rel}\mu^{eff\,2}}. \tag{17}$$

In the case of periodicity boundary conditions, the choice $(\varepsilon_{rel}, n) = (1\%, 10)$ gives a minimal volume corresponding to $V = 445$. For $n = 100$ successive computations, this volume reduces to 45.

5 Conclusions

A computational homogenization methodology was applied to the determination of RVE sizes for the isotropic linear elastic behaviour of copper polycrystals. For a given precision of 1% in the estimation of the effective property, and a number of affordable computations ranging from 10 to 100, RVE sizes remain of the order of 40 to 400 grains, provided that periodicity boundary conditions are applied to the polycrystalline aggregates. The convergence of apparent properties obtained using homogeneous boundary conditions towards the effective modulus is significantly slower than for periodicity conditions. The asymptotic shear modulus can be accurately estimated by a small number of huge computations or by a large number of small-scale computations, looking at the ensemble average of the apparent properties.

The integral range A_3 defined and identified in this work is a well-suited parameter to compare RVE sizes for different properties and morphologies. It characterizes the rate of decrease in the dispersion of apparent properties for increasing volume sizes, according to Equation (14). It depends on the investigated property (volume fraction, elasticity moduli, thermal conductivity, etc.). The value calculated in this work can be compared to the integral range found for a two-phase elastic material with a contrast in Young's moduli of 100 and 50% volume fraction of hard phase, namely $A_3 = 1.64$ (Kanit et al., 2003). The relatively high value found in the present work in spite of the relatively small contrast in properties between different orientations can be attributed to the multi-phase character of polycrystals, each crystal orientation being regarded as an individual phase. The volume must be large enough to contain enough individual orientations.

Acknowledgements

The authors are indebted to P. Featonby and J.-C. Leroy (SAGEM) for stimulating discussions during this work. The authors thank SAGEM and DGA for their financial and scientific support.

References

Barbe, F., Decher, L., Jeulin, D. and Cailletaud, G. (2001) Intergranular and intragranular behavior of polycrystalline aggregates. Part 1: F.E. model, *Int. J. Plasticity* **17**, 513–536.

Cailletaud, G., Forest, S., Jeulin, D., Feyel, F., Galliet, I., Mounoury, V. and Quilici, S. (2003) Some elements of microstructural mechanics, *Comput. Mater. Sci.* **27**, 351–374.

Drugan, W.J. and Willis, J.R. (1996) A micromechanics-based nonlocal constitutive equation and estimates of representative volume element size for elastic composites, *J. Mech. Phys. Solids* **44**, 497–524.

Feyel, F. (1999) Multiscale FE^2 elastoviscoplastic analysis of composite structures, *Comput. Mater. Sci.* **16**, 344–354.

Gairola, B.K.D. and Kroner, E. (1981) A simple formula for calculating the bounds and the self-consistent value of the shear modulus of polycrystalline aggregates of cubic crystals, *Int. Engng. Sci.* **19**, 865–869.

Huet, C. (1990) Application of variational concepts to size effects in elastic heterogeneous bodies, *J. Mech. Phys. Solids* **38**, 813–841.

Kanit, T., Forest, S., Galliet, I., Mounoury, V. and Jeulin, D. (2003) Determination of the size of the representative volume element for random composites: Statistical and numerical approach, *Int. J. Solids Struct.* **40**, 3647–3679.

Lippmann, N., Steinkopf, T., Schmauder, S. and Gumbsch P. (1997) 3D-Finite-Element modeling of microstructures with the method of multiphase elements, *Comput. Mater. Sci.* **9**, 28–35.

Matheron, G. (1971) *The Theory of Regionalized Variables and Its Applications*, Paris school of Mines Publications.

Nygårds, M. (2003) Number of grains necessary to homogenize elastic materials with cubic symmetry, *Mech. Mater.* **35**, 1049–1057.

Ren, Z.-Y. and Zheng, Q.-S. (2002) A quantitative study of minimum sizes of representative volume elements of cubic polycrystals-numerical experiments, *J. Mech. Phys. Solids* **50**, 881–893.

Sab, K. (1992) On the homogenization and the simulation of random materials, *Eur. J. Mech. Solids*, **11**, 585–607. Sanchez-Palencia, E. and Zaoui, A. (1987) *Homogenization Techniques for Composite Media*, Lecture Notes in Physics No. 272, Springer, Berlin.

Z-set package (2001) www.nwnumerics.com, www.mat.ensmp.fr.

Atomistic Corroboration of a Multiscale Approach for the Analysis of Dislocation Nucleation at a Surface Step

G. Xu*, D.E. Segall and C. Li

*Department of Mechanical Engineering, University of California at Riverside, Riverside, CA 92521, U.S.A.; *E-mail: gxu@engr.ucr.edu*

Abstract. The variational boundary integral formulation of the Peierls–Nabarro dislocation model has recently become one of the most effective multiscale approaches for the analysis of dislocation nucleation problems. By representing the structure of a dislocation as the relative displacement between two adjacent atomic layers along the slip plane, the model allows for the convenient incorporation of atomic information to treat the deformation of the dislocation core as continuous deformation, therefore eliminating the uncertain core cutoff parameter associated with the singularity of continuum elastic dislocation theory. By reducing many atomic degrees of freedom to fewer, yet more physically intuitive, degrees of freedom in this multiscale approach, one may gain a greater understanding of relevant physical processes in larger systems with more realistic geometries. Application of this approach requires the understanding of the reliability of this approach, or at least, it correlation to that of all atom calculations. Using nucleation of a $\langle 111 \rangle$ screw dislocation at a step from a $\{112\}$ surface of tantalum as an example, this paper provides an atomistic corroborative study of this multiscale approach. The results show the critical stresses for dislocation nucleation in this configuration obtained by the multiscale approach are in good agreement with all atom calculations.

Key words: dislocation nucleation, surface step, atomistic simulation, multiscale modeling.

1 Introduction

The recent increasing interest in energetics of dislocation nucleation in various configurations such as cracks (Schoeck and Pueschl, 1991; Rice, 1992; Xu et al., 1995; 1997; Zhu et al., 2004), voids (Rudd, 2002), interfaces (Spearot et al., 2005), surface steps (Godet et al., 2004) has been largely driven by the desire to understand the physics of mechanical behavior of crystalline materials, particularly those nanostructured polycrystalline materials, as well as to grow the high quality of thin layer semiconductor structures for advanced electronic and optoelectronic applications. Here we shall point out that dislocation nucleation means the creation of dislocations from perfect lattice structures but not dislocation generation or multiplication from existing dislocations. The fundamental physical processes for these two phenomena

Yilong Bai et al. (eds), IUTAM Symposium on Mechanical Behavior and Micro-Mechanics of Nanostructured Materials, 181–189.

are quite different. The local stress required to nucleate a dislocation is comparable to the theoretical stress, while the local stress required to generate or multiply dislocations from existing dislocations is generally comparable to the yield stress of the crystal, which is typically several orders lower than the theoretical stress. In general, dislocation nucleation is a complex atomic scale process that involves the formation of a curved dislocation in a configuration of the length scale of few nanometers, somewhat one order larger than the size of dislocation core structure. Hence, it is well recognized that the energetics of such a process cannot be reliably obtained based on continuum elastic dislocation theory because it ignores atomic information by using an uncertain cutoff parameter associated with the dislocation core region. On the other hand, direct atomistic simulation is also limited to the problems of relatively small size because of the difficulties to enforce the boundary conditions as well as to determine the unstable saddle point configurations of curved dislocations in large atomic systems.

Alternatively, the multiscale approach based on the variational boundary integral formulation of the Peierls–Nabarro dislocation model has proven to be an effective approach to study dislocation nucleation problems in complicated configurations (Xu et al., 1997; Xu and Argon, 2000; Xu, 2002; Xu and Zhang, 2002; Zhang and Xu, 2004). By representing the structure of a dislocation as the relative displacement between two adjacent atomic layers along the slip plane, this model treats the deformation of the dislocation core in a continuous fashion, therefore eliminating the uncertain core cutoff parameter associated with continuum elastic dislocation theory. In addition, appropriate atomic potentials based on *ab initio* calculations can also be conveniently incorporated along the adjacent atomic layers so that the effects of atomic properties on dislocation nucleation can be ascertained. Furthermore, the variational boundary integral method (Xu and Ortiz, 1993; Xu, 2000) can be effectively utilized to solve the structure of unstable saddle point configurations of embryonic dislocations in general three dimensional configurations by controlling slip based on Lagrange multiplier method (Xu et al., 1995).

It should be noted that the Peierls–Nabarro dislocation model is a continuum based multiscale model with minimum commitment to the atomistic information. Although this method appears to be far superior to continuum elastic dislocation theory, the reliability of this method for the analysis of dislocation nucleation processes is still not well understood. The ultimate solution of this problem based on *ab initio* approach such as density functional theory with local approximation is out of the question in the foreseeable future. It appears, however, there is still a need to understand how this model compares to all atom methods based on empirical potentials. Since the variational boundary integral method provides superior control over the system of study, such corroboration establishes the confidence to use this multiscale approach to explore dislocation nucleation in realistic structures of technological importance. Moreover, the saddle point configurations of the curved dislocation obtained by this method may also be used as the initial configurations for further direct atomic refinement using the nudged elastic band method (Zhu et al., 2003) if such refinement is deemed necessary for certain applications.

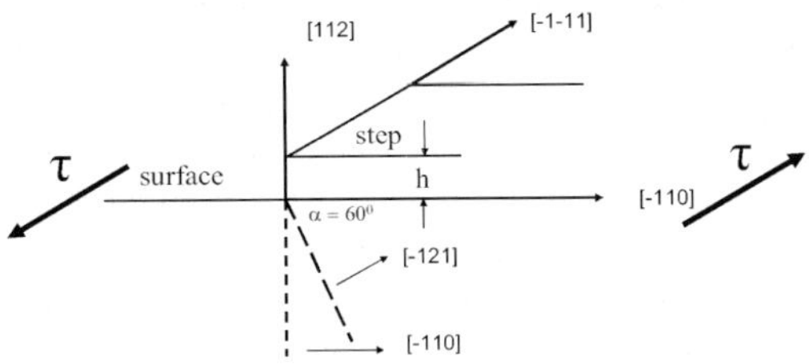

Fig. 1. Configuration of a surface step of tantalum. The slip direction is assumed to be parallel to the intersection of the step surface and the crystal surface. The remote shear stress is assumed to be parallel the slip direction.

In this paper, we summarize our recent comparative study of dislocation nucleation at a surface step using both the multiscale method and direct atomic calculations. The focus is to understand how these two methods correlate with each other. To make the comparison as exact as possible, we start with the same atomic potential of tantalum which is modeled based on a host of *ab initio* simulation. Tantalum has a number of interesting properties which make it a good system to study. It is part of the bcc family of metals whose low temperature plasticity is controlled by long, straight, low mobile $\langle 111 \rangle$ screw dislocations and has many active slip planes. These interesting features have inspired numerous atomic level investigations of the motion of such screw dislocations under stress using empirical potential and *ab initio* methods [7]. In our study, all elastic constants used in the multiscale method are calculated directly from this potential. The interatomic layer potential between the adjacent atomic layer along the possible slip planes, namely (110) and (211), are modeled based on the atomic calculation of the generalized stacking fault energy, or the so-called γ surface. The critical stress for dislocation nucleation at a step of tantalum surface obtained with both methods will be compared.

2 Configuration for Dislocation Nucleation from a Surface Step in Tantalum

Figure 1 shows the configuration of a surface step at a tantalum surface. The symbol denotes the height of the step. Dislocation nucleation is driven by the simple remote shear stress which is parallel to the surface of the step. Nucleation of a screw dislocation should occur on the most energetically favorable slip plane such as the slip plane (-110) or (-121). Such configuration is selected also for its simplicity because of no involvement of the effect of normal stress across the slip plane and the surface production or elimination from the step.

As the shear stress increases from zero, the stress concentration on the slip plane near the step induces a stable slip until the shear stress reaches a critical value which is governed by the periodic shear resistance between the adjacent interatomic layers. Under the stress level below this critical value, a curved embryonic dislocation may also form as a consequence of a localized outward protrusion of slip into an unstable saddle point configuration from the stable slip configuration through the assistance of thermal motion. Thus, there are two solutions associated with the same level of stress below this critical value, one is stable and the other is unstable. The energy difference between these two configurations is the activation energy E_{act} required for thermally assisted dislocation nucleation. Note that this activation energy is stress-dependent and approaches zero as the stress approaches the critical stress for athermal dislocation nucleation.

3 Variational Boundary Integral Formulation of the Peierls–Nabarro Dislocation Model

In the variational boundary integral formulation of the Peierls–Nabarro dislocation model, the structure of a dislocation along the slip plane, corresponding to the relative displacement between two atomic layers, is modeled as continuous distribution of infinitesimal dislocations. Between the adjacent atomic layers is assumed an interatomic layer potential along the slip plane, while the surrounding crystal is treated as a linear elastic solid. The structure of the dislocation can then be determined by the variational boundary integral method (Xu et al., 1997). To account for the surface effect, a straightforward approach is adopted to allow the use of this method for general half space problems. In this approach, the surface with the step is modeled as part of a very large crack surface embedded in the infinite elastic solid. As the size of this crack is selected to be much larger than any relevant size of the dislocation and step configuration, the problem can then be solved in an equivalent system that is essentially composed of the slip plane and the crack embedded in the infinite solid (Xu and Zhang, 2003).

Let $\Delta(\mathbf{x})$ denote the displacement along the slip plane and the opening displacement of the crack. The total energy $\Pi[\Delta(\mathbf{x})]$ of the system can be expressed by

$$\Pi[\Delta(\mathbf{x})] = W[\Delta(\mathbf{x})] + V[\Delta(\mathbf{x})] - P[\Delta(\mathbf{x})], \tag{1}$$

where W is the elastic strain energy, V is the interatomic layer potential energy, and P is the work of external force. The elastic strain energy W for general anisotropic solids is obtained by Xu (2004). The interatomic layer potential energy is given by

$$V[\Delta(\mathbf{x}) = \int_S \Phi[\Delta(\mathbf{x})]dS, \tag{2}$$

where $\Phi[\Delta(x)]$ is the interatomic layer potential per unit area of the slip plane, which we model based on the generalized stacking fault energy calculated by our all atom method.

The minimization of the potential energy $\Pi[\Delta(\mathbf{x})]$ with respect to the displacement $\Delta(\mathbf{x})$ leads to nonlinear integral equations. A finite element method with the six-node triangular elements is used to discretize the slip plane and the crack. The resulting nonlinear equations are solved by the Newton–Raphson method. The more detailed numerical methodology is described in Xu et al. (1997).

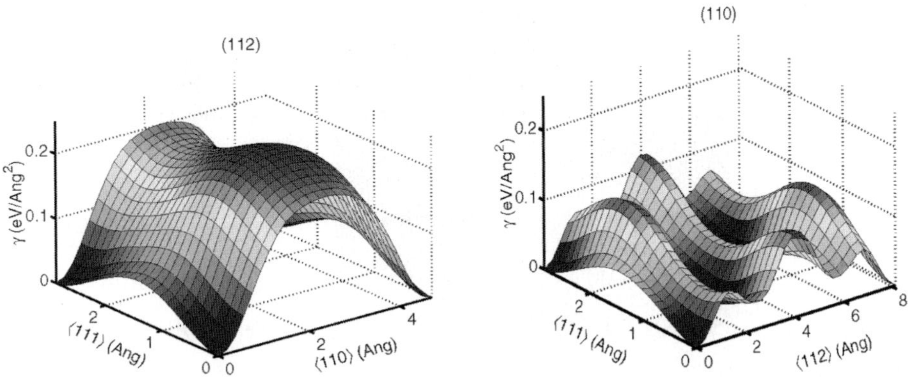

Fig. 2. The generalized stacking fault energy for slip planes (112) and (110) of tantalum.

4 Atomic Information and Calculations

The major advantage of our multiscale model is its capability to reduce many atomic degrees of freedom to fewer, yet more physically intuitive, degrees of freedom. This allows us not only to gain a greater understanding of the relevant physical process controlling the system of study but also to study much larger systems with more realistic geometries. However, an overriding problem with any multiscale approach is the ability to know *a priori* if all of the relevant atomic information has been properly taken into account in the multiscale model. One clear test to determine the reliability of such methods, or at least to help gain confidence in it, is the ability for the multiscale approach to reproduce results from that of all atom calculations. The goal of this section is to calculate the atomic level information (generalized stacking fault energy) which is incorporated into the variational boundary integral method and to provide results from all atom simulations in order to make direct comparison with our multiscale approach.

We represent the atomic scale interactions by employing a first-principles-based embedded atom method (qEAM) (Strachan et al., 2003). This empirical potential has been fitted to reproduce a host of *ab initio* calculations. The form of the potential is as follows:

$$U(\{\vec{r}_i\}) = \sum_i F(\rho_i) + \frac{1}{2}\sum_{i,j} \varphi(r_{ij}), \tag{3}$$

where $F(\rho_i)$ is the embedding energy, ρ_i is the so-called electron density, $\varphi(r_{ij})$ is a two-body term and $r_{ij} = |\vec{r}_i - \vec{r}_j|$. Details of the functional form for each term can be found in Strachan et al. (2005). Calculations for the generalized stacking fault energy are carried out in periodic boundary conditions. Atoms are fully relaxed in the direction perpendicular to the slip plane, while lattice vectors are not relaxed under dilation. Figure 2 shows the profile of the generalized stacking fault energy for both slip planes (112) and (110).

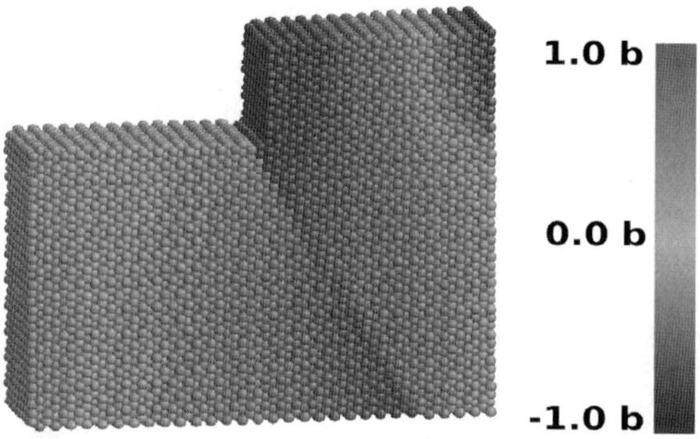

Fig. 3. Shear displacement of a dislocation in a typical simulation cell.

All atom calculations are performed to obtain the critical stress required for nucleation of a dislocation from a surface step. Figure 3 shows a typical simulation cell. Periodic conditions are enforced. To determine the stress needed for the nucleation of a dislocation we simply apply a shear strain to the system and calculate the stress from the relation

$$\sigma_{xz} = -\frac{1}{V_c}\frac{\partial E}{\partial \varepsilon_{xz}}, \tag{4}$$

where E is the energy of the cell and V_c is its volume. At each particular strain we minimize the energy functional, with respect to atomic positions, by the method of conjugate gradients. Only motion along the slip direction $[-1-11]$ is taken into account. At a given strain the atomic configuration is relaxed such that the maximum force on an atom along the slip direction is less than 6×10^{-5} eV/Å. At the instability point a dislocation nucleates and our conjugate gradient method becomes unstable. The atoms are then relaxed by a steepest decent method. To find the critical stress τ we gradually increase the strain to that slightly beyond the critical point. Figure 3 also displays the shear displacement of the cell when a dislocation first becomes nucleated on the slip plane (-121). The critical shear loading as a function of the height of the step is plotted in Figure 4.

5 Dislocation Nucleation at a Surface Step by the Multiscale Approach

The critical stress required for nucleation of a straight screw dislocation in the $[-1-11]$ direction on the (-110) or (-121) slip plane is also obtained by the multiscale approach based on the variational boundary integral formulation of the Peierls–Nabarro dislocation model. The ineratomic potentials on each slip plane are modeled based on the obtained generalized stacking fault energy illustrated in Figure 2. The results are plotted in Figure 4 in comparison with the results obtained

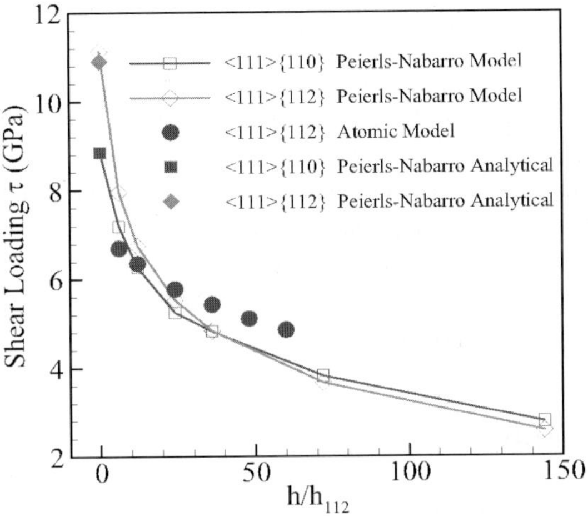

Fig. 4. Comparison of the critical stress for nucleation of a screw dislocation as a function of step height between analytic solution, atomic calculation, and multiscale approach.

by the all atom calculations, together with the analytical solutions of the Peierls–Nabarro model for dislocation nucleation along a smooth surface. The close agreement between the analytic solutions and the multiscale calculation validates the numerical accuracy of the multiscale calculation as well as the treatment of the surface as a very large crack. The multiscale approach also compares quite well with the all atom calculations as long as the finite size of the simulation box does not affect the result significantly. In all cases the multiscale approach is within 15% of the all atom solution. The slight discrepancy between the multiscale approach and the all atom solution in predicting the slip system, (−110) vs. (−121), probably results from the fact that the difference in the shear resistance profile of both slip systems. Despite this discrepancy, the multiscale approach accurately predicts the critical stress and the influence of surface steps. Moreover, in the multiscale approach the competition between the two slip systems is evident, a fact that would be extremely difficult to determine from an all atom calculation. The multiscale solution demonstrates that the presence of stress concentration at surface steps can significantly facilitate dislocation nucleation; the critical stress is reduced by over a factor of three for steps as small as 20 nm in height.

6 Summary and Conclusions

Nucleation of a ⟨111⟩ screw dislocation at a step from a {112} surface of tantalum has been studied by both an all atom method and a multiscale approach. In the multiscale approach, the interatomic layer potential, modeled based on atomic calculations of generalized stacking fault energies, is assumed along the slip plane. The structure of the dislocation, represented by the relative displacement between two adjacent atomic layers along the slip plane, is solved by the variational boundary integral method. The results show that atomic scale steps can reduce the critical stress required for athermal nucleation of a dislocation by nearly an order of magnitude. Comparison of the results of the multiscale approach with atomic calculations demonstrates that the multiscale approach correlates well with the all atom method for the study of the dislocation nucleation problem. Since the variational boundary integral method provides superior control over the system of study and effective way to solve saddle point configurations, such correlation further establishes the confidence on this multiscale approach to explore dislocation nucleation in realistic geometries of technological importance.

Acknowledgements

We gratefully acknowledge the support of the Lawrence Livermore National Laboratory and the National Science Foundation through the CAREER (0103257) and NIRT (0134446) awards.

References

Beltz, G.E. and Freund, L.B. (1993) On the nucleation of dislocations at a crystal surface, *Phys. Stat. Sol. B* **180**, 303–313.

Godet, J., Pizzagalli, L., Brochard, S. and Beauchamp, P. (2004) Theoretical study of dislocation nucleation from simple surface defects in semiconductors, *Phys. Rev. B* **70**, 054109-1-8.

Rice, J.R. (1992) Dislocation nucleation from a crack tip: An analysis based on the Peierls concept, *J. Mech. Phys. Solids* **40**, 239–271.

Rudd, R.E. and Belak, J.F. (2002) Void nucleation and associated plasticity in dynamic fracture of polycrystalline copper: An atomistic simulation, *Comput. Mater. Sci.* **24**(1–2), 148–153.

Schoeck, G. and Pueschl, W. (1991) The formation of dislocation loops at crack tips in three dimensions, *Phil. Mag. A* **64**, 931–49.

Segall, D.E., Strachan, A., Goddard, W.A., Ismail-Beigi, S. and Arias, T.A. (2004) Ab initio and finite-temperature molecular dynamics studies of lattice resistance in tantalum, *Phys. Rev. B* **68**, 014104.

Spearot, D., Jacob, K. and McDowell, D. (2005) Molecular dynamics simulations of dislocation nucleation from bicrystal interfaces in FCC metals, in *Proceedings of IMECE2005 2005 ASME International Mechanical Engineering Congress and Exposition*, November 5–11, 2005, Orlando, FL, USA.

Strachan, A., Cagin, T., Gulseren, O., Mukherjee, S., Cohen, R.E. and Goddard, W.A. (2004) First principles force field for metallic tantalum, *Mod. Sim. Mat. Sci. Eng.* **12**, s445–s459.

Xu, G. (2000) A variational boundary integral method for the analysis of 3D cracks of arbitrary geometry in anisotropic elastic solids, *J. Appl. Mech.* **67**, 403–408.

Xu, G., Argon, A.S. and Ortiz, M. (1995) Nucleation of dislocations from crack tips under mixed mode loading: Implications for brittle against ductile behavior of crystals, *Phil. Mag. A* **72**, 415–451.

Xu, G. and Argon, A.S. (2000) Homogeneous nucleation of dislocation loops under stress in perfect crystals, *Phil. Mag. Lett.* **80**, 605–611.

Xu, G. and Ortiz, M. (1993) A variational boundary integral method for the analysis of 3-D cracks of arbitrary geometry modeled as continuous distributions of dislocation loops, *Int. J. Num. Meth. Engng.* **36**, 3675–3701.

Xu, G. and Zhang, C. (2003) Analysis of dislocation nucleation from a crystal surface based on the Peierls–Nabarro dislocation model, *J. Mech. Phys. Solids* **51**, 1371.

Zhang, C. and Xu, G. (2005) Energetics of dislocation nucleation under a nanoindenter, *Mat. Sci. Eng. A* **400–401**, 471–475.

Zhu, T., Li, J. and Yip, S. (2004) Atomistic study of dislocation loop emission from a crack tip, *Phys. Rev. Lett.* **93**, 025503-1-4.

Indenter Tip Radius and Micro-Indentation Hardness

C.J. Tao, T.C. Wang*, X.Y. Feng and S.H. Chen

LNM, Institute of Mechanics, Chinese Academy of Sciences, Beijing 100080, China;
**E-mail: tcwang@imech.ac.cn*

Abstract. Using finite element method with the conventional J_2 theory and strain gradient theory respectively, the effect of the indenter tip radius on the micro-indentation hardness is investigated in the present paper. It is found that the former can not predict the size effect even considering the indenter tip radius, while the latter gives a good agreement to the experimentally measured micro-indentation hardness, which confirms that the size effect of micro-indentation hardness does exist due to the factor of the strain gradient effect.

Key words: finite element method, size effect, micro-indentation hardness, indenter tip radius, conventional J_2 theory, strain gradient theory.

1 Introduction

When a material or a structure possesses a micrometer scale, some mechanical characters are different totally from those in macro scale. Size effect is a main phenomenon found in the micro-scale, especially for the micro-indentation hardness [1–3]. The micro-indentation hardness increases as the indent depth decreases, which can not be explained by the conventional plasticity theory due to no length scale is included. Strain gradient theories are proposed as an effectively theoretical tool to understand the size effect, see for example [4–8].

Using the strain gradient theories, the size effect in micro-indentation hardness has been studied [9–11] with an assumption that the indenter tip is perfect without a curvature and the theoretically predicted results are consistent well with the experimentally measured data.

As for the size effect in micro-indentation hardness, some researchers thought that many experimental factors, such as the effect of the surface layer, the friction between the indenter and the indented material, the indenter tip curvature, should have a great influence on the measured hardness, which could cause an increasing indentation hardness for a decreasing indent depth [12]. However, recent experiments carried out by Swadener et al. [13] on a fine-grained polycrystalline iridium using spherical indenters with different radius show a different trend from [12]: at a fixed

Yilong Bai et al. (eds), IUTAM Symposium on Mechanical Behavior and Micro-Mechanics of Nanostructured Materials, 191–199.

indenter radius, the hardness decreases with the indentation depth decreasing for the case of $a/R < 0.2$. One question arises from [12] and [13]: which result is right? Of course, experiment is a believable truth, which is always adopted to check the correctness of any theory.

In the present paper, the micro-indentation hardness tests will be analyzed numerically using finite element method with the conventional J_2 theory and strain gradient theory proposed by Chen and Wang [7, 8] respectively. A conical indenter with a round tip is considered. The effect of the indenter tip radius on the micro-indentation hardness is emphasized.

The strain gradient theory is briefly reviewed in Section 2. The finite element analysis in the present paper is given in Section 3. In Section 4, numerical results are compared to the existing experimental data for several materials.

2 Review of Strain Gradient Theory

The incremental constitutive relations of the strain gradient theory [7, 8] are

$$\begin{cases} \dot{\sigma}_{ij} = 2\mu\dot{\varepsilon}'_{ij} + K\dot{\varepsilon}_m\delta_{ij}, \\ \dot{m}_{ij} = 2\mu l_{cs}^2\dot{\chi}'_{ij} + K_1 l_{cs}^2 \cdot \chi_m\delta_{ij}, \end{cases} \quad \Sigma_e < \sigma_Y, \tag{1}$$

$$\begin{cases} \dot{\sigma}_{ij} = \dfrac{2\Sigma_e}{3E_e}\dot{\varepsilon}'_{ij} + \dfrac{2\dot{\Sigma}_e}{3E_e}\dot{\varepsilon}_{ij} - \dfrac{2\Sigma_e}{3E_e^2}\dot{\varepsilon}_{ij}\dot{E}_e + K\dot{\varepsilon}_m\delta_{ij}, \\ \dot{m}_{ij} = \dfrac{2\Sigma_e}{3E_e}l_{cs}^2\dot{\chi}'_{ij} + \dfrac{2\dot{\Sigma}_e}{3E_e}l_{cs}^2\dot{\chi}'_{ij} - \dfrac{2\Sigma_e}{3E_e^2}l_{cs}^2\chi'_{ij}\dot{E}_e + K_1 l_{cs}^2\dot{\chi}_m\delta_{ij}, \end{cases} \quad \Sigma_e \geq \sigma_Y, \tag{2}$$

where the generalized effective strain and the generalized effective stress are defined as

$$\begin{cases} E_e^2 = \varepsilon_e^2 + l_{cs}^2\chi_e^2, \ \Sigma_e = (\sigma_e^2 + l_{cs}^{-2}m_e^2)^{1/2}, \\ \sigma_e^2 = \dfrac{3}{2}s_{ij}s_{ij}, \qquad m_e^2 = \dfrac{3}{2}m'_{ij}m'_{ij}. \end{cases} \tag{3}$$

Here ε_e is the effective strain, χ_e the effective rotation gradient and η_1 the effective stretch gradient, defined by Fleck and Hutchinson [4].

$$\varepsilon_e = \sqrt{\frac{2}{3}\varepsilon'_{ij}\varepsilon'_{ij}}, \quad \chi_e = \sqrt{\frac{3}{2}\chi_{ij}\chi_{ij}}, \quad \eta_1 = \sqrt{\eta_{ijk}^{(1)}\eta_{ijk}^{(1)}}. \tag{4}$$

The rotation gradient χ_{ij} is defined as a curvature tensor related to the micro-rotation ω_i,

$$\chi_{ij} = \omega_{i,j}. \tag{5}$$

l_{cs} is an intrinsic material length scale required on dimensional grounds. K is the volumetric modulus, K_1 the bend-torsion volumetric modulus. The influence of stretch gradient is introduced by the following hardening law,

$$\begin{cases} \dot{\Sigma}_e = A'(E_e)\left(1+\dfrac{l_1\eta_1}{E_e}\right)^{1/2}\dot{E}_e = B(E_e, l_1\eta_1)\dot{E}_e, & \Sigma_e \geq \sigma_Y, \\ \dot{\Sigma}_e = 3\mu\dot{E}_e, & \Sigma_e < \sigma_Y, \end{cases} \tag{6}$$

where $B(E_e, l_1\eta_1)$ is the hardening function; σ_Y is the yield stress and μ is the shear modulus; For power law hardening material, $A(E_e) = \sigma_Y E_e^n$; l_1 is the second intrinsic material length associated with the stretch gradient. The strain gradient theory [7, 8] reduces to the conventional J_2 deformation theory in the absence of strain gradient effects.

3 Finite Element Analysis

The principal of virtual work and the detailed formulas of the finite element method for the strain gradient theory [7, 8] can be found in [10].

3.1 The Coordinate System

It is convenient to express the field quantities in terms of cylindrical coordinate system (r, θ, z). Both the geometry of the indented solid and loading are axis-symmetric. The displacement field of the indented solid is

$$u_r = u_r(r, z), \quad u_\theta = 0, \quad u_z = u_z(r, z) \tag{7}$$

and the micro-rotation field is

$$\omega_\theta = \omega_\theta(r, z), \quad \omega_r = \omega_z = 0. \tag{8}$$

3.2 The Blunt Conical Indenter

In order to simulate the micro-indentation test, the axis-symmetric model is adopted in this paper. The blunt conical indenter and the axis-symmetric model are shown in Figure 1. The contact depth can be expressed as:

$$\begin{cases} \delta(r) = \dfrac{r}{\tan\beta} - \xi, & r_0 \leq r \leq a, \\ \delta(r) = R - (R^2 - r^2)^{1/2}, & r \leq r_0, \end{cases} \tag{9}$$

where $r_0 = R\cos\beta$, $\xi = R/\sin\beta - R$.

The assumptions in the numerical simulations and the boundary conditions can be found in [10]. The indented body is taken to be a circular cylinder. The size of the indented body is much larger than the depth of the indentation. On the whole surface of the cylinder, the torque tractions are taken to be zero, which yields $\omega_i = 0$, $m_{ij} = 0$ so that the influence of l_{cs} can be ignored.

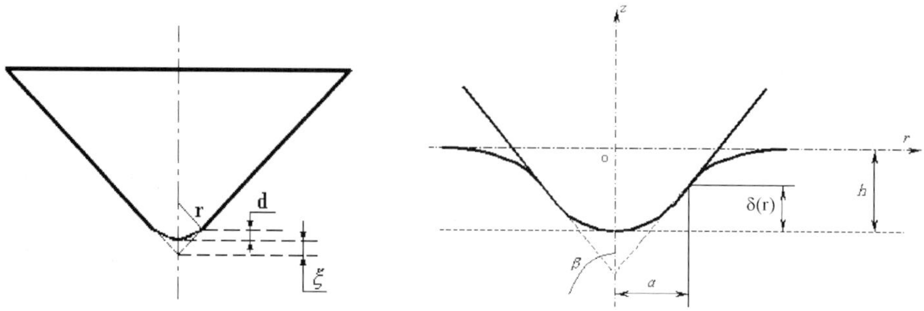

a. Blunt conical indenter $(R \neq 0)$ b. Axis-symmetric indentation model

Fig. 1. Blunted conical indenter and axis-symmetric model.

3.3 The Material Parameters

Young's modulus and Poisson ratio can be obtained from the existing literature – but the yield stress, power-law hardening exponent and intrinsic length scale need to be determined by the numerical fitting method based on the simulation for the experiment data. Following is the detailed steps:

1. Choose three values of depths h_0, h_1, h_2 at large depth where the indentation hardness nearly keeps constant, the corresponding loads $P(h_0)$, $P(h_1)$, $P(h_2)$ can be obtained from the experiment data.
2. First, the values of n and l_1 are prescribed and the initial value σ_Y is chosen as $\sigma_y = \sigma$, one can get $P^*(h_0, \sigma, n, l_1)$, $P^*(h_1, \sigma, n, l_1)$ and $P^*(h_2, \sigma, n, l_1)$ by finite element calculations.
3. Introducing

$$F(\sigma, n, l_1) = \sum_{i=0}^{2} (P^*(h_i, \sigma, n, l_1) - P(h_i))^2,$$

one can get the proper σ^* by the quasi-Newton method [14]:

$$\sigma_{k+1} = \sigma_k - \frac{F(\sigma_k)}{F'(\sigma_k)}, \quad F'(\sigma_k) = \frac{F(\sigma_k) - F(\sigma_{k-1})}{\sigma_k - \sigma_{k-1}},$$

which makes the functional F to be minimum $\min_\sigma F(\sigma, n, l_1) = G(n, l_1)$. Obviously, σ^* depends on n and l_1 , that is, $\sigma^* = \sigma^*(n, l_1)$.
4. Keeping l_1 as a constant, for a given value n $(0 \leq n \leq 1)$, carrying out the above calculations to get each $G(n, l_1)$, then from the curve $G(n, l_1)$ versus n, one can get the proper value n^*, which makes the functional G to be minimum $T(l_1) = \min_n G(n, l_1) = G(n^*, l_1)$.
5. Finally, for a given value l_1 $(0 < l_1 < 1\ \mu\text{m})$, carrying out above calculations to get each $T(l_1)$, then from the curve $T(l_1)$ versus l_1 to make $\min_{l_1} T(l_1) = T(l_1^*)$. The final parameters σ^{n}, n^* and l_1^* are the needed yield stress σ_Y, the power-law hardening exponent n and the intrinsic length scales l_1.

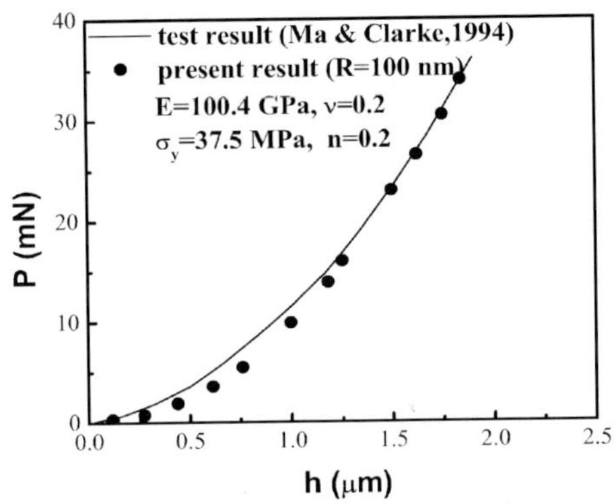

a. for single crystal silver

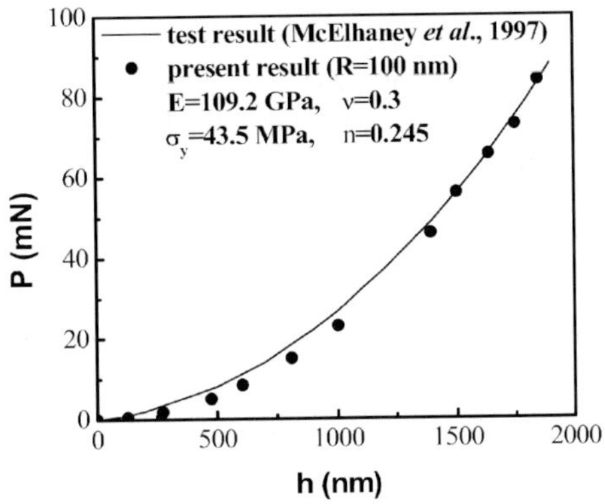

b. for polycrystal copper

Fig. 2. The indentation load P versus depth h, the calculation results are based on the classical J_2 plasticity theory.

4 Calculation Results and Comparison with the Test Data

The calculation results are shown in this section. Comparisons between the calculation results and the experimental indentation data are emphasized. The indenter tip radius is taken to be $R = 100$ nm.

4.1 Calculation Results and Comparison with the Test Data

Figure 2a shows the results of the indentation load P versus indentation depth h for single crystal silver. The solid line is the experimental results given by Ma and Clarke [2] and the full circles are the present calculation results. Young's modulus $E = 100.4$ GPa and Poisson's ratio $\nu = 0.2$ were given by Ma and Clarke [2]. The yield stress $\sigma_Y = 37.5$ MPa and power-law hardening exponent $n = 0.2$ are obtained by the fitting method of Section 3.3. It can be seen that the calculation results agree well with the test results at deep depth, but lower than the test results at the shallower depth.

The nominal indentation hardness H^* versus the indentation depth h for single crystal silver is shown in Figure 3a. Here $H^* = P/(24.56h^2)$, which is defined by Ma and Clarke [2]. From Figure 3a, one can see that the calculation results are much lower than the test results for shallower depth.

The calculation results on polycrystalline copper are shown in Figures 2b and 3b respectively, which have the same trends as Figures 2a and 3a.

Figure 4a shows the results of the indentation hardness H versus indentation depth h for single crystal silver. The indentation hardness H is defined as $H = P/\pi a^2$.

Since one cannot directly measure the contact area, the existed experimental data did not include any measured values of contact area. We only use the calculated values of the contact area instead the measured values of contact area. From Figure 4a,

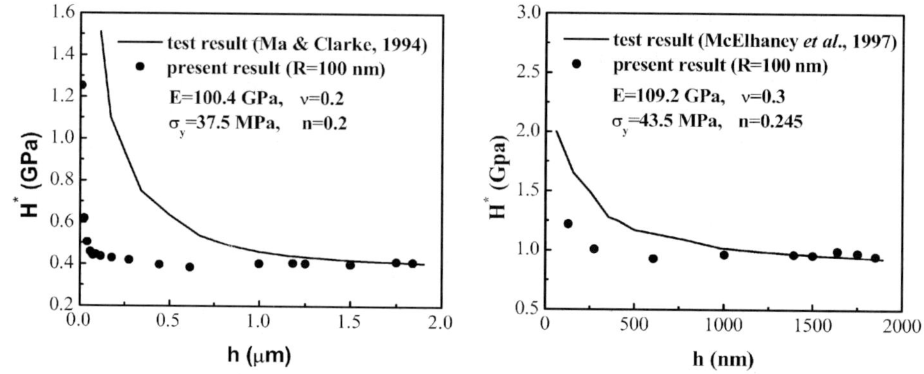

a. for single crystal silver

b. for polycrystal copper

Fig. 3. The nominal indentation hardness H^* versus depth h, the calculation results are based on the classical J_2 plasticity theory.

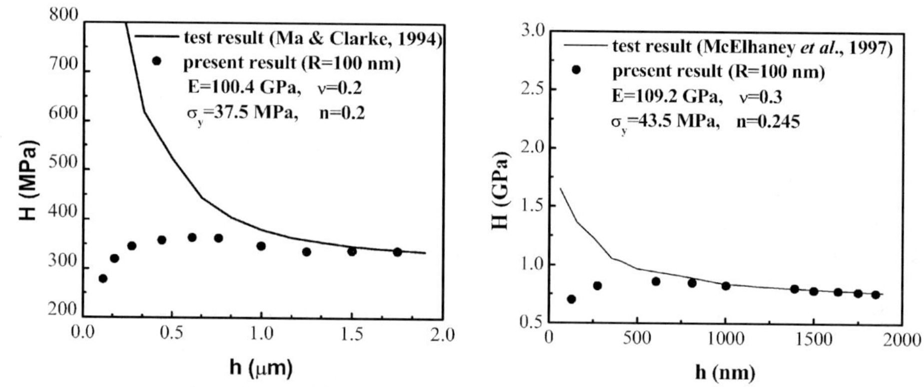

a. for single crystal silver

b. for polycrystal copper

Fig. 4. The indentation hardness H versus depth h, the calculation results are based on the classical J_2 plasticity theory.

one can see that the calculation results are not only much lower than the test results, but the hardness decreases with the decreasing depth when the depth is lower than 500 nm. It means that the predicted load and hardness based on the conventional J_2 theory cannot agree well with the experimentally measured micro-indentation load and hardness over a wide range of the indentation depth, even considering the effect of indenter tip radius, which reveals that the indenter tip radius cannot explain the size effect.

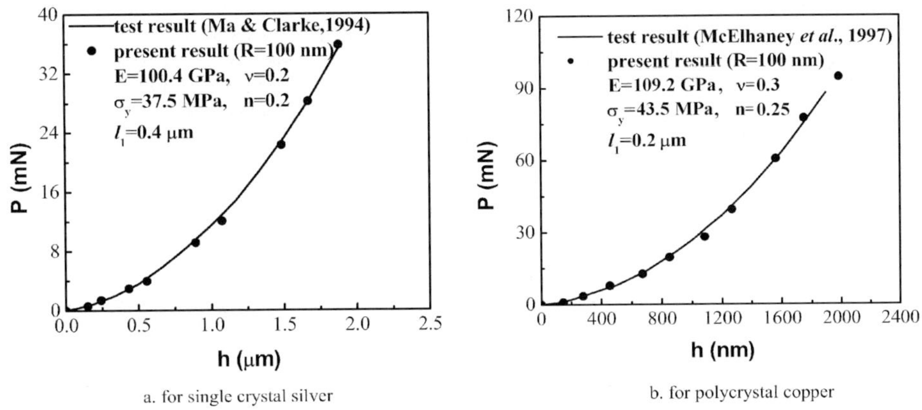

Fig. 5. The indentation load P versus depth h, the calculation results are based on the strain gradient theory [7, 8].

4.2 Calculation Results for Strain Gradient Theory

The above results clearly show that the effect of the indenter tip radius cannot explain the size effect, so we carry out the calculations based on Chen and Wang's strain gradient theory [7, 8] with considering the influence of the indenter tip radius.

Figure 5a shows the results of the indentation load P versus indention depth h for single crystal silver. The solid line is the experimental results given by Ma and Clarke [2] and the full circles are the present calculation results. Young's modulus $E = 100.4$ GPa and Poisson's ratio $\nu = 0.2$ were given by Ma and Clarke [2]. The yield stress $\sigma_Y = 37.5$ MPa, power-law hardening exponent $n = 0.2$ and the intrinsic length scale $l_1 = 0.4$ μm are obtained by the fitting method of Section 3.3.

From Figure 5a, one can see that the predicted load agree very well with the experimentally measured micro-indentation load over the whole range of the indentation depth, which provides a validation of Chen and Wang's strain gradient theory [7, 8].

The nominal indentation hardness H^* versus the indention depth h for single crystal silver is shown in Figure 6a.

From Figure 6a, one can see that the predicted nominal indentation hardness agree very well with the experimentally measured nominal indentation hardness over the whole range of the indentation depth. Similar calculations are carried out on polycrystalline copper and shown in Figures 5b and 6b respectively, which have the same trends as Figure 5a and 6a.

5 Conclusions

The predicted load and hardness based on the conventional J_2 theory cannot agree well with the experimentally measured micro-indentation load and hardness at micron scales even considering the indenter tip radius, which means that indenter tip

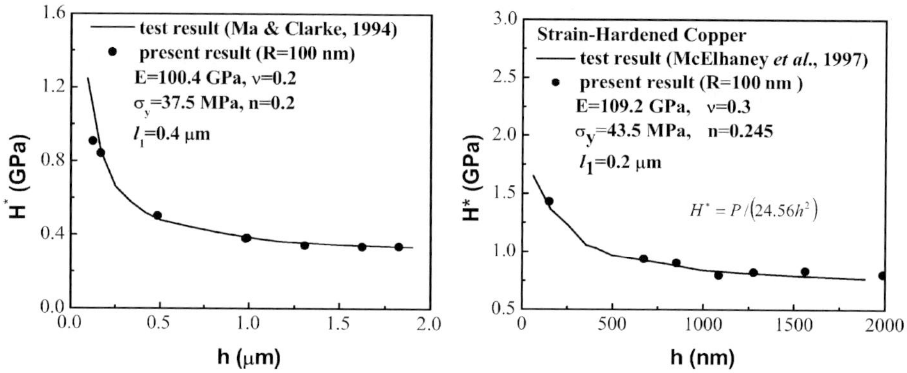

a. for single crystal silver

b. for polycrystal copper

Fig. 6. The nominal indentation hardness H^* versus depth h, the calculation results are based on the strain gradient theory [7, 8]

radius is not the reason for size effect as some researchers said. While based on the strain gradient theory proposed by Chen and Wang [7, 8], the predicted load and hardness agree very well with the experimentally measured micro-indentation load and hardness over a whole range of the indentation depth.

Acknowledgments

This work is supported by the National Natural Science Foundation of China (10272103) and (10202023).

References

1. Stelmashenko, N.A., Walls, M.G., Brown, L.M. and Milman, Y.V., Microindentations on W and Mo oriented single crystals: An STM study, *Acta Metall. Mater.* **41**(10), 1993, 2855–2865.
2. Ma, Q. and Clarke, D.R., Size dependent hardness in silver single crystals, *J. Mater. Res.* **10**, 1995, 853.
3. McElhaney, K.W., Vlassak J.J. and Nix, W.D., Determination of indenter tip geometry and indentation contact area for depth-sensing indentation experiments, *J. Mater. Res.* **13**, 1998, 1300.
4. Fleck, N.A. and Hutchinson, J.W., Strain gradient plasticity. In J.W. Hutchinson and T.Y. Wu (eds.), *Advances in Applied Mechanics*, Vol. 33, Academic Press, New York, 1997, p. 295.
5. Gao, H., Huang, Y., Nix, W.D. and Hutchinson, J.W., Mechanism-based strain gradient plasticity – I: Theory, *J. Mech. Phys. Solids* **47**, 1999, 1239.
6. Huang, Y., Gao, H., Nix, W.D. and Hutchinson J.W., Mechanism-based strain gradient plasticity – II: Analysis, *J. Mech. Phys. Solids* **48**, 2000, 99.

7. Chen, S.H. and Wang, T.C., A new hardening law for strain gradient plasticity, *Acta Mater.* **48**, 2000, 3997–4005.
8. Chen, S.H. and Wang, T.C., A new deformation theory for strain gradient effects, *Int. J. Plasticity* **18**, 2002, 971–995.
9. Huang, Y., Xue, Z., Gao, H., Nix, W.D. and Xia, Z.C., A study of micro-indentation hardness tests by mechanism-based strain gradient plasticity, *J. Mater. Res.* **15**, 2000, 1786–1796.
10. Chen, S.H., Tao, C.J. and Wang, T.C., A study of micro-indentation with size effects, *Acta Mech.* **167**(1–2), 2004, 57.
11. Chen, S.H., Liu, L. and Wang, T.C., Strain gradient effects in nano-indentation of film-substrate systems, *Acta Mater.* **52**(5), 2004, 1089.
12. Li Min, Liang Naigang, Zhang Taihua, et al., 3D finite element simulation of the nanoindentation process, *Acta Mech. Sinica* **35**(3), 2003, 257 [in Chinese].
13. Swadener, J.G., George, E.P. and Pharr, G.M., The correlation of the indentation size effect measured with indenter of various shapes, *J. Mech. Phys. Solids* **50**, 2002, 681–694.
14. Dennis, Jr J.E. and More, J.J., Quasi-Newton methods – Motivation and theory, *SIAM Rev.* **19**, 1977, 46–89.

Part 6

Mechanical Behaviors of Other Nano-Materials

The Phase Angle of an Interface Crack Induced by Indentation Delamination with Buckling

Tong-Yi Zhang[1,*], Bin Huang[1] and Ming-Hao Zhao[2]

[1] *Department of Mechanical Engineering, Hong Kong University of Science and Technology, Clear Water Bay, Kowloon, Hong Kong, China; *E-mail: mezhangt@ust.hk*
[2] *Department of Engineering Mechanics, Zhengzhou University, Zhengzhou, Henan 450002, China*

Abstract. In the indentation delamination test, an indentation load generates a compressive stress field to delaminate a film from its substrate and the interfacial fracture is always of a mixed mode. The interfacial fracture toughness depends on the phase angle of the fracture mixed mode. If the delaminated film buckles, the buckling induces a tensile stress field and thus changes the phase angle and the measured value of interfacial fracture toughness. A simple equation was derived to calculate the phase angle. Based on the simple equation and the empirical formula, which describes the interfacial fracture toughness as a function of the phase angle, relations among interfacial fracture toughness, phase angle and indentation load/crack-length are discussed in details.

Key words: phase angle, interfacial fracture toughness, indentation, delamination, buckling.

1 Introduction

Thin films on substrates have been attracting intensive interests from academic and industrial researchers due to their wide applications in the fields such as microelectronics, optoelectronics and microelectromechanics, etc. The durability and reliability of devices composed of thin films on substrates depend greatly upon the mechanical properties of film/substrate systems. One of the most important mechanical properties is adhesion between a film and its substrate. When the adhesion is evaluated by fracture, it is more accurate to call it the interfacial fracture toughness. Due to the small dimension of film thickness, it is a challenging task to accurately measure the interfacial fracture toughness between a film and its substrate. Researchers are developing new techniques to characterize the interfacial fracture toughness in film/substrate systems and six developed testing methods are briefly summarized in a recent overview [1]. Among the six testing methods, the indentation-induced delamination method is the most attractive approach because of its easiness and simplicity. Furthermore, indentation tests may be more appropriate than other methods when sample size and/or testing region become very small. In the indentation-induced delamination test, stresses near an interface crack tip are always comprised

Yilong Bai et al. (eds), IUTAM Symposium on Mechanical Behavior and Micro-Mechanics of Nanostructured Materials, 203–209.

of both mode I and mode II components, thereby leading to a mixed mode fracture. The mode mixity is characterized by a phase angle, ψ. Measured interfacial fracture toughness depends on the phase angle. Jensen and Thouless [2] proposed the following empirical formula,

$$\Gamma(\psi) = \Gamma_I^c[1 + (1 - \lambda)\tan^2\psi], \tag{1}$$

where $\Gamma(\psi)$ is the interfacial fracture toughness at a phase angle, ψ, $\Gamma_I^c(\psi = 0)$ is the pure mode I interfacial fracture toughness, and λ is a fitting parameter. Since $\Gamma(\psi)$ changes with the phase angle, Γ_I^c is the fundamental property to represent the adhesion strength between a film and its substrate. To accurately determine the value of Γ_I^c from measured $\Gamma(\psi)$, it is necessary to precisely determine the phase angle. Huang et al. [3] investigated the phase angle for the indentation delamination test and applied the theoretical results in the assessment of Γ_I^c between ZnO and Si [4]. The present work briefly reports the results published in [3] and discusses in detail the relations among interfacial fracture toughness, phase angle and indentation load/crack-length.

2 Energy Release Rate

Marshall and Evans [5] and Evans and Hutchinson [6] developed the fundamentals for measuring interfacial fracture energy from the indentation-induced delamination test. They modeled the delamination region as a clamped circular plate with a delamination radius, a, on a rigid substrate. During indentation testing on a film/substrate system, an indenter penetrates and displaces a volume, V_I, into the film, which induces a compressive stress, σ_I, in the film,

$$\sigma_I = \frac{V_I E_f}{2\pi h_f a^2(1 - \nu_f)}, \tag{2a}$$

where E_f, ν_f and h_f are Young's modulus, Poisson's ratio, and thickness of the film, respectively. The indentation-induced compressive stress is the driving force to delaminate the film from the substrate and the corresponding strain energy release rate, G, for the delamination is:

$$G = \frac{h_f\sigma_I^2(1 - \nu_f^2)}{2E_f} + (1-\alpha)\frac{h_f\sigma_R^2(1 - \nu_f)}{E_f} - (1-\alpha)\frac{h_f(\sigma_I - \sigma_B)^2(1 - \nu_f)}{E_f}, \tag{2b}$$

where σ_B is the critical buckling stress given by

$$\sigma_B = \frac{\mu^2 E_f}{12(1 - \nu_f^2)}\left(\frac{h_f}{a}\right)^2, \tag{2c}$$

with $\mu = 3.8317$ for single buckling, σ_R denotes residual stress of the film, and the parameter α is given by

$$\alpha = 1 - \frac{1}{1 + 0.902(1 - \nu_f)}, \quad \text{for } \sigma_I + \sigma_R > \sigma_B \text{ (with buckling)}. \tag{2d}$$

3 Phase Angle

Based on the thin film approximation that the film thickness is much smaller than the substrate thickness and considering only circular buckling, Huang et al. [3] derived the phase angle, which takes the explicit form:

$$\tan\psi = \frac{\left[\frac{\sigma_I}{\sigma_B} - \frac{3.6304(1+\nu_f)}{\mu^2}\xi^2\right]\sin\varpi - \sqrt{12}\frac{4.2156}{\mu^2}\xi\cos\varpi}{\left[\frac{\sigma_I}{\sigma_B} - \frac{3.6304(1+\nu_f)}{\mu^2}\xi^2\right]\cos\varpi + \sqrt{12}\frac{4.2156}{\mu^2}\xi\sin\varpi}, \tag{3}$$

where ξ is the ratio of the buckling amplitude at the center of the circular buckled film to the film thickness, and ϖ is a dimensionless function of Dundurs' elastic mismatch parameters. Under the thin film approximation, Suo and Hutchinson [7] calculated and tabulated the value of ϖ and $\varpi = 52.1°$ if the film has the same elastic constants as the substrate.

4 Relations among Interfacial Fracture Toughness, Phase Angle and Load/Crack-Length

In the present work, we shall take the empirical formula, Equation (1), and make assumptions that the interfacial fracture toughness is independent of the film thickness and the film residual stress is a constant, etc in order to plot and illustrate the relationship among the indentation load, the phase angle, and the energy release rate in the indentation delamination tests with buckling. For simplicity, we re-write the phase angle, Equation (3), the energy release rate, Equation (2), and the empirical equation, Equation (1) in dimensionless forms:

$$\tan\psi = \tag{4}$$

$$\frac{\left[\tilde{V}_I - \frac{3.6304(1+\nu_f)}{\mu^2 c}(\tilde{V}_I + \tilde{a}\tilde{\sigma}_R - 1)\right]\sin\varpi - \sqrt{12}\frac{4.2156}{\mu^2}\left[\frac{1}{c}(\tilde{V}_I + \tilde{a}\tilde{\sigma}_R - 1)\right]^{1/2}\cos\varpi}{\left[\tilde{V}_I - \frac{3.6304(1+\nu_f)}{\mu^2 c}(\tilde{V}_I + \tilde{a}\tilde{\sigma}_R - 1)\right]\cos\varpi + \sqrt{12}\frac{4.2156}{\mu^2}\left[\frac{1}{c}(\tilde{V}_I + \tilde{a}\tilde{\sigma}_R - 1)\right]^{1/2}\sin\varpi},$$

$$\tilde{G} = (1-\alpha)\tilde{\sigma}_R^2 + [(1+\nu_f)\tilde{V}_I^2/2 - (1-\alpha)(\tilde{V}_I - 1)^2]/\tilde{a}^2, \tag{5}$$

$$\tilde{\Gamma} = \tilde{\Gamma}_I^c[1 + (1-\lambda)\tan^2\psi], \tag{6}$$

where

$$c = 0.2473(1+\nu_f) + 0.2231(1-\nu_f^2), \quad \tilde{V}_I = 6(1+\nu_f)V_I/(\pi\mu^2 h_f^3),$$
$$\tilde{a} = 12(1-\nu_f^2)a^2/(\mu^2 h_f^2), \quad \tilde{\sigma}_R = \sigma_R/E_f, \quad \tilde{G} = G/[(1-\nu_f)E_f h_f],$$
$$\tilde{\Gamma} = \Gamma/[(1-\nu_f)E_f h_f], \quad \tilde{\Gamma}_I^c = \Gamma_I^c/[(1-\nu_f)E_f h_f].$$

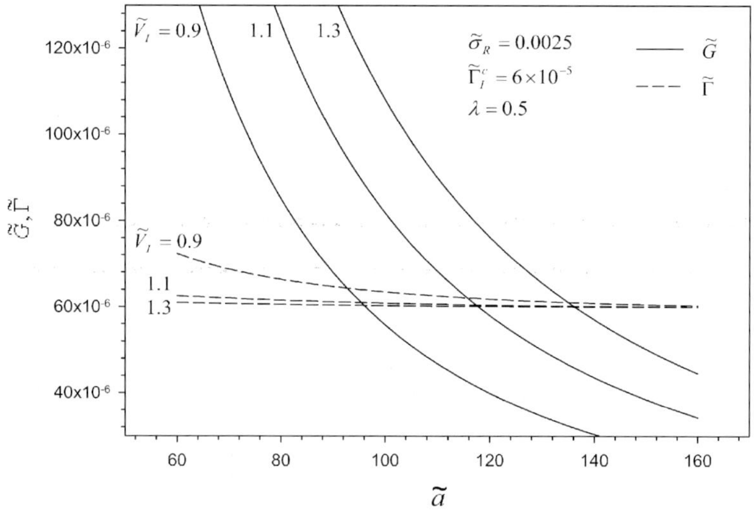

Fig. 1a. Strain energy release rate and interfacial fracture toughness as a function of interface crack length for indentation loads of $\tilde{V}_I = 0.9$, 1.1, and 1.3.

In the dimensionless expression, the indentation volume represents the indentation load. Equation (5) indicates that for a given residual stress, the energy release rate is determined by the indentation volume and the crack length. On the other hand, the indentation volume and the crack length determine also the phase angle if the residual stress and the value of ϖ are known. Once the phase angle, the mode I interfacial fracture toughness, and the λ fitting parameter are known, the empirical formula gives the interfacial fracture toughness. In following calculations, we took $\varpi = 52.1°$, $\tilde{\sigma}_R = 0.0025$, $\tilde{\Gamma}_I^c = 6 \times 10^{-5}$, and $\lambda = 0.5$ for simplicity.

Figure 1 shows plots of $\tilde{G}$ and $\tilde{\Gamma}$ versus $\tilde{a}$ for different indentation loads of $\tilde{V}_I$, where Figures 1a and 1b are gross and fine views, respectively. For a given indentation load, both $\tilde{G}$ and $\tilde{\Gamma}$ drop with the extension of $\tilde{a}$, but $\tilde{G}$ drops much faster than $\tilde{\Gamma}$. Since $\tilde{G}$ represents the driving force for the delamination, a high indentation load will give a large value of $\tilde{G}$ for a given delamination radius. At a given value of $\tilde{G}$, a high indentation load yields a large delamination radius. On the other hand, $\tilde{\Gamma}$ represents the resistance of a film/substrate system against delamination. Figure 2 illustrates that a high indentation load will give a small value of $\tilde{\Gamma}$ for a given delamination radius. Under a given indentation load, the delamination radius increases if the driving force is higher than the resistance. The delamination growths until that the driving force is balanced by the resistance, which means that an indentation load yields a delamination radius in a film/substrate system, as shown by the intercepts of A, B, C between $\tilde{G}$ and $\tilde{\Gamma}$ curves under the three indentation loads in Figure 1b. In addition, the intercepts give out the interfacial fracture toughnesses under the three indentation loads. When the indentation load is increased, the equilib-

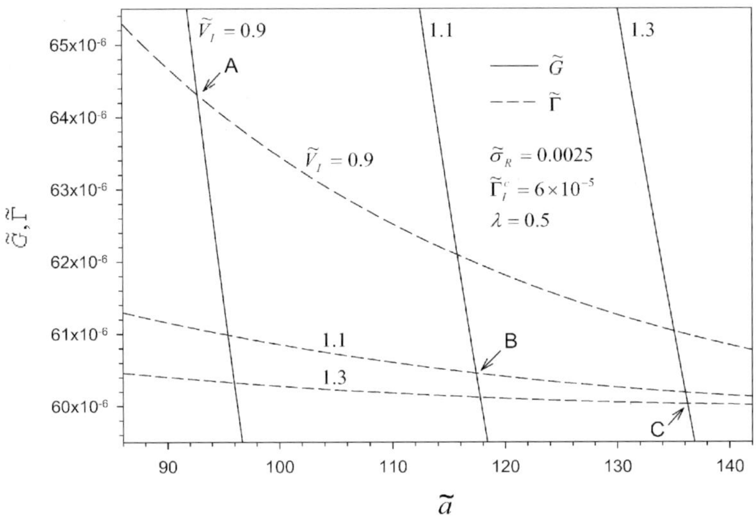

Fig. 1b. Fine view of strain energy release rate and interfacial fracture toughness. The intercept of two lines with same load gives out the equilibrium crack length and corresponding interfacial fracture toughness at that load.

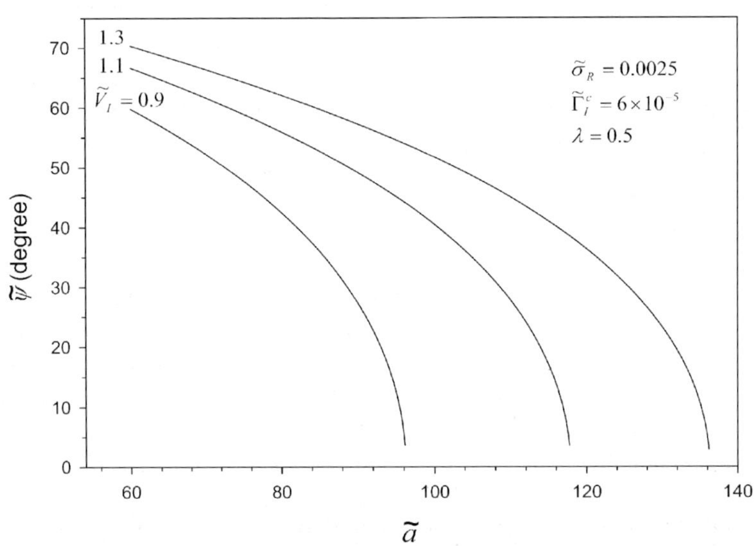

Fig. 2. Phase angle given by Equation (8) as a function of the crack length.

rium crack length increases and the interfacial fracture toughness decreases, which is consistent with the experimental results [4, 8].

Alternately, we may determine the phase angle by combining the empirical formula and the energy release rate based on the argument that at equilibrium, the en-

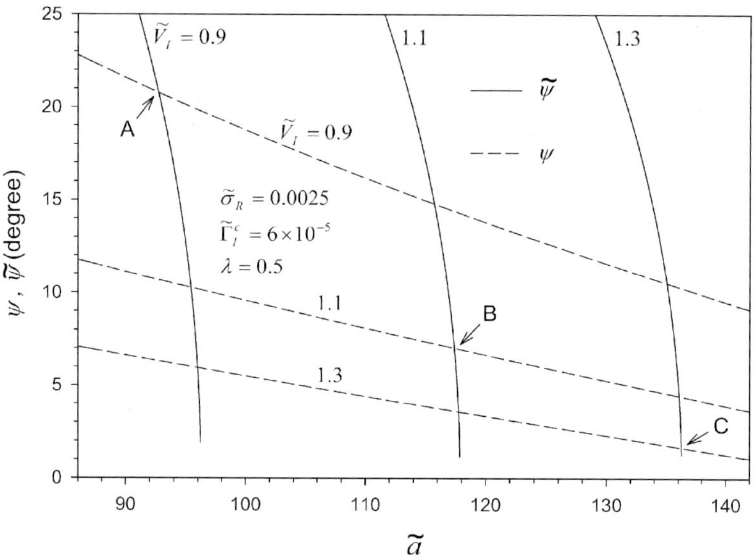

Fig. 3. Superposition plots of phase angles of ψ and $\tilde{\psi}$. The intercept of two lines with the same indentation load determines the equilibrium crack length and corresponding phase angle under that load.

ergy release rate should be balanced by the fracture resistance. In this sense, we may assume that the empirical Equation (6) is also valid for crack extension force, i.e.,

$$\tilde{G} = \tilde{\Gamma}_I^c[1 + (1-\lambda)\tan^2\psi]. \tag{7}$$

Combining Equation (7) with Equation (5) produces a phase angle as follows:

$$\tan^2\tilde{\psi} = \frac{\frac{(1-\alpha)\tilde{\sigma}_R^2}{\tilde{\Gamma}_I^c} + \frac{(1+\nu_f)\tilde{V}_I^2/2-(1-\alpha)(\tilde{V}_I-1)^2}{\tilde{\alpha}^2\tilde{\Gamma}_I^c} - 1}{(1-\lambda)}. \tag{8}$$

Since Equation (5) has been discussed and plotted in [3], we plot Equation (8) in Figure 2 with only positive values of the phase angle for clarity. Figure 2 shows that for a given indentation load of $\tilde{V}_I$, the phase angle decreases as the crack length gets longer. For a given crack length, the phase angle increases as the indentation load is higher. To determine the phase angle at the equilibrium state, we plot both Equations (8) and (5) in Figure 3. The intercepts under same indentation loads, as labeled with A, B, C in Figure 3, determine the equilibrium crack lengths and phase angles for those loads. The trend of variation of crack length and phase angle with the load is the same as that in Figure 1b. Furthermore, comparison of Figure 3 with Figure 1b demonstrates that the equilibrium crack lengths are the same by the two approaches.

5 Concluding Remarks

The determination of the phase angle of an interface crack with buckling in the indentation delamination test could be crucial in the assessment of the pure mode I interfacial fracture toughness. In our previous work [3], we reported a simple formula to calculate the phase angle of indentation delamination with buckling. The results show that with the increase of buckling, the phase angle deviates from that of unbuckle state and decreases smoothly from a positive value to a negative value, reflecting the cooperation of indentation compressive stress and buckling induced tensile stress. In the present work, we discuss the relations among interfacial fracture toughness, phase angle and indentation load/crack-length. These discussions are based on the empirical formula for interface fracture toughness. To demonstrate these relationships, we have numerically calculated the phase angle, the energy release rate, and the interface fracture toughness. Although the calculations are based on some assumptions with certain values of the parameters, the general approach described in the present work should be applicable to general cases.

Acknowledgement

This work was supported by a grant from the Research Grants Council of the Hong Kong Special Administrative Region, China.

References

1. Volinsky, A.A., Moody, N.R. and Gerberich, W.W., Interfacial toughness measurements for thin films on substrates. *Acta Mater.* **50** (2002) 441–466.
2. Jensen, H.M. and Thouless, M.D., Effects of residual stresses in the blister test. *Int. J. Solids Structures* **30** (1993) 779–795.
3. Huang, B., Zhao, M.H., Gao, C.F. and Zhang, T.Y., Phase angle in indentation-induced delamination with buckling. *Scripta Mater.* **50** (2004) 607–611.
4. Huang, B., Zhao, M.H. and Zhang, T.Y., Indentation fracture and indentation delamination in ZnO film/Si substrate systems. *Philos. Magazine A* **84** (2004) 1233–1256.
5. Marshall, D.B. and Evans, A.G., Measurement of adherence of residually stresses thin films by indentation. I. Mechnical of interface delamination. *J. Appl. Phys.* **56** (1984) 2632–2638.
6. Evans, A.G. and Hutchinson, J.W., On the mechanics of delamination and spalling in compressed films. *Int. J. Solids Structures* **20** (1984) 455–466.
7. Suo, Z. and Hutchinson, J.W., Interface crack between two elastic layers. *Int. J. Fract.* **43** (1990) 1–18.
8. Zheng, X.J., Zhou, Y.C. and Li, J.Y., Nano-indentation fracture test of $Pb(Zr_{0.52}Ti_{0.48})O_3$ ferroelectric thin films. *Acta Mater.* **51** (2003) 3985–3997.

Microscopic Shape Memory and Superelastic Effects and Their Novel Tribological Applications

Yang-Tse Cheng[1,*], Wangyang Ni[1,2], Yijun Zhang[1,2] and David S. Grummon[2]

[1]*Materials and Processes Laboratory, General Motors Research and Development Center, Warren, MI 48090, U.S.A.; *E-mail: yang.t.cheng@gm.com*
[2]*Department of Chemical Engineering and Materials Science, Michigan State University, East Lansing, MI 48824, U.S.A.*

Abstract. We provide an overview of our recent work on microscopic shape memory and superelastic effects and their tribological applications.

Key words: shape memory materials, friction, wear, tribology, indentation, adhesion.

1 Introduction

Since the discovery of shape memory effect in NiTi in the 1960s, there have been extensive efforts on understanding and applications of the shape memory and superelastic effects [1]. Recent years have seen an increasing interest in its tribological applications. For example, shape memory alloys have desirable wear properties under cavitation erosion [2–4] and dry sliding wear conditions [5–12]. In this paper, we provide an overview of our recent studies which established the microscopic shape memory and superelastic effects under complex loading conditions. We will also discuss several novel tribological applications of shape memory and superelastic effects: (i) the use of shape memory NiTi alloys as self-healing surfaces, and (ii) the use of the superelastic NiTi as an interlayer between a hard coating and a soft substrate to improve interfacial adhesion, decrease friction coefficient, and improve wear resistance.

2 Microscopic Shape Memory Effect and Self-Healing Surfaces

Contact induced surface damages are often unavoidable. It is thus highly desirable to have a tribological system that can detect and heal such damages automatically. The shape memory effect offers a possibility of self-healing in tribological applications. However, it was unknown whether shape memory effect existed at microscopic length scales and under complex loading conditions such as indentation and sliding,

Yilong Bai et al. (eds), IUTAM Symposium on Mechanical Behavior and Micro-Mechanics of Nanostructured Materials, 211–217.

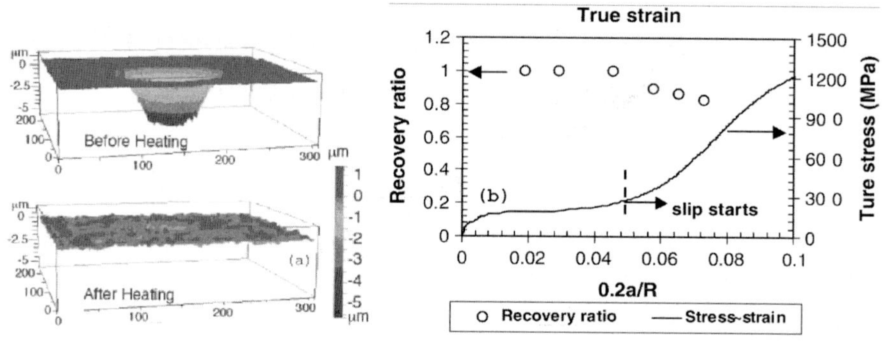

(a) (b)

Fig. 1. Shape memory NiTi: (a) recovery of spherical indents, and (b) relationship between thermally activated recovery ratio and representative strain ($0.2a/R$), and the relationship between the true stress and true strain.

although the shape memory effect was well established at macroscopic length scales and under simple loading conditions such as tensile, compression, or shear. To establish the basis for using shape memory materials as self-healing tribological surfaces, we have conducted a systematic investigation of the magnitude of shape recovery on the surfaces of a martensitic NiTi alloy using indentation and scratch tests. The indentation and scratch experiments can simulate an asperity contact under normal and tangential loading, respectively.

Figure 1a shows the 3-D profiles of a spherical indent on the martensitic NiTi before and after recovery [13]. The recovery of the indent was induced by heating the specimen to a temperature above the austenite finish temperature. The specimen almost completely recovered its original shape after being heated. This self-healing capability of indents can be quantitatively characterized by defining a thermal-induced recovery ratio, $\delta = (h_f - h'_f)/h_f$, where h_f is the residual indentation depth recorded immediately after unloading and h'_f is the final indentation depth after the completion of thermal-induced recovery [13, 14]. Our experimental study shows that for spherical indentation δ depends on both the indenter radius and indentation depth. The results can be rationalized using the concept of the representative strain defined as, $\varepsilon_r = 0.2a/R$, where a and R are the contact radius and indenter radius, respectively. Figure 1b shows the relationship between the thermo-induced recovery ratio and the representative strain together with the stress-strain relationship [14]. It shows that self-healing remains constant and almost complete until a critical strain is reached. This critical strain, which coincides with the end of the stress plateau in stress-strain curve, is the maximum recoverable strain.

The self-healing effect also exists in sliding contact [15]. Figure 2 shows the profile of a scratch before and after recovery. The scratch was generated by a spherical indenter with tip radius 213.4 μm using a progressive load from 0 to 10 N. It clearly demonstrates that the scratch scar can be healed when the temperature is raised above

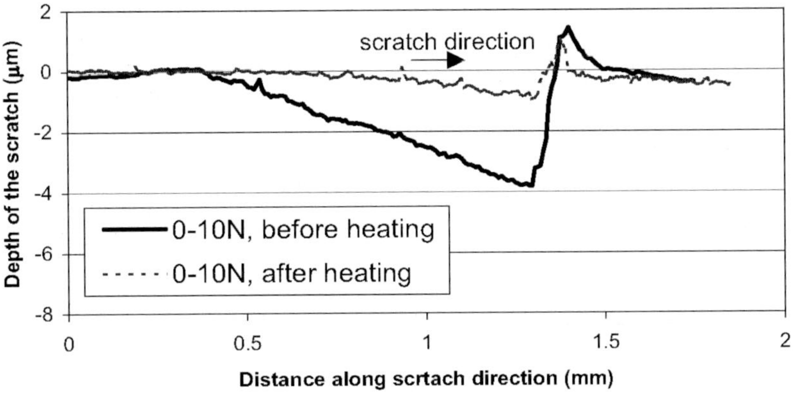

Fig. 2. Self-healing of a scratch scar on a shape memory alloy.

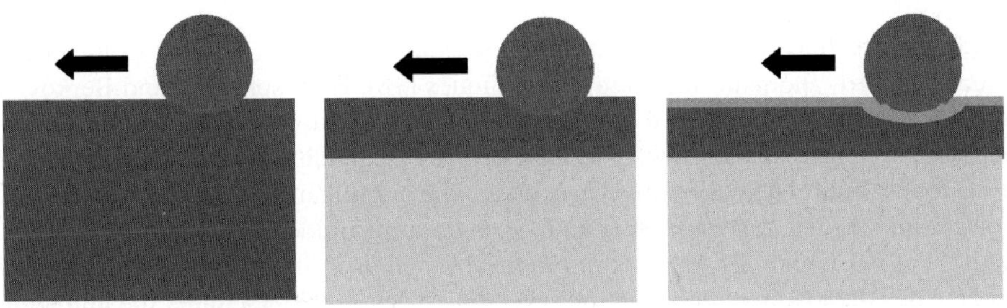

Fig. 3. Three concepts for self-healing tribological surfaces.

its austenite finish temperature. The recovery is almost complete when the scratch scar is shallow. Recently, we have explored the synergetic effects of combining hard coatings and shape memory alloys as wear resistant self-healing tribological surfaces. Specifically, we investigated the indentation and scratch behavior of a NiTi shape memory alloy with and without a thin CrN coating layer. We found that a CrN layer can significantly improve the wear resistance of the NiTi shape memory alloy, while maintaining the recovery ability of indents and scratches upon heating [16]. Figure 3 illustrates three concepts for self-healing tribological surfaces using shape memory materials [17].

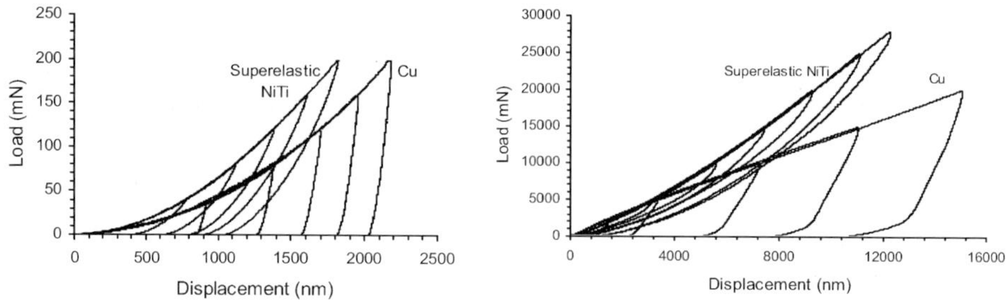

(a) (b)

Fig. 4. Load-displacement curves of Berkovich indents (a) and spherical indents (b) made in an austenite NiTi alloy and Cu.

3 Microscopic Superelastic Effects and Strong Interfaces

Recently, we have also established microscopic superelastic effects using instrumented micro- and nano-indentation techniques [18]. Both spherical and Berkovich diamond indenters were used for the investigation. Figures 4a and 4b are the typical load-displacement curves for the respective Berkovich and spherical indentation in a superelastic NiTi alloy and in copper. The significant difference between the load-displacement curves in NiTi and Cu is the magnitude of recoverable work relative to total work. The total work required to move an indenter into a solid is the area under the loading curve and the reversible work is the area under the unloading curve. The area between the loading and unloading curves is the dissipated energy. For elastic-plastic solids such as copper, the reversible and irreversible work result from elastic and plastic deformation, respectively. In contrast to copper, a significant amount of the total work of indentation in the superelastic NiTi alloy is reversible. This indentation induced superelastic effect is much more pronounced under spherical indentation than under Berkovich indentation.

The large difference in the magnitude of indentation-induced superelasticity is the result of stress distribution under the two types of indenters. The stress at the tip of a perfectly sharp pyramidal or conical indenter rises to a theoretically infinite value at the apex unless plastic deformation occurs. A large volume of material directly below the pyramidal indenter is therefore highly strained that significant deformation occurred by dislocation motion rather than indentation induced phase transformation. This volume would not, therefore, contribute to the superelastic behavior. In contrast, the maximum stress under the spherical indenter remains finite. It would thus appear that the strain caused by the stress distribution under the spherical indenter was largely accommodated by stress induced martensite formation, leading to superelastic strain recovery upon unloading.

As an application of the microscopic superelastic effect, we compare the stress-strain relationships, schematically shown in Figure 5, for superelastic NiTi alloy and that for elastomeric polymer adhesives [1, 19]. Both of them have large recoverable strain. However, the elastic modulus and strength of the superelastic NiTi alloy are

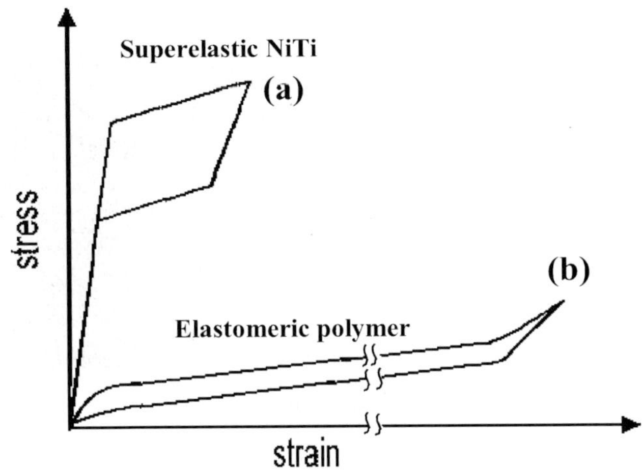

Fig. 5. Schematic illustration of the stress-strain curves of superelastic NiTi (a), and elastomeric polymer (b).

several orders of magnitude greater than that of the polymeric adhesives. In addition, a hysterisis loop, which arises from the internal friction, is associated with the stress-strain curve of the superelastic NiTi. A part of the energy is thus consumed due to the impedance to the movement of the phase boundaries between austenite and martensite phases [1]. Considering the similarity between the stress strain relationships, we have suggested that the superelastic NiTi material can act as a high strength metallic adhesive for bonding ceramic coatings to ductile substrates by its large recoverable strain, large stain tolerance, and energy dissipation [20, 21].

To test the idea of using superelastic interlayers to improve adhesion, we have prepared, using magnetron sputtering techniques, thin film structures of CrN hard coatings on aluminum substrates either with or without superelastic NiTi interlayers. Temperature-controlled scratch tests were performed with a 100 micron radius diamond indenter on a CSM Instruments Micro-Scratch Tester which also recorded acoustic emission events. Scratch load was increased from zero to 5 N over the 1 mm length of the scratch.

Results from the scratch tests are shown in Figure 6. Acoustic emission from the CrN-only specimen indicated that film delamination commenced at loads near 2–2.5 N, decreasing slightly at higher temperatures between 298 and 398 K. In contrast, the CrN:NiTi:Al specimen exhibited a critical delamination load that increased with temperature, from about 2 N at 298 K to about 4.4 N at 398 K. A substantial jump in critical load is seen to occur at temperatures near the austenite finish temperature (A_f), indicating that the improved performance is associated with temperatures allowing superelastic response in the interlayer. Figure 6 also includes SEM micrographs of the scratch terminations (with the full scratch shown in insets), indicating that at temperatures above Af adhesion performance is greatly enhanced. In addition,

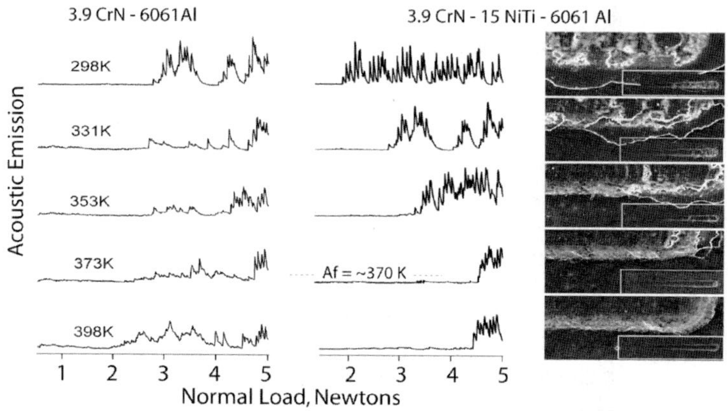

Fig. 6. Results of temperature-controlled scratch tests. The micrographs at the right are scratch terminations from the specimen with a 3.9 μm CrN hard coating on a 15 μm NiTi interlayer on 6061 aluminum. Scratches were made at several temperatures as indicated. The AE scans at the far left are for CrN coated aluminum without the NiTi interlayer, whereas the plots (and micrographs) on the right are for the specimen with the NiTi interlayer.

we have shown that friction and wear characteristics can also be improved using a superelastic interlayer [21, 22].

4 Conclusions

We have shown that microscopic shape memory and superelastic effects exist under indentation and sliding contact conditions. The shape memory effect can be exploited for making self-healing tribological surfaces. By using the superelastic NiTi alloy as an interlayer between a hard coating and a soft substrate, it is possible to improve interfacial adhesion, thus providing protection of very soft substrates using hard coatings for which mechanical property mismatches are large. We hope the results presented in this paper will stimulate future research on fundamentals and applications of shape memory materials by the mechanics-materials community.

Acknowledgments

The authors would like to thank Michael Lukitsch, Anita M. Weiner, Leo C. Lev, Curtis Wong, Robert K. Kubic, Michael P. Balogh, Yue Qi, Louis G. Hector, Thomas A. Perry, and Mark W. Verbrugge technical assistance and valuable discussions.

References

1. K. Otsuka and C.M. Wayman, *Shape Memory Alloys*, Cambridge University Press, Cambridge, 1998.
2. R.H. Richman and W.P. McNaughton, *J. Mat. Eng. and Performance* **6**, 1997, 633–641.
3. T. Zhang and D.Y. Li, *Mat. Sci. Eng.* **A293**, 2000, 208–214.
4. H. Hiraga, T. Inoue, H. Shimura and A. Matsunawa, *Wear* **231**, 1999, 272–278.
5. J. Jin and H. Wang, *Acta Metall. Sinica* **24**, 1988, A66–A70.
6. P. Clayton, *Wear* **162–164**, 1993, 202–210.
7. Y.N. Liang, S.Z. Li, Y.B. Jin, W. Jin and S. Li, *Wear* **198**, 1996, 236–241.
8. R. Liu and D.Y. Li, *Mat. Sci. Tech.* **16**, 2000, 328–332.
9. J. Singh and A.T. Alpas, *Wear* **181–183**, 1995, 302–311.
10. H.C. Lin, H.M. Liao, J.L. He, K.C. Chen and K.M. Lin, *Metall. Mat. Tran.* **28A**, 1997, 1871–1877.
11. R. Liu and D. Li, *Wear* **250–251**, 2001, 956–964.
12. F.T. Cheng, P. Shi and H.C. Man, *Scripta Materialia* **45**, 2001, 1089.
13. W. Ni, Y.-T. Cheng and D.S. Grummon, *Appl. Phys. Lett.* **80**, 2002, 3310–3312.
14. W. Ni, Y.T. Cheng and D.S. Grummon, *Surf. Coatings Tech.* **177–178**, 2004, 512–517.
15. W. Ni, Y.-T. Cheng and D.S. Grummon, *J. de Phys. IV* **112**, 2003, 1147–1150.
16. W. Ni, Y.-T. Cheng and D.S. Grummon, *Surface and Coatings Technology*, submitted.
17. Y.-T. Cheng, W. Ni, M.J. Lukitsch, A.M. Weiner and D.S. Grummon, US patent application publication 20040202888 A1 (October 14, 2004).
18. W. Ni, Y.-T. Cheng and D.S. Grummon, *Appl. Phys. Lett.* **82**, 2003, 2881.
19. D. Brandon and W.D. Daplan, *Joining Process: An Introduction*, John Wiley & Sons, New York, 1997.
20. Y.-T. Cheng, W. Ni, L.C. Lev, M.J. Lukitsch, D.S. Grummon and A.M. Weiner, US patent 6,866,730 (March 15, 2005).
21. W. Ni, Y.-T. Cheng, M. Lukitsch, A.M. Weiner, L.C. Lev and D.S. Grummon, *Wear* **259**, 2005, 842.
22. W. Ni, Y.-T. Cheng, M.J. Lukitsch, A.M. Weiner, L.C. Lev and D.S. Grummon, *Appl. Phys. Lett.* **85**, 2004, 4028–4030.

Dynamics of Self-Organized Epitaxial Island Formation under Controlled Annealing

Y. Ni[1,2], A.K. Soh[1,*] and L.H. He[2]

[1] *Department of Mechanical Engineering, The University of Hong Kong, Pokfulam Road, Hong Kong, China; *E-mail: aksoh@hkucc.hku.hk*
[2] *CAS Key Laboratory of Mechanical Behavior and Design of Materials, University of Science and Technology of China, Hefei, Anhui 230026, People's Republic of China*

Abstract. It is a well known fact that strain-driven self-assembly via Stranski–Krastonov growth is a promising way to fabricate ordered quantum dot array. However, control of the morphology remains to be a critical issue. One approach towards controlled self-assembly is, but not limited to, epitaxial growth on patterned substrates or patterned epilayers. The possibility of controlling the growth morphology of quantum dots upon patterned substrates and patterned epilayers is explored by numerical studies of three-dimensional phase field simulation. The results indicate that, by creating appropriate patterns, such as topographical pattern created in the substrate or epilayer, and periodically strained substrate, etc, the initial strain distributions on the surfaces of the substrate or epilayer can be altered, and thus the subsequent evolution path of surface morphology under annealing can be controlled efficiently. This may lead to highly ordered quantum dot array.

Key words: epitaxy, surface morphology, phase field method, pattern formation.

1 Introduction

In recent years there has been considerable experimental and theoretical interest in self-assembly of three-dimensional quantum dot array due to their important applications in the development of fabricating novel photoelectronic and magnetic devices [1–3]. One promising way is to employ the strain-driven self-organization process during heteroepitaxial growth of thin films, where a lattice parameter mismatch between the film and the substrate causes internal stresses in the thin film due to the constraint of the substrate, driving the self-organization of surface morphology. It turns out, however, that merely exploiting self-organized growth is far from sufficient either for the physics or for the applications. The experimental observations of heteroepitaxial growth have shown that in some cases, the quantum dots can be increasingly uniform and regular, while in other cases, they are less ordered or even disordered (refer to [1–3] and other cited references). This is because that spontaneous formation of the regular surface morphology is not uncommon and arises due

Yilong Bai et al. (eds), IUTAM Symposium on Mechanical Behavior and Micro-Mechanics of Nanostructured Materials, 219–228.

to the competing long- and short-range interactions, thereby they will not have perfect long range order. To date, controlling these quantum dot arrays in both size and position, especially the degree of lateral ordering still becomes more of an issue. One way to overcome this problem is controlled self-organized growth [4]. A typical approach is tuning the self-organized growth through strain management, i.e. manipulating the surface strain field of the substrate. It is widely accepted that the periodic strain field can be induced by the subsurface island array, subsurface dislocation network or the inclusions in the substrate etc., and the elastic energy distribution exhibits pronounced minima and maxima in the lateral directions. This leads to a diffusional bias of the deposited atoms and to a favorable surface nucleation site of the islands, which are linked to the top island positions. Hence, the top islands may become increasingly uniform, regular and vertically aligned due to these elastic interactions, which are observed in many experiments [5–7].

Although many experimental results show that self-organized growth through strain management, for example, through pre-patterned substrate or pre-patterned epilayer technique to manipulate the growth kinetics of strain-driven quantum dot formation during heteroepitaxial growth, is an effective means of guiding quantum dots ordering. However, related theories are not well understood due to its complexity. The present study is through three-dimensional phase field simulation to investigate the effect of manipulating the surface strain field of the substrate in the dynamics of self-organized epitaixal island formation under annealing, such as the dynamics of film breaking up into islands and coarsening of islands. The surface strain field of the substrate can be created by two ways, one is pre-patterning of the substrate template, such as composition-modulated substrate, the substrate with a buried topographic pattern, or with buried voids or inclusions, and the other is topographically patterned epilayer. This paper is organized as follows. In Section 2, a three-dimensional continuum phase field model to simulate the dynamics of self-organized epitaixal island formation under annealing is briefly outlined. In Section 3, several numerical examples on the possibility of controlling the growth morphology of quantum dots upon patterned substrates and patterned epilayers will be provided; some conclusions will be drawn in Section 4.

2 The Phase Field Model

Consider a thin film of cubic form with thickness h deposited on a substrate with thickness H ($H \gg h$), as illustrated schematically in Figure 1.

Assume that both the film and substrate are linear elastic cubic crystals with identical elastic constants. The epitaxial orientation is along the crystallographic plane $(h_3\ k_3\ l_3)$, which is parallel to the substrate. A current orthogonal coordinate system is introduced such that the z-axis is normal to the film-substrate interface, on which its x- and y-axes are $[h_1\ k_1\ l_1]$ and $[h_2\ k_2\ l_2]$, respectively, denoted by means of the Miller indices. Here, the inhomogeneities buried in the substrate induced by the pre-patterned substrate technique are characterized by a homogeneous but spatially varying mismatch strain, $\varepsilon_{xx}^{\text{misfit}}(\mathbf{r})$ and $\varepsilon_{yy}^{\text{misfit}}(\mathbf{r})$ in the x- and y-directions,

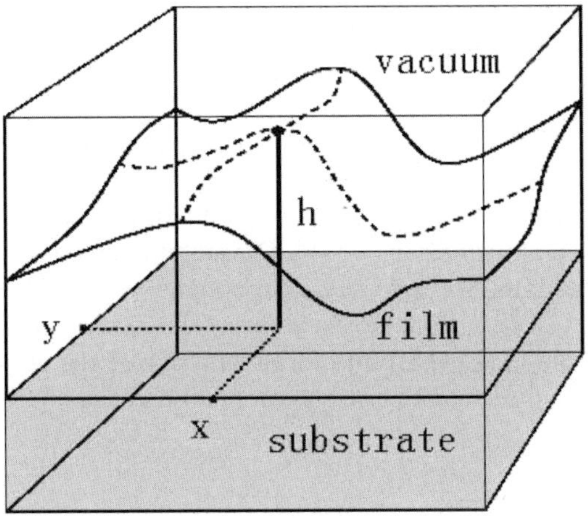

Fig. 1. Sketch of a thin film on a lattice mismatched substrate.

respectively, while, the pre-patterned epilayer technique manipulates the distribution of total stress field of the film/substrate system via modifying the initial surface configuration. Once the strained film lose the flat surface shape under arbitrary small fluctuation, and the roughness amplitude increases, eventually the film breaks into islands caused by the strain-mediated surface diffusion. A phase field method is used to handle this free boundary problem coupled with the solution of the complex stress field with arbitrarily-shaped free surface and embedded inhomogeneity in the substrate. A long-range order parameter $\eta(\mathbf{r}, t)$ was defined to denote the film-substrate system, with $\mathbf{r}$ being the position vector of a point, and the total free energy was enabled to be a function of this order parameter, e.g. $\eta = 0$ in the vacuum and $\eta = 1$ in the solid phase. Hence, the position dependent elastic modulus $C_{ijkl}(\mathbf{r})$ changed its value from 0 to C^0_{ijkl} for a transition from vacuum to solid. In the computation, however, during evolution the value of the long-range order parameter might deviate from 0 and 1 in the vacuum and solid, respectively, and the emergence of small negative values of $C_{ijkl}(\mathbf{r})$ would cause mechanical instability. To avoid this problem of instability of the computation procedure, a scaled order parameter, which was termed as the materials density distribution $\rho(\mathbf{r}(t) = 1/2[1+\tanh(2\eta(\mathbf{r}, t) - 1)/(2\Delta)]$, was introduced in [8], where Δ was a parameter chosen to control the interface thickness between the solid and the vacuum. The elastic modulus $C_{ijkl}(\mathbf{r}, t)$ is then described as a linear function of $\rho(\mathbf{r}, t)$, i.e., $C_{ijkl}(\mathbf{r}) = C^0_{ijkl}\rho(\mathbf{r})$. For this heterogeneous system, the total free energy can be expressed as

$$E^{\text{total}} = \int_V \{f[\eta(\mathbf{r})] + \lambda_0[\nabla\eta(\mathbf{r})]^2 + f_{\text{el}}\}d^3r, \tag{1}$$

where the first term in the integral is the bulk free energy in the absence of stress, which is defined as a double-well function $a\eta^2(1-\eta)^2$; and the second term is the gradient energy corresponding to the interfacial energy between the vacuum and the solid phase, ∇ is the gradient operator, while a and λ_0 are constants. The last term is the strain energy density which is given by

$$f_{\mathrm{el}} = \frac{1}{2} C_{ijkl}(\mathbf{r})(\varepsilon_{ij} - \varepsilon_{ij}^{\mathrm{misfit}}(\mathbf{r}))(\varepsilon_{kl} - \varepsilon_{kl}^{\mathrm{misfit}}(\mathbf{r})), \tag{2}$$

where ε_{ij} are the strain components to be determined later, and $\varepsilon_{ij}^{\mathrm{misfit}}(\mathbf{r})$ is the spatially varying epitaxial misfit strain tensor. Throughout this paper, the usual summation convention is adopted for repeated indices, where Latin indices take the value of 1, 2 or 3, while Greek ones take the value of 1 or 2. Since surface diffusion is typically the dominant mass transport mechanism during the annealing process, the mass conservation requires the kinetic equation to be a Cahn–Hilliard equation

$$\frac{\partial \eta}{\partial t} = \nabla \left[\frac{D(\eta)}{\Lambda k_B T} \nabla \frac{\delta E^{\mathrm{total}}}{\delta \eta} \right], \tag{3}$$

where Λ is the number of atomic sites per unit volume, k_B is Boltzmann's constant, T is the absolute temperature and $D(\eta)$ is the diffusion coefficient. Note that $D(\eta)$ is assumed to be a constant D in the vacuum and film and the difference between the bulk and the surface diffusion coefficient is neglected for simplification, without losing main physics; while $D(\eta) = 0$ in the substrate since it does not evolve during film surface roughening. In the framework of phase field methods, the derived strain energy density should also be a function of the long-range order parameter. To determine the stress field of this heterogeneous system, the phase field microelasticity method proposed in [9] is adopted. Through a variational approach, this theory has proven that the elastic strain and the strain energy of the heterogeneous system with inhomogeneous elastic modulus can be calculated by establishing an equivalent system with a homogeneous modulus C_{ijkl}^0 and a distributed pre-unknown effective strain ε_{ij}^0 defined by

$$C_{ijkl}^0[\varepsilon_{kl}(\mathbf{r}) - \varepsilon_{kl}^0(\mathbf{r})) = C_{ijkl}(\mathbf{r})[\varepsilon_{kl}(\mathbf{r}) - \varepsilon_{kl}^{\mathrm{misfit}}(\mathbf{r})], \tag{4}$$

where the pre-unknown effective strain ε_{ij}^0 are obtained by solving the following phase field microelasticity (PFM) kinetic equation [9].

$$\frac{\partial \varepsilon_{ij}^0}{\partial t} = -K_{ijkl} \frac{\partial E_{\mathrm{elas}}^{\mathrm{equiv}}}{\partial \varepsilon_{kl}^0}. \tag{5}$$

Once the effective strain ε_{ij}^0 has been obtained, the elastic strains and stresses can be computed as follows:

$$\varepsilon_{ij}(\mathbf{r}) = \bar{\varepsilon}_{ij} \frac{1}{2} \int \frac{d^3\xi}{(2\pi)^3} (\xi_i \tilde{G}_{jk} + \xi_j \tilde{G}_{ik}) \tilde{\sigma}_{kl}^0(\boldsymbol{\xi})^* \xi_l e^{i\xi \cdot \mathbf{r}}, \tag{6}$$

$$\sigma_{ij}(\mathbf{r}) = C_{ijkl}^0[\varepsilon_{kl}(\mathbf{r}) - \varepsilon_{kl}^0(\mathbf{r})], \tag{7}$$

where the integral $\int$ in the Fourier space excludes the point $\boldsymbol{\xi} = 0$, the elastic constants and Green's function tensor in the current coordinate system are obtained through coordinate transformation, respectively [10]

$$C^0_{ijkl} = \Omega_{im}\Omega_{jn}\Omega_{kp}\Omega_{lq}C'_{mnpq}, \tag{8}$$

$$\tilde{G}_{ij}(\boldsymbol{\xi}) = \Omega_{ik}\Omega_{jl}\tilde{G}'_{kl}(\zeta), \tag{9}$$

where $\boldsymbol{\Omega}$ is the coordinate transformation matrix defined as $r_i = \Omega_{ik}r'_k$, $\zeta_i = \Omega_{ji}\xi_j$, $\tilde{G}'_{im}(\zeta) = \int_V G'_{im}(\mathbf{r}')e^{-i\zeta\cdot\mathbf{r}'}d^3r'$, C'_{ijkl}, $\tilde{G}'_{kl}(\zeta)$ and $\tilde{G}_{ij}(\boldsymbol{\xi})$, C^0_{ijkl} are the corresponding elastic constants and Green's function tensor in the natural x', y', z'-coordinate system and in the current coordinate system, respectively.

After scaling the length and time such that $x \to x/l_0$ and $t \to t/\pi$, where $l_0 = \sqrt{\lambda_0/a}$ and $\tau = (\lambda_0/Da)$ are the characteristic length and time scales, respectively, a dimensionless parameter, $\omega = l/l_0$, is defined with $l = \gamma/\mu(\varepsilon^{\text{misfit}})^2$. In our preliminary computation, the periodic boundary condition is imposed and the simulation is restricted to a cuboid cell of size $64l_0 \times 64l_0 \times 32l_0$, where $\omega = 2$. The total thickness of the system is $32l_0$, which consists of 21 layers of the substrate, 3 layers of the film and 8 layers of the vacuum phase. The real grid size and the discrete time step are taken as $\Delta x = 1.0l_0$ and $\Delta t = 0.02\tau$, respectively. In the computation, the scaled kinetic coefficient is taken as $K^* = 10$ and $\nu = 1/3$ is assumed. The initial random small fluctuation of surface morphology is added to allow the elastic misfit strain to destabilize the film, which is described as the superposition of m static plane waves with amplitude β_m, wave numbers k_m and random phase shift angles ϕ_m, φ_m.

$$h(x, y) = h_0 + \sum_{m=1}^{32} \beta_m \sin(k_m x + \phi_m)\sin(k_m y + \varphi_m), \tag{10}$$

where $h(x, y)$ is the film thickness; $\beta_m = 0.02$ and $k_m = 2\pi m/(64l_0)$ are assumed. The scaled Equations (3) and (5) can be solved numerically using a semi-implicit Fourier-spectrum method and an explicit forward Euler method combined with the FFT algorithm, respectively.

3 Results and Discussion

For our purpose is to investigate the effect of the pre-patterned substrate and epilayer, we did not consider the effect of elastic anisotropy here. Namely, we adopted an isotropic elasticity in the following numerical simulation. The first numerical results for the case of unpatterned substrate are shown in Figure 2.

In the beginning of the annealing process, the amplitudes of some dominated surface undulation increase, and the surface quickly evolves into random ripples. Once the ripples break up to become islands, the coarsening process proceeds, and the larger islands expand at the expense of smaller islands till a partly self-organized island array emerges due to the long-range elastic interaction between

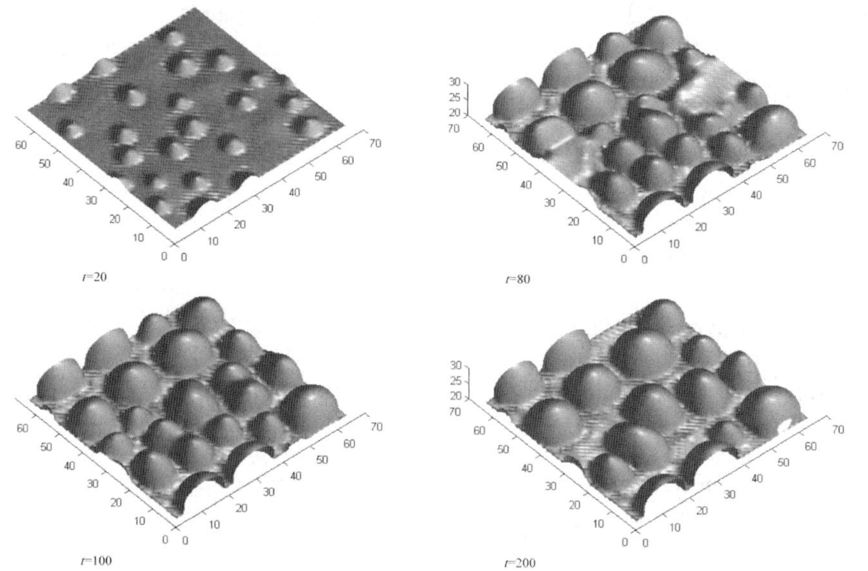

Fig. 2. Typical evolving surface morphology on an unpatterned substrate during the annealing process at a given time.

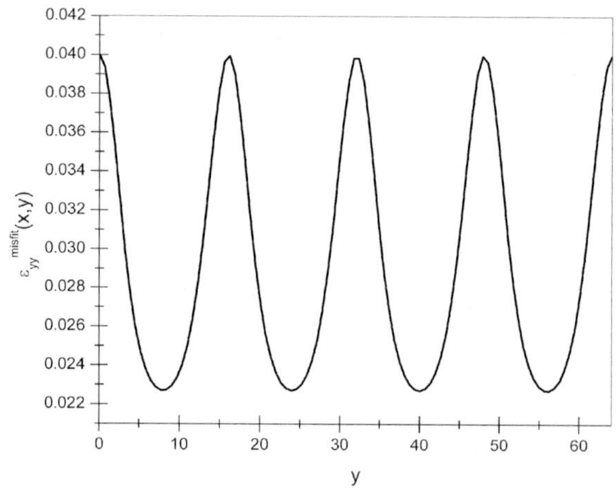

Fig. 3. Plot of the unidirectional mismatch strain modulation $\varepsilon_{yy}^{\text{misfit}}(x, y) = \varepsilon^{\text{misfit}} (1 + e^{-2(\sin(\pi y/16)^2)})$ where $\varepsilon^{\text{misfit}} = 0.02$.

islands. For the pre-patterned substrate, the embedded inhomogeneity of the substrate can be characterized by those resulted from an equivalent homogenous but spatially varying mismatch strain. We consider first the effect of unidirectional mismatch strain modulation on the dynamics of self-organized epitaixal island formation under annealing. The distribution of mismatch strain is assumed to be of the form

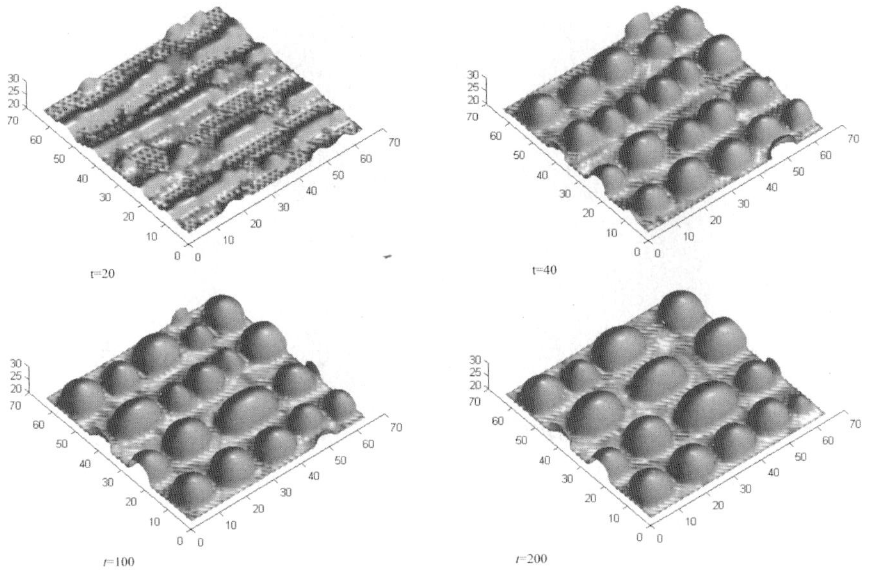

Fig. 4. Evolving surface morphology on the substrate with unidirectional mismatch strain modulation during the annealing process at a given time.

$\varepsilon_{yy}^{\text{misfit}}(x, y) = \varepsilon^{\text{misfit}}(1 + e^{-2(\sin(\pi y/16)^2)})$, $\varepsilon_{xx}^{\text{misfit}} = \varepsilon^{\text{misfit}}$ with $\varepsilon^{\text{misfit}} = 0.02$. The plot of $\varepsilon_{yy}^{\text{misfit}}(x, y)$ is shown in Figure 3.

Because the strain field in the film and substrate are strongly dependent on both the misfit strain modulation and the details of surface profile, which can be exactly solved based on phase field microelasticity model. The corresponding evolution path of surface morphology is shown in Figure 4.

We found that regions of lower mismatch strain are clearly served as preferred areas for island formation thus leading to ordering in the direction, while in the direction, ordering is poorer since the mismatch strain in this direction is constant. By comparison between Figures 2 and 4, we demonstrate that the distribution of periodically mismatch strain leads to directed quantum dot ordering. Further, we consider the effect of two-dimensional distribution of mismatch strain modulation. The contour plot of assumed mismatch strains is shown in Figure 5.

One can imagine that the evolving surface morphology of the epilayer will be ordering in both the x and the y directions. From the simulation results presented in Figure 6, it can be seen that the quantum dot array really is to show ordering in both the x and the y directions, thereby leading to more regular distribution than for a unidirectional mismatch strain modulation. In the last numerical example, we note that the pre-patterned surface profile of the film could also result a periodic strain distributions on the surfaces of the substrate. Figure 7 shows the evolving sequence of surface morphology.

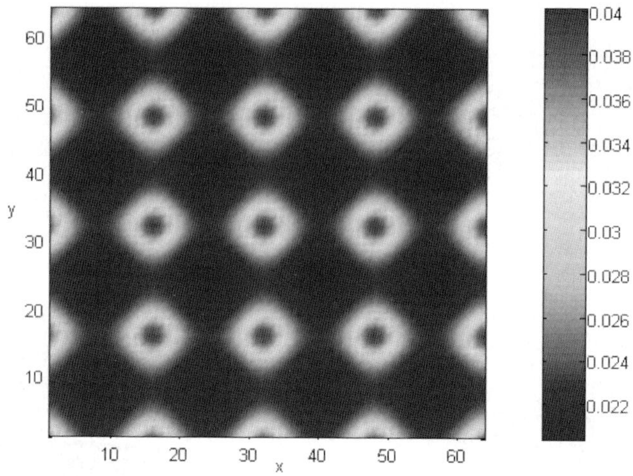

Fig. 5. Contour plot the two-dimensional mismatch strain modulation $\varepsilon_{xx}^{\text{misfit}} = \varepsilon_{yy}^{\text{misfit}} = \varepsilon^{\text{misfit}}(1 + e^{-2(\sin(\pi x/16)^2 + \sin(\pi y/16)^2)})$ where $\varepsilon^{\text{misfit}} = 0.02$.

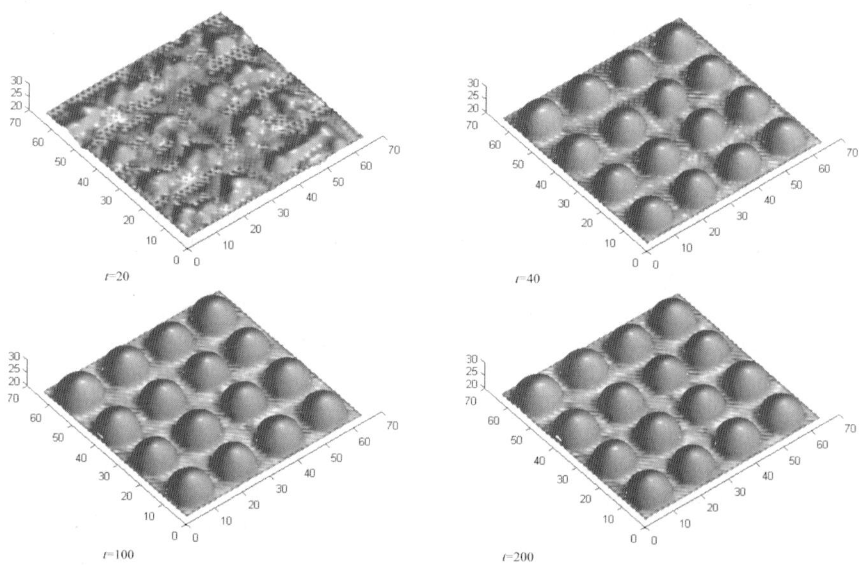

Fig. 6. Evolving surface morphology on the substrate with two-dimensional mismatch strain modulation during the annealing process at a given time.

An interesting observation is that the surface morphology evolution could be modulated by initial surface configuration. Namely, the surface patterns depend not only on the competition between the surface energy and the strain energy, but also

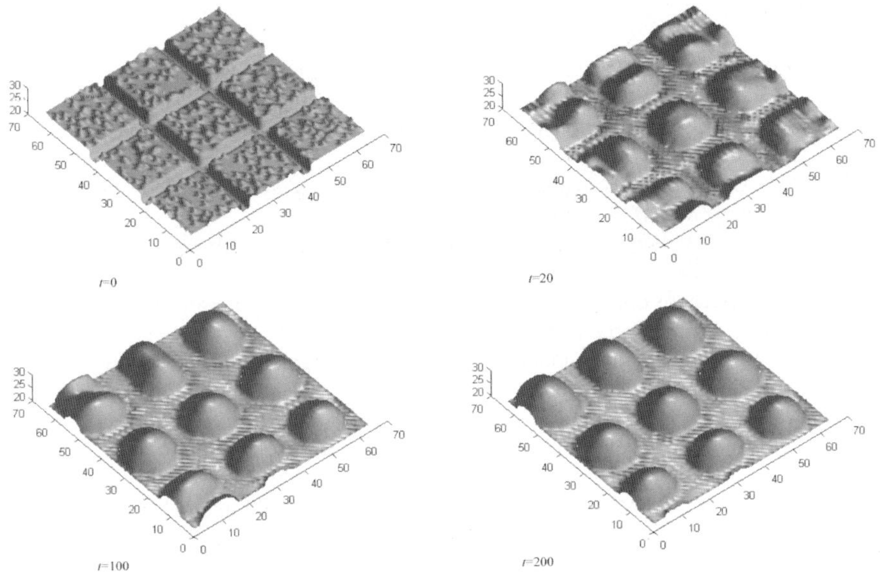

Fig. 7. Evolving surface morphology on an unpatterned substrate with patterning surface of the epilayer during the annealing process at a given time.

on the initial surface configuration. The surface evolution under annealing is strongly history dependent.

4 Conclusion

In summary, the nonlinear surface evolution of the strained thin film under controlled annealing is investigated by means of phase field model. Our preliminary numerical results show that the evolving surface morphologies are strongly dependent on the details of pre-patterned substrate and pre-patterned epilayer, which provide an effective means of guiding quantum dots ordering. We wish it to be realizable by combining lithography, AFM and self-organized growth in the real fabrication process.

Acknowledgements

This research is supported from the National Natural Science Foundation of China (Grant No. 10172079) and the Ph. D. Programs Foundation of Ministry of Education of China. Support from the Research Grants Council of the Hong Kong Special Administrative Region, China (Project No. HKU 7203/03E) is also acknowledged. The authors acknowledge productive discussions with Y.M.M Jin.

References

1. Shchukin, V.A. and Dimberg, D., Spontaneous ordering of nanostructures on crystal surfaces, *Rev. Mod. Phys.* **71**, 1999, 1125–1171.
2. Teichert, C., Self-organization of nanostructures in semiconductor heteroepitaxy, *Phys. Rep.* **365**, 2000, 335–432.
3. Stangl, J., Holy, V. and Bauer, G., Structural properties of self-organized semiconductor nanostructures, *Rev. Mod. Phys.* **76**, 2004, 725–783.
4. Shchukin, V.A., Ledentsov, N.N. and Bimberg, D., *Epitaxy of Nanostructures*, Springer-Verlag, Berlin, 2004.
5. Tersoff, J., Teichert, C. and Lagally, M.G., Self-organization in growth of quantum dot superlattices, *Phys. Rev. Lett.* **76**, 1996, 1675–1678.
6. Springholz, G., Three-dimentional stacking of self-assembled quantum dots in multilayer structures, *C.R. Physique* **6**, 2005, 89–103.
7. Rousset, S., Croset, B., Girard, Y., Prévot, G., Repain, V. and Rohart, S., Self-organized epitaxial growth on spontaneously nano-patterned templates, *C.R. Physique* **6**, 2005, 33–46.
8. Wang, Y.U., Jin, Y.M.M. and Khachaturyan, A.G., Phase field microelasticity modeling of surface instability of heteroepitaxial thin films, *Acta Materialia* **52**, 2004, 81–92.
9. Wang, Y.U., Jin, Y.M.M. and Khachaturyan, A.G., Phase field microelasticity theory and modeling of elastically and structurally inhomogeneous solid, *J. Appl. Phys.* **92**, 2002, 1351–1360.
10. Ni, Y., He, L.H. and Soh, A.K., Three-dimensional phase field simulations for surface roughening of heteroepitaxial films with elastic anisotropy, *J. Crystal Growth* **284**, 2006, 281–292.

Studying Visco-Plasticity of Amorphous Polymers by Indentation Tests

C.Y. Zhang[1], Y.W. Zhang[1,*], K.Y. Zeng[2] and L. Shen[3]

[1]*Department of Materials Science and Engineering, National University of Singapore, Singapore 119260; *E-mail: msezyw@nus.edu.sg*
[2]*Department of Mechanical Engineering, National University of Singapore, Singapore 119260*
[3]*Institute of Materials Research and Engineering, Singapore 117602*

Abstract. A constitutive model for thermoplastic polymeric materials and its finite element implementation are presented. A five-step indentation scheme was formulated to extract a complete list of the parameters in the constitutive model. Indentation tests on polymethylmethacrylate (PMMA), following the five-step and other schemes, were performed. The experimental data using the five-step scheme were used to extract all the parameters in the model. The extracted parameters were then used to predict the other experimental results. Good agreements between the experimental results and model prediction indicate that the five-step indentation scheme is a practical approach to determine the mechanical properties of thermoplastic polymers.

Key words: polymer, nanoindentation, viscoelastic, plasticity, constitutive model.

1 Introduction

The widely used nanoindentation techniques face some challenging issues when applied to polymeric materials. Not only time-independent elasto-plastic deformation but also time-dependent viscoelastic/viscoplastic deformations are present during indentation tests. To characterize the mechanical properties of polymeric materials, two issues should be addressed: The first is to construct a proper constitutive model which can capture not only time-independent elasto-plastic but also time-dependent viscoelastic/viscoplastic deformations of polymers; and the other one is to accurately and efficiently extract the material parameters in the constitutive model from the load-depth curves obtained by indentation tests.

In the present paper, we first construct a phenomenological model to describe all four types of deformations of polymeric materials, that is, elasticity, plasticity, viscoelasticity and viscoplasticity by combining the mechanical elements of elastic, plastic, and viscoelastic/viscoplastic units. Under the assumption that the effect of loading rate is negligible and the strain hardening is absent, a five-step loading

Yilong Bai et al. (eds), IUTAM Symposium on Mechanical Behavior and Micro-Mechanics of Nanostructured Materials, 229–238.

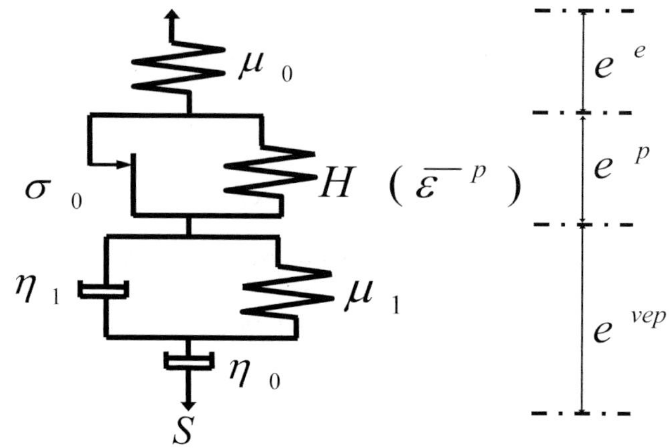

Fig. 1. The constitutive model used to describe the shear deformation of polymeric materials. μ_0 is the shear modulus of the elastic component; σ_0 is the yield strength under pure shear deformation; $H(\bar{\varepsilon}^p)$ defines the hardening rule; μ_1 is the shear modulus of the spring in Voigt–Kelvin unit and η_1 is the viscosity of the dashpot; η_0 is the viscosity coefficient of the separate dashpot. All the shear modulus are connected with the Young's modulus by $\mu_i = E_i/2(1+\nu)$ $(i = 0, 1)$ where ν is the Poisson's ratio.

scheme is proposed to determine all the elastic, plastic and viscoelastic/plastic material parameters in the model. Indentation experiments on polymethylmethacrylate (PMMA) following the five-step scheme are performed. All materials parameters in the model are extracted by fitting to the experimental results. The numerical predictions using these materials parameters are in good agreement with experimental results.

2 Constitutive Model and Numerical Procedures

Four main kinds of deformations exist during indentation on amorphous polymeric materials, namely, elastic, plastic, viscoelastic and viscoplastic deformations. To describe these features, a phenomenological model shown in Figure 1 is proposed.

The following two assumptions have been made:

(I) The volumetric response of polymeric materials is elastic under moderate hydrostatic pressure [1–5],

$$\Delta\sigma_{kk} = 3K_0\Delta\varepsilon_{kk}, \tag{1}$$

where K_0 is the bulk modulus.

(II) The shear strain can be decomposed into elastic, plastic and viscoelastic/viscoplastic components,

$$\Delta e_{ij} = \Delta e^e_{ij} + \Delta e^p_{ij} + \Delta e^{vep}_{ij}, \tag{2}$$

where

$$\sigma_{ij} = s_{ij} + \frac{1}{3}\sigma_{kk}\delta_{ij}, \quad \varepsilon_{ij} = e_{ij} + \frac{1}{3}\varepsilon_{kk}\delta_{ij},$$

δ_{ij} is the Kronecker symbol; s_{ij} is the deviatoric part of the stress tensor σ_{ij}; and e_{ij} is the deviatoric part of the strain tensor ε_{ij}. Here the Einstein summation convention is used.

The stress-strain relation of the elastic component follows a linear equation,

$$\Delta e_{ij}^{e} = \frac{\Delta s_{ij}}{2\mu_0}. \tag{3}$$

The stress-strain relation of the viscoelastic/viscoplastic component in the model shown in Figure 1 takes the following differential form,

$$2e_{ij}^{vep} = \frac{s_{ij}}{\{\mu_1 + \eta_1 \partial t\}} + \frac{s_{ij}}{\{\eta_0 \partial t\}}. \tag{4}$$

For the plastic component, considering the yielding behavior of polymers may be sensitive to the hydrostatic pressure and also influenced by the strain-rate, a modified von Mises yield surface is usually adopted,

$$F = q - kP - [\sigma_0 + H(\bar{\varepsilon}^p) + \alpha \log(\dot{\bar{\varepsilon}})], \tag{5}$$

where

$$q = \sqrt{\frac{3}{2}s_{ij}s_{ij}}$$

is the effective stress; k is the coefficient of internal friction, which reflects the effect of hydrostatic pressure on the yield function; $P \equiv -\sigma_{kk}/3$ is the hydrostatic pressure; σ_0 is the yield strength under pure shear deformation; $H(\bar{\varepsilon}^p)$ defines the hardening rule in terms of the effective plastic strain

$$\bar{\varepsilon}^p = \bar{e}^p \equiv \int_0^t \sqrt{\frac{2}{3}\dot{e}_{ij}\dot{e}_{ij}}\, dt;$$

the last term $\alpha \log(\dot{\bar{\varepsilon}})$ introduces the effect of the strain rate using the logarithmic function proposed by Eyring [6] in terms of the effective strain rate

$$\dot{\bar{\varepsilon}} \equiv \sqrt{\frac{1}{2}\dot{\varepsilon}_{ij}\dot{\varepsilon}_{ij}}.$$

The plastic components of the strain tensor are assumed to be proportional to the derivative of the potential function with respect to the components of the stress tensor by the associative flow rule [7],

$$\Delta e_{ij}^{p} = \frac{3}{2}\Delta\bar{e}^p \frac{s_{ij}|_{t+\Delta t}}{\sqrt{\frac{3}{2}s_{ij}|_{t+\Delta t}s_{ij}|_{t+\Delta t}}}. \tag{6}$$

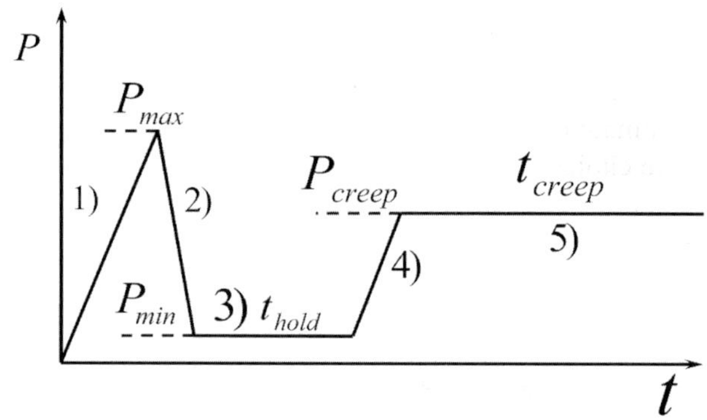

Fig. 2. Schematic illustration of the five-step test scheme.

The updated stress can now be calculated as,

$$s_{ij}|_{t+\Delta t} = 2\mu_0(e^e_{ij}|_t + \Delta e_{ij} - \Delta e^{vep}_{ij} - \Delta e^p_{ij}). \tag{7}$$

The consistency rule of the theory of plasticity which requires that $F = \dot{F} = 0$ can determine the scalable variable $\Delta\bar{e}^p$.

The constitutive model was implemented in the general-purpose finite element program ABAQUS/Standard [8] by writing a user-defined material subroutine UMAT. Finite deformation was considered.

3 A Five-Step Indentation Scheme

3.1 Separation of Time-Independent Plastic Deformation

It is known that the loading curves of elasto-plastic materials can be well fitted by a power law,

$$P = Ch^n \tag{8}$$

and for conical or pyramidal indenters, the power n is equal to 2 and C is a time-independent constant depending only on the elasto-plastic properties of the material. However, if the power law was applied to fit the loading curves of the present visco-elastic-plastic material, it was found that C increases with the reduction of the loading time when the maximum load was held fixed. Gradually, C approaches to the upper limiting value, which corresponds to the extreme case of the model excluding viscoelasticity/viscoplasticity.

Based upon our above analysis, a five-step test scheme (Figure 2) may be used to study elastic-viscoelastic deformation and plastic deformation separately: (1) a fast loading step to the maximum load $P_{\max}$, (2) a fast unloading step to a very small load $P_{\min}$ at 1% of $P_{\max}$, (3) an holding step under $P_{\min}$ for a period of t_{hold}, (4) a fast reloading step to the creep test load P_{creep}, and (5) a final holding step for a period of t_{creep}. The steps are based on the assumption that: the elastic-plastic deformation is dominant during the fast loading/unloading steps (Steps 1 and 2), and only negligible viscoelastic/viscoplastic deformation is induced; the viscoelastic/viscoplastic deformation is dominant during Steps 4 and 5, and negligible instantaneous plasticity is induced. The choice of $P_{creep} < P_{max}$ is to prevent further time-independent plastic deformation and ensure the dominance of the viscoelastic deformation during the reloading and creeping steps. Hence the time-independent plastic deformation can be isolated using the five-step loading scheme.

3.2 Formulating Time-Independent Plastic Deformation

To reduce the complexity of the plastic constitutive model while retaining the important features for describing the deformations of polymers, the following two simplifications were made:

(I) Strain hardening is absent during the indentation test, i.e. $H(\bar{\varepsilon}^p) = 0$. Two facts may justify the simplification: firstly, strain hardening only occurs at large strains when the polymer chains are oriented in long extensions [9]; secondly, large plastic strains are only localized near the indenter tip during indentation tests.

(II) The effect of the strain rate on instantaneous plastic deformation is negligible, i.e., $\alpha \approx 0$ [10, 11]. This may be justified by the fact that the indentation tests are normally performed at a relatively lower speed.

With these two simplifications, the plastic properties of a polymeric materials can be fully described by only two material parameters: the yield strength under pure shear deformation, σ_0, and the internal friction coefficient, k.

Simulations were conducted to investigate the influence of the internal friction coefficient k by using the current material model with very short loading times (that is, to minimize the time-dependent deformation). It is found that a simple linear equation exists between k and C:

$$C = a + b \cdot k, \tag{9}$$

where a and b are material-dependent parameters. For the elasto-plastic material whose yield strength is insensitive to the hydrostatic pressure, i.e. $k = 0$, Dao et al. [12] have proposed the following empirical relation,

$$C = a = N_1 \sigma_{0.29} \left(1 + \frac{\sigma_y}{\sigma_{0.29}}\right) \left[N_2 + \ln\left(\frac{E^*}{\sigma_{0.29}}\right)\right], \tag{10}$$

where N_1, N_2 are computationally derived dimensionless constants which depend only on indenter geometry; σ_y is the yield strength and $\sigma_{0.29}$ is the stress when the plastic strain reaches 0.29 under a tensile test; $E^* = E/(1 - \nu^2)$ is defined as the reduced modulus where E, ν are the Young's modulus and the Poisson's ratio of the sample material, respectively. For elastic-perfectly-plastic materials, $\sigma_y = \sigma_{0.29} = \sigma_0$, thus Equation (10) can be rewritten as,

$$C = a = 2N_1\sigma_0\left[N_2 + \ln\left(\frac{E^*}{\sigma_0}\right)\right]. \tag{11}$$

For coefficient b in Equation (9), our parametric studies show that there is an exponential function between b and σ_0 when the reduced modulus is fixed. On the other hand, an exponential relation between b/σ_0 and $1/E^*$ is also observed when σ_0 is fixed. Further comprehensive parametric studies combining the effects of both σ_0 and E^* indicate that the coefficient b can be well described by the following equation,

$$b = \sigma_0\left[M_1 + M_2 \exp\left(-M_3\frac{\sigma_0}{E^*}\right)\right], \tag{12}$$

where M_1, M_2 and M_3 are also computationally derived dimensionless constants depending only on indenter geometry. Substituting Equations (12) and (11) into Equation (9), one obtains,

$$C = 2N_1\sigma_0\left[N_2 + \ln\left(\frac{E^*}{\sigma_0}\right)\right] + \sigma_0\left[M_1 + M_2 \exp\left(-M_3\frac{\sigma_0}{E^*}\right)\right]k. \tag{13}$$

If the reduced modulus E^* can be determined by the creep steps of the five-step scheme, only two independent indentations are required to determine the two unknowns, that is, the yield strength under pure shear deformation, σ_0, and the internal friction coefficient, k. Here we propose to use two indenters with different geometries to get the equations,

$$\begin{aligned} C_1 &= 2N_1^1\sigma_0\left[N_2^1 + \ln\left(\frac{E^*}{\sigma_0}\right)\right] + \sigma_0\left[M_1^1 + M_2^1 \exp\left(-M_3^1\frac{\sigma_0}{E^*}\right)\right]k, \\ C_2 &= 2N_1^2\sigma_0\left[N_2^2 + \ln\left(\frac{E^*}{\sigma_0}\right)\right] + \sigma_0\left[M_1^2 + M_2^2 \exp\left(-M_3^2\frac{\sigma_0}{E^*}\right)\right]k. \end{aligned} \tag{14}$$

3.3 Formulating Elastic, Visco-Elastic-Plastic Deformation

Steps 4 and 5 were used to extract the elastic and visco-elastic-plastic properties of polymeric materials using a sharp indenter. Analytical solutions to the elastic-viscoelastic deformation based on the concept of "effective indenters" proposed by both Pharr and Bolshakov [13] and Sakai [14] were derived by applying Radok's method [15]. The materials parameters related to elastic, visco-elastic-plastic deformation in the model can be extracted by fitting the equations with the deformation data obtained from Steps 4 and 5 in the five-step indentation scheme. The formulations and procedures for extracting the elastic, visco-elastic-plastic properties can be found in our recent paper [16], thus will not be discussed in details here.

4 Experiments Using the Five-Step Scheme

Depth-sensing indentation experiments were performed on polymethylmethacrylate (PMMA, $E = 2.4 \sim 3.3$ GPa, tensile strength 0.085 GPa, Goodfellow Ltd, Huntingdon, UK) with the MTS Nano Indenter XP (MTS Cooperation, Nano Instruments

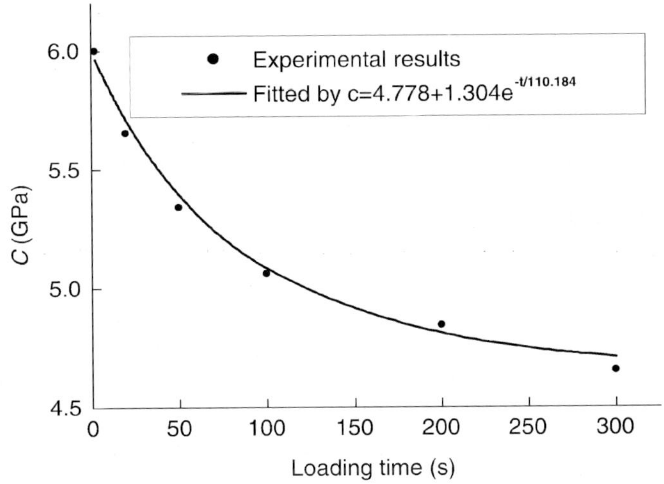

Fig. 3. The coefficient C with the exponent n fixed at 2.

Innovation Center, Oak Ridge, TN, USA) under the load-control mode. Two different indenters were used: a Berkovich indenter and a conical indenter of which the half-included angle is 45°. The X-Ray Diffraction (XRD) test shows the bulk PMMA is amorphous.

Three groups of tests were conducted. The first group was carried out to determine the elastic-viscoelastic parameters using the five-step scheme with the Berkovich indenter. The second group focusing on the loading curves was carried out to find the coefficient C under the extreme condition with negligible viscoelastic influence using a triangle-wave loading history with different loading time. Both indenters were used. The maximum load was fixed at 10 mN for the Berkovich indenter and fixed at 2 mN for the conical indenter. The last group was carried out using the Berkovich indenter to check the predictive capability of the present model. Both the triangle-wave loading history and the three-segment loading history with arbitrary combinations of the loading time, the holding time and the unloading time were used.

5 Material Parameter Extraction and Comparison

5.1 Determination of the Material Parameters

One can obtain the elastic, visco-elastic-plastic parameters of PMMA by following the procedures proposed by Zhang et al. [16]: $E_0 = 3.1$ GPa, $\upsilon = 0.4$, $E_1 = 6.5$ GPa, $\eta_1 = 855.5$ GPa.s, $\eta_0 = 800.0$ GPa.s. It can be seen that the values of these parameters are consistent with literature values [17].

Figure 3 shows the influence of the loading time on the coefficient C , where the values of C were obtained by fitting the loading curves with $P = Ch^2$. To

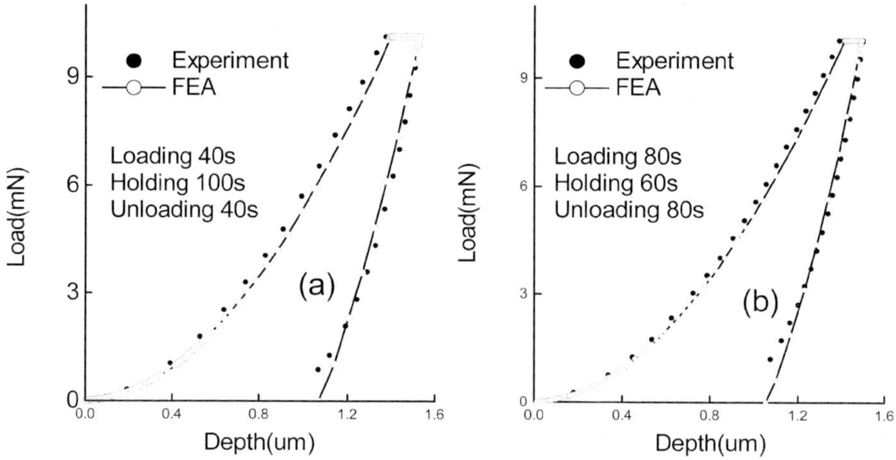

Fig. 4. Comparison between experiment results and simulated results under a three-segment loading history.

obtain a more accurate value of C, that is, without the influence of viscoelasticity/viscoplasticity, curve fitting and extrapolation was used to determine the value of C as shown in Figure 3. It is found that $C = 6.0132$ GPa for the Berkovich indenter, and $C = 1.3131$ GPa for the conical indenter.

The indenter geometry-dependent constants in Equation (14) can be found by running a series of finite element simulations. It was found that $N_1^1 = 13.05$, $N_2^1 = -1.24$, $M_1^1 = 3.90$, $M_2^1 = 65.98$, $M_3^1 = 64.44$ for the Berkovich indenter, and $N_1^2 = 1.33$, $N_2^2 = 0.25$, $M_1^2 = 2.82$, $M_2^2 = 16.68$, $M_3^2 = 84.71$ for the conical indenter. By solving Equation (14), one obtains $\sigma_0 = 0.12$GPa and $k = 0.3$. The theoretic tensile strength, which can be calculated by $\sigma_T = \sigma_0/(1 + k_3) = 0.108$ GPa, is close to the experimental values 0.085 GPa. In addition, the internal friction coefficient k is also comparable to the literature value of 0.25 [18]. These minor differences may be attributed to the difference in either materials, or experimental measurements or the constitutive model.

5.2 Predictive Performance of the Present Model

Once all the material parameters in the constitutive model are fully determined by the five-step indentation scheme, they can be used to predict the indentation behavior of PMMA under other test conditions. Figures 4a and 4b show the comparison between the experimental results and the simulation results under a three-segment loading history, while Figures 5a, 5b and 5c show the comparison between the experimental results and the simulation results under a triangle-wave loading history. Good agreement between experiment results and model predictions implies that the model provides a reasonable good description for thermoplastic polymers and the five-step

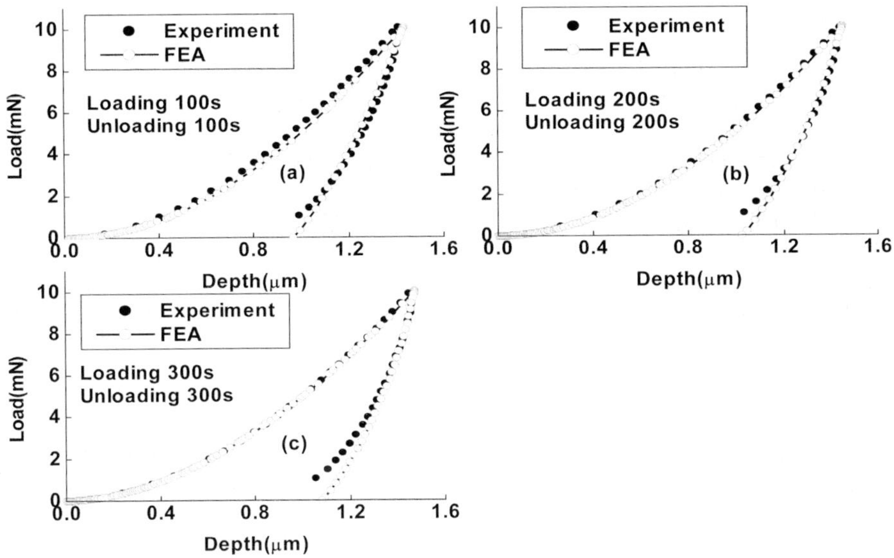

Fig. 5. Comparison between experimental results and simulated results under a triangle-loading history.

indentation scheme is a useful and practical approach to extract all the parameters in the proposed model.

6 Summary

A constitutive model was proposed to describe the elastic, plastic and viscoelastic/viscoplastic deformations for polymeric materials. The model predictions were compared with and verified by experimental results. A five-step indentation scheme was proposed to extract all the parameters in the constitutive model. Indentation experiments on polymethylmethacrylate (PMMA) following the five-step scheme were performed. To reduce the complexity of the model while retaining important features of mechanical responses, both the strain hardening and the strain rate effect were neglected. The predictions using the extracted values of the elastic-visco-elastic-plastic parameters are in good agreement with experimental results.

References

1. S.G. Bardenhagen, M.G. Stout and G.T. Gray, *Mech. Mater.* **25**, 1997, 235.
2. L. Cheng, X. Xia, W. Yu, L.E. Scriven and W.W. Gerberich, *J. Poly. Sci. B* **38**, 2000, 10.
3. C.Y. Zhang, Y.W. Zhang and K.Y. Zeng, *J. Mater. Res.* **19**, 2004, 3053.
4. L. Cheng, X. Xia, L.E. Scriven and W.W. Gerberich, *Mech. Mater.* **37**, 2005, 213.

5. E. Riande, R. Díaz-Calleja, M.G. Prolongo, R.M. Masegosa and C. Salom, *Polymer Viscoelasticity: Stress and Strain in Practice*, Marcel Dekker, New York, 2000, pp. 704.
6. H. Eyring, *J. Chem. Phys.* **4**, 1936, 283.
7. R.K. Goldberg and G.D. Roberts, NASA/TM-2003-212382, National Aeronautics and Space Administration, Washington, DC, 2003.
8. ABAQUS version 6.4, Hibbitt, Karlsson and Sorensen, Inc., Pawtucket, RI.
9. E. Riande, R. Díaz-Calleja, M.G. Prolongo, R.M. Masegosa and C. Salom, *Polymer Viscoelasticity: Stress and Strain in Practice*, Marcel Dekker, New York, 2000, pp. 586–602.
10. A.J. Lesser and R.S. Kody, *J. Polym. Sci. B/Polym. Phys.* **35**, 1997, 1611.
11. G. Dean and B. Read, *Polym. Test* **20**, 2001, 677.
12. M. Dao, N. Chollacoop, K.J. Van Vliet, T.A. Venkatesh and S. Suresh, *Acta. Mater.* **49**, 2001, 3899.
13. G. M. Pharr and A. Bolshakov, *J. Mater. Res.* **17**, 2002, 2660.
14. M. Sakai, *J. Mater. Res.* **18**, 2003, 1631.
15. J.R.M. Radok, *Q. Appl. Math.* **15**, 1957, 198.
16. C.Y. Zhang, Y.W. Zhang, K.Y. Zeng and L. Shen, *J. Mater. Res.* **20**, 2005, 1597.
17. H. Lu, B. Wang, J. Ma, G. Huang and H. Viswanathan, *Mech. Time-Depend. Mater.* **7**, 2003, 189.
18. R. Quinson, J. Perez, M. Pink and A. Pavan, *J. Mater. Sci.* **32**, 1997, 1371.

Phase Transitions of Carbon Materials under High Pressure

Wanlin Guo*, Yitao Dai and Bin Zhang

*Institute of Nano Science, Nanjing University of Aeronautics and Astronautics, Nanjing 210016, China; *E-mail: wlguo@nuaa.edu.cn*

Abstract. Paradoxical experimental observations are explained by studying the high-pressure physical mechanics processes of graphite and carbon nanotubes (CNTs) and the macroscopic mechanics behaviors in the experiments of diamond anvil cells (DAC). The stress concentration on the graphite sample under non-hydrostatics compression in DAC experiments can produce a new phase that is hard enough to crack the superhard diamond. Those soft to hard phase transitions occur at the pressure of about 17 GPa for both graphite and CNTs, independent of the shape and the size of the indenter and the amount of the graphite layers. And a theoretical route is provided to industrially produce diamond and high strength CNTs-bundles composite at room temperature by using of high-pressure technology. Physical mechanics of nanomaterials in particular environment is also discussed.

Key words: graphite, carbon nanotubes, ultrahigh pressure, bond switching, physical mechanics.

1 Introduction

Carbon is the fourth most abundant element in the solar system, which can exist in diverse polymorphs such as diamond, graphite, amorphous carbon, carbines, carbon nanotubes (CNTs), fullerences (C60, for example), and nanofoam [1]. These polymorphs have great differences both in physical and mechanical properties. In graphite, fullerences and CNTs, most of the carbon atoms are arranged in a structure of hexagonal rings with hybridized sp^2 bonds. In amorphous carbon, triangle, pentagon, or heptagon structures coexist. And in diamond, carbon atoms form sp^3 bonds with the four nearest neighbors, creating the superhard pyramidal structure. Graphite is very soft and opaque and also chemically inert, it can be used as lubricant, electronic and thermal conductors, and high temperature materials. CNT is a new material with high strength (the tensile strength of CNTs is approximately 100 times greater than that of steel), low specific gravity (the specific gravity of CNTs is 1/6 that of steel) and excellent electronic properties, and it can be a conductor or semiconductor. C60 is a spherical molecule, and it is a new kind of semiconductor

Yilong Bai et al. (eds), IUTAM Symposium on Mechanical Behavior and Micro-Mechanics of Nanostructured Materials, 239–249.

and can also be a potential high temperature superconductor. Diamond is one of the hardest materials with inert chemical properties and high compressive strength, and it can be used as thermal conductors, cutting and wearable materials, and is also a kind of expensive adornment. Can we have the magic to "Touch graphite and turn it into diamond" just as the fairy tale says? It is indicated by experiments that those polymorphs can transform into each other under certain conditions, such as high pressure.

Ultrahigh pressure can change the atomic structure of matters, and it is an important method to find new phenomena and obtain new phases of materials. With the development of the DAC technique, the high pressure beyond 400 GPa has been obtained and widely used to study the physical and mechanical properties of materials under compression. It has been found in experiments that at the temperature of 20–700°C, a high pressure of 12.5 GPa can convert fullerenes into polycrystalline superhard amorphous carbon with the hardness as high as that of the single-crystal diamond [2]. On the other hand, in graphite subjected to high pressure up to 65 GPa at room temperature, no super-hardness is found after the pressure is removed [3]. But recently, a new phase hard enough to crack diamond anvils was obtained from graphite undergoing a transition of sp^2 to sp^3 bonding at a pressure of about 17 GPa [4].

CNTs can be one-dimensional conductor or semiconductor, depending on the chiral structure and its atomic structure [5]. Its conductivity can be modulated by mechanical deformations [6], and it can even be tuned as a single electron transistor [7]. For the existence of unpaired electrons, CNTs can form chemical bonding with other matters, so it is sensitive to the surroundings and can be used as gas ionization sensors [8]. Therefore, controlling the bond switching and phase transition of CNTs by compression may have significant potential in nanotechnology. The physical mechanics properties of CNT bundles under high pressure have been widely studied [9–11]. Though the pressure treatment of single-wall CNTs to 62 GPa did not produce an after-pressure-release superhard carbon phase [12], and intertube or interlayer sp^3 bonding in single-walled CNT bundles was not observed up to a stress level of 20 GPa in first-principles calculations [13], the nanoindentation of single-walled CNTs can produce force-depth curves comparable to that of diamond in some experiments [14]. However, the condition and mechanism of the bond formation, bond broken and phase transition are paradoxical. To explain these incompatible experiments, more study about the physical, chemical and mechanical properties of carbon matters under high pressure is needed.

The physical mechanics, which can describe the recombination of the bonds of carbon, should be developed to study the microscopic and macroscopic properties of carbon structure under ultrahigh pressure. In that theory, the microscopic and macroscopic properties of carbon matter are predicted by study the microscopic mechanism of the variety of the physical and mechanical properties of carbon structure induced by the lattice recombination under ultrahigh pressure. That methodology can also help us to find new door for synthesizing new carbon materials and improve the design for better high-pressure experimental setup.

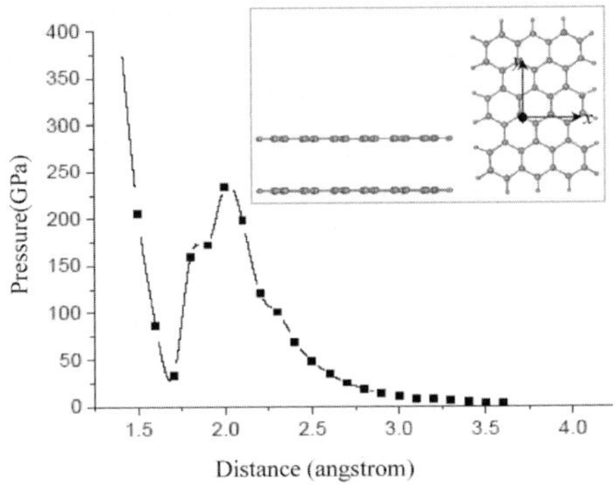

Fig. 1. Variation of interlayer pressure with interlayer interatomic distance from the QM simulations. The transition region is between 2.0 and 1.7 Å. The inset is the bi-layer graphite model we used, in which all the boundary carbon-bonds are closed with H atoms. The atom-on-atom stack is used to show more directly the relation between the pressure and the interatomic distance [18].

Forming new structures of materials under ultrahigh pressure at room temperature is a practical and convenient alternative compared with the chemical methods. In order to study the physical mechanical and the macroscopic mechanical behavior of the carbon structures under ultrahigh pressure, we should combine the quantum mechanics (QM), the molecular dynamics (MD) and the finite element (FE) analysis together. In that scheme, the electronic processes, such as the electron transmission, the band structure and the quantum effect can be described by the QM method, the atomic and molecular processes, such as bonding formations, bonding breakings can be described by the MD method, and the macroscopic behaviors such as stress and strain can be described by the FE analysis [15–17]. In this work, we give a primary study of the physical mechanical properties of carbon structures (we take graphite and CNTs as examples) and carry out atomic simulations of the nanoindentation of graphite and CNTs by the MD method, and a macroscopic mechanical analysis of the DAC experiments of graphite by the FE method.

To describe the bonding formation of the carbon atoms of different interlayer, a QM simulation is performed. In the QM simulation, a bi-layer graphite model, in which each layer contains 60 atoms and all the boundary carbon-bonds are closed with H atoms, is used (as shown in the inset of Figure 1). The interlayer distance was reduced gradually to apply compressive stress. The atom-on-atom stack is used to show more directly the relation between the stress and the interatomic distance. All of the QM simulations of the compression of graphite sheet were performed using the Roothaan-Hall form of Hartree–Fock formula, and the PM3 method [19]. It is found that when the distance of the two graphite layers approaches 2 Å, the nominal compressive stress increases sharply with the decreasing of the interlayer distance and forms a peak at about 2 Å. When the distance is further compressed below 2 Å, suddenly drop in stress occurs till the distance down to 1.7 Å. Within 1.7 Å, the stress

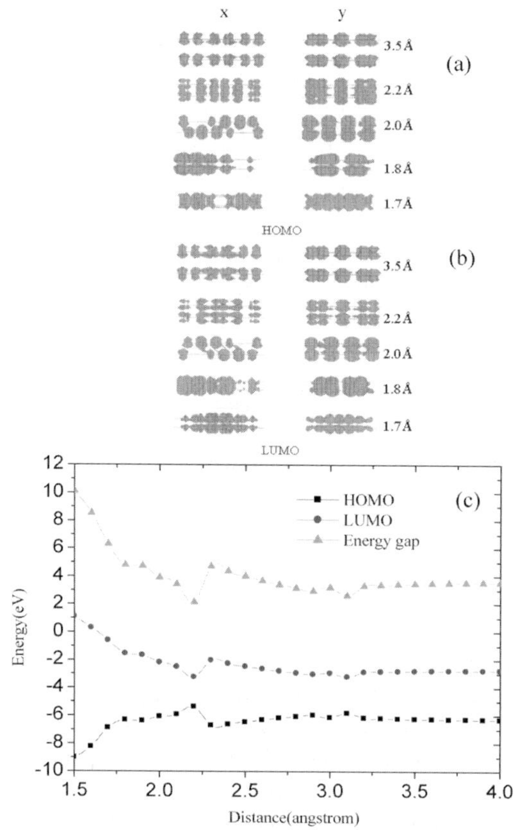

Fig. 2. Transition of electronic structure occurs with decreasing interlayer distance of graphite sheet under uniform compressing. (a) and (b) show the density distribution of the HOMO and LUMO, respectively, and (c) shows the corresponding energy level of those molecular orbital and there gap. The atomic model of the graphite sheet is the same as shown by the inset of Figure 1.

increases steeply with the decreasing of the distance (Figure 1). This result indicates that the bonding switching happens when the interlayer distance is about 2 Å∼1.7 Å. To give a perspicuous explanation of the bonding switching, the density distribution of the highest occupied molecular orbital (HOMO) and the lowest unoccupied molecular orbital (LUMO) are shown in Figures 2a and 2b respectively. When the inter-atomic distance decreases from 3.5 to 2.2 Å, both distributions of the HOMO and LUMO remain similar shapes, but when the distance decreases further from 2.2 to 2.0 Å, interlayer overlap of the pattern start to develop and the widening of the energy gap occurs, as shown by Figure 2c. When the distance goes further down to 1.8 Å, essential changes occur, such that the interlayer HOMO pattern first separates and then merges at 1.7 Å. The corresponding transition in the pattern shapes of the LUMO density is also remarkable around 2.0 Å, and the energy gap widens with

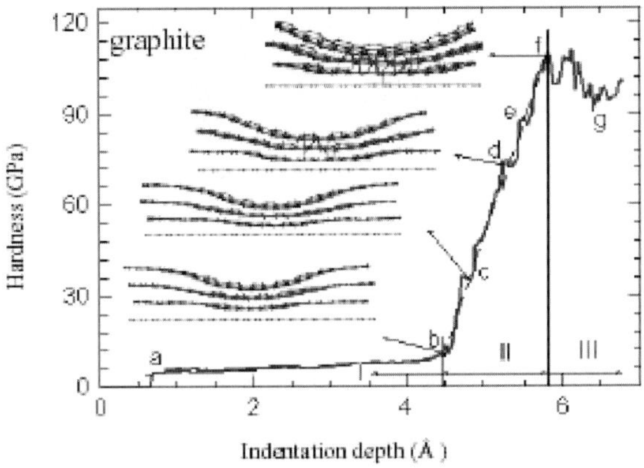

Fig. 3. Hardness-indentation depth curve of a graphite sample. Insets at points b, c, d, f show the atomic geometrical structures and the interlayer sp^3 bonding at the points (modified from [25, figure 1]).

a higher speed. These show that the sp^2–sp^3 transition occurs when the interatom distance is between 2.0 and 1.7 Å. This result is consistent with the MD simulations in the following section.

In the MD simulation, the second generation C-C potential developed by Brenner [20, 21] is used to describe the bonds interaction, bonding switching and breaking of carbon atoms, and the nonbonded long-range interaction is modeled with the Lennard–Jones 6–12 potential [22]. To simulate the process of nanoindentation, the virtual nanoindenter is used. In this method, a repulsive potential in the form of $V(r) = A\theta(R - r)(R - r)^3$ is used for both the indenter and the substrate to avoid the interlinking [23] with the carbon atoms. Here A is a force constant, $\theta(x)$ is the step function, R is radius of the indenter, and r is the distance from the carbon atom to the center of the indenter sphere. In all of our simulations, the indenter is pressed at the center of the samples and penetrates with a speed of 5 m/s, and a constant temperature system (NVT) with the temperature of 300 K is used. The method of the temperature controlling is the Berendsen scheme [24], and the time step is 1 fs.

Figure 3 shows the nano-hardness (defined as the indentation load divided by the contact area) of graphite as a function of the indentation depth. It is found that there are three distinctive phases of graphite in this simulation: the soft phase (phase I, from a to b, where the nearest distance between the different layers is about 2 Å), the hard phase (phase II, from b to f, the modulus is about 730 GPa, which is comparable to the corresponding value of diamond [26]) and the unstable phase (phase III, from f to g, the compressive load begin to drop beyond f where the nano-hardness is about 109 GPa). These phenomena are caused by the interlayer sp^2–sp^3 bonding switching [25]. The hard phase can further be divided into two different sub-stages (IIa (from b to d) and IIb (from d to f)) though the compressive slopes are the same. In stage IIa , the interlayer sp^3 bonding of the graphite sample is recoverable once unloading, showing the super-hard elastic property similar to that of CNTs [27], while entering

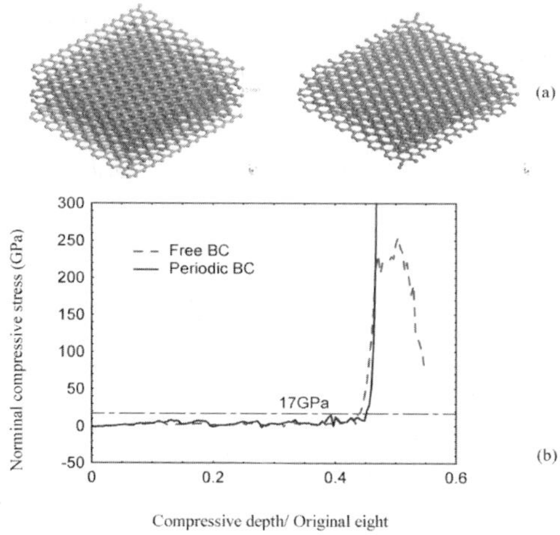

Fig. 4. Nano-hardness of graphite sample under planar compression under free and periodical boundary conditions. (a) The atomic model which contains five layers of graphite. The top and the bottom layers are fixed in plane, served as the grip gear. (b) The nano-hardness of the graphite sample under planar compression with free and periodical boundary conditions.

in stage IIb, the interlayer sp^3 bonding of the sample becomes permanent even when the compressive loading is removed. The bond type analysis indicates that at point c, where a drop in nano-hardness occurs, the first sp^3 bond form; at point d, where the nano-hardness is about 75 GPa, residual sp^3 bonds exist after the loading is removed (this is proved by a latest experiment [28]), and beyond point f, where the hardness become unstable, an amorphous phase with sp^2 and sp^3 bonds coexisting is found. These results are consistent with experimental observations.

To remove the effect of the shape of the indenter on the nano-hardness, calculations of planar compression are carried out (Figure 4). It is found that the limiting stress in the planar compression case with periodic boundary condition (BC), where no shear stress can occur, can smoothly increase to more than 800 GPa before the first large stress drop occurs, which is much higher than $\sim$100 GPa in the indentation case. When the free BC is applied, the planar compression leads to the nominal pressure strength of about 250 GPa. However, the soft-hard phase transition still occurs at about 17 GPa under both the free BC and the periodic BC, which is similar to the indentation case.

Calculations of compressions of different layers are also performed to investigate the influence of the amount of the layers on the hardness (Figure 5). In these cases, although the limiting hardness of the graphite sheet may be influenced by the number of graphite layers, but the soft-hard transition is little affected. Under free boundary condition, samples with more layers become unstable under compression.

Nano-indentation of CNTs gives similar results with those of graphite. The nano-hardness of the $(6, 6)/(11, 11)$ bi-walled CNT is shown in Figure 6. Spheric indenters with different radius (2.5 nm and 5 nm) are used to depict the effect of indenter size on the nano-hardness. The soft-hard transition also occurs when the

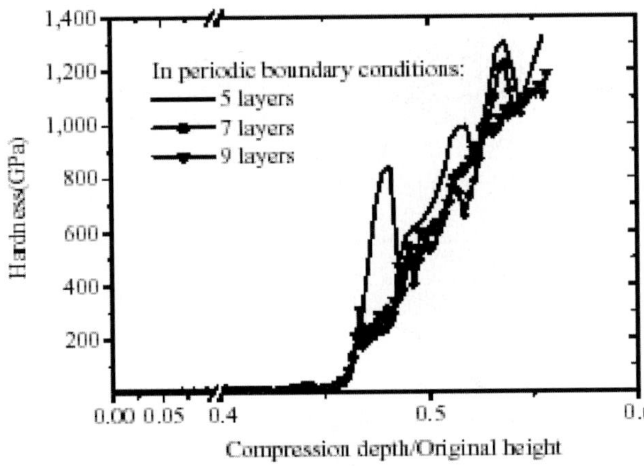

Fig. 5. The influence of number of graphite layers on the hardness. The model is similar to that shown in Figure 4, but with 5, 7 and 9 layers of graphite, and the periodic boundary condition is used.

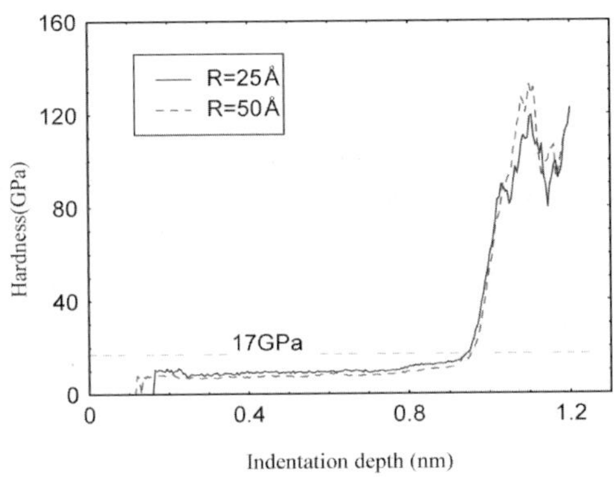

Fig. 6. Nano-hardness of the (6,6)/(11,11) bi-walled CNTs with different indenter radius. The model is the same as that in [25].

compressive stress is about 17 GPa. Increasing the tip radius from 2.5 to 5 nm produces little changes on the results.

The maximal hardness of the graphite shows close relationship to the shear stress induced by the non-hydrostatics compression, and the hardness can reach up to 1 TPa under ideal hydrostatic compression. According to the elasticity theory, under non-hydrostatics compression, the stress concentration on the boundary surface exists. To describe this macroscopic mechanical behavior during the process of the DAC experiment of graphite samples, FE calculations are carried out to examine the stress distributions along the contact boundary using the code ANSYS.

In the FE model (Figure 7), eight-node axis-symmetry elements are adopted for the spatial axial symmetry and horizontal. The interfacial friction and nonlinear con-

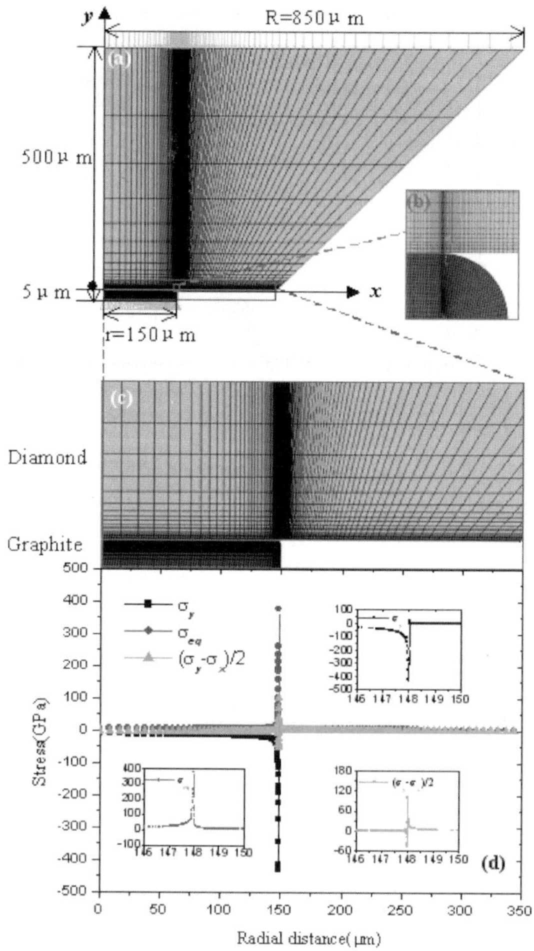

Fig. 7. The finite element model and the radial distributions of the stress components on the diamond interface. (a) The spatial axis-symmetry FE model of the diamond culet contacted with the disk-shape graphite sample, where the radii of the table R and the sample r are 850 μm and 150 μm respectively. (b) The snapshot of the mesh refinement at the boundary corner, where the radius of the sample blunt edge is 2 μm. (c) The snapshot of the interface between the diamond and the graphite. (d) The stress components are plotted as functions of the radial distance when the average stress in the mid plane of the graphite sample is 17 GPa. The compressive stress σ_y, the Von-Mises equivalent stress σ_e, and the shear stress $(\sigma_y - \sigma_x)/2$ curves near the boundary are plotted in the insets.

tact between the diamond and graphite are included using contact pair elements. The modulus of the hard-phase graphite is 730 GPa, which is obtained from the above MD simulations, the modulus of diamond is 1000 GPa, the friction coefficient is 0.1, and the Poisson ratio is 0.2. The absolute values of the stresses all increase sharply near the contact boundary, and can exceed 100 GPa, which is the compressive strength of diamond, in certain regions around the edge when the average pressure in the graphite sample is 17 GPa as the conditions in the experiment of Mao et al. [4]. These results can explain why diamond anvil is cracked by graphite sample in the DAC experiment at nominal pressure of 17 GPa. Changes in the friction coefficient from 0.05 to 0.1 and the Poisson ratio from 0.1 to 0.29 do not change the general trend.

In conclusion, both graphite and CNTs present soft-hard transformation at the pressure of about 17 GPa during the nanoindentation or uniform compression. Due to the formations of interlayer sp^3 bonds under high pressure, the modulus of graphite can exceed 730 GPa in the hard phase. This phase transition is independent of the shape and the size of the indenter and the amount of the graphite layers though they may affect the nano-hardness and the strength. The phase transition between the soft and hard phases is recoverable when the compressive stress is lower than a certain value, for graphite the value is about 74 GPa; beyond that value some of the interlayer sp^3 become permanent even after unloading. The nano-hardness can reach 109 GPa in the graphite sample, which exceeds the strength of diamond of 100 GPa. And in the DAC experiments of graphite under non-hydrostatics compression, the maximal concentrated stress can exceed 100 GPa when the average pressure is 17 GPa in the sample. These results can explain the seemingly paradoxical experimental observations that graphite sheets can be hard enough to crack the super-hard diamond anvil during a compressive process at 17$\sim$23 GPa in the DAC experiments, while no hard phase can be detected in graphite and CNTs after unloading from high pressure up to 65 GPa.

Under particular conditions such as high pressure, confined system, electric field, magnetic field, iron and laser irradiation, and high temperature, the electronic and atom structure of matter can be changed, and new phase, new materials and new properties may appear. By use of physical mechanics, we can describe different phenomena in different scales, which can lead to the synthesis of new materials in more ordinary and convenient ways.

Acknowledgements

This work is supported by National NSF (No. 10372044), Jiangsu Province NSF, the Innovation Team Programme, the Cultivation Fund of the Key Scientific and Technical Innovation Project of the Ministry of Education of China (No. 705021).

References

1. Henning, T.h. and Salama, F. (1998) Carbon in the universe, *Science* **282**, 2204; Haggerty, S.E. (1999) A diamond trilogy: Superplumes, supercontinents, and supernovae, *Science* **285**, 851.

2. Kozlov, M.E., Hirabayashi, M. and Nozaki K. et al. (1995) Transformation of C60 fullerenes into a superhard form of carbon at moderate pressure, *Appl. Phys. Lett.* **66**, 1199; Hirai, H., Kondo. K. and Kim, M. et al. (1997) Transparent nanocrystalline diamond ceramics fabricated from C60 fullerene by shock compression, *Appl. Phys. Lett.* **71**, 3016; Lyapin, G., Brazhkin, V.V. and Gromnitskaya, E.L. et al. (2000) Hardening of fullerite C60 during temperature-induced polymerization and amorphization under pressure, *Appl. Phys. Lett.* **76**, 712.
3. Patterson, J.R., Catledge, S.A. and Vohra, Y.K. et al. (2000) Electrical and mechanical properties of C70 fullerene and graphite under high pressures studied using designer diamond anvils, *Phys. Rev. Lett.* **85**, 5364.
4. Mao, W.L., Mao, H.K. and Eng, P.J. et al. (2003) Bonding changes in compressed superhard graphite, *Science* **302**, 425.
5. Tombler, T.W., Zhou, C. and Alexseyev, L. et al. (2000) Reversible electromechanical characteristics of carbon nanotubes under local-probe manipulation, *Nature* **405**, 769.
6. Postma, H.W., Teepen, T. and Yao, Z. et al. (2001) Carbon nanotube single-electron transistors at room temperature, *Science* **293**, 76.
7. Odom, T.W., Huang, J.L. and Kim, P. et al. (1998) Atomic structure and electronic properties of single-walled carbon nanotubes, *Nature* **391**, 62.
8. Modi, A., Koratkar, N. and Lass, E. et al. (2003) Miniaturized gas ionization sensors using carbon nanotubes, *Nature* **424**, 171.
9. Tang, J., Qin, L. and Sasaki, T. et al. (2002) Revealing properties of single-walled carbon nanotubes under high pressure, *J. Phys. Condensed Matter.* **14**, 10575.
10. Venkateswaran, U.D., Rao, A.M. and Richter, E. et al. (1999) Probing the single-wall carbon nanotube bundle: Raman scattering under high pressure, *Phys. Rev. B* **59**, 10928.
11. Tang, J., Qin, L. and Sasaki, T. et al. (2000) Compressibility and polygonization of single-walled carbon nanotubes under hydrostatic pressure, *Phys. Rev. Lett.* **85**, 1887.
12. Patterson, J.R., Vohra, Y.K. and Weir, S.T. et al. (2001) Single-wall carbon nanotubes under high pressures to 62 GPa studied using designer diamond anvils, *J. Nanosci. Nanotech.* **1**, 143.
13. Chan, S., Yim, W. and Gong, X. et al. (2003) Carbon nanotube bundles under high pressure: Transformation to low-symmetry structures, *Phys. Rev. B* **68**, 075404.
14. Popov, M., Kyotani, M. and Koga, Y. (2003) Superhard phase of single wall carbon nanotube: Comparison with fullerite C60 and diamond, *Diamond and Related Materials* **12**, 833.
15. Chacham, H. and Kleinman, L. (2000) Instabilities in diamond under high shear stress, *Phys. Rev. Lett.* **85**, 4904.
16. Richter, A., Ries, R. and Smith, R. et al. (2000) Nanoindentation of diamond, graphite and fullerene films, *Diamond and Related Materials* **9**, 170.
17. Merkel, S., Hemley, R.J. and Mao, H.K. (1999) Finite-element modeling of diamond deformation at multimegabar pressures, *Appl. Phys. Lett.* **74**, 656.
18. Zhang, B., Guo, W. and Dai, Y.T. (2005) "Touch graphite and turn it into diamond" – Physical mechanics of carbon under ultrahigh pressure, *Physics* **34**(7), 410 [in Chinese].
19. Guo, Y. and Guo, W. (2003) Coupled mechanical and electronic properties of single-wall carbon nanotubes by QM/MD, *J. Phys. D* **36**, 805.
20. Brenner, D.W. (1990) Empirical potential for hydrocarbons for use in simulating the chemical vapor deposition of diamond films, *Phys. Rev. B* **42**, 9458.
21. Brenner, D.W., Shenderova, O.A. and Harrison, J.A. et al. (2002) Second generation reactive empirical bond order (REBO) potential energy expression for hydrocarbons, *J. Phys.: Condens. Matter* **14**, 783.

22. Web site of Brenner's software. http://www.engr.ncsu.edu/mat/CompMatSci/projects.html.
23. Kelchner, C.L., Plimpton, S.J. and Hamilton, J.C. (1998) Dislocation nucleation and defect structure during surface indentation, *Phys. Rev. B* **58**, 11085.
24. Berendsen, H.J.C., Postma, J.P.M. and van Gunsteren, W.F. et al. (1984) Molecular dynamics with coupling to an external bath, *J. Chem. Phys.* **81**, 3684.
25. Guo, W., Zhu, C.Z., Yu, T.X., Woo, C.H., Zhang, B. and Dai, Y.T. (2003) Formation of sp^3 bonding in nanoindented carbon nanotubes and graphite, *Phys. Rev. Lett.* **93**, 245502.
26. MatWeb, http://www.matweb.com/.
27. Zhu, C.Z., Guo, W., Yu, T.X. and Woo, C.H. (2005) Radialcompression of carbon nanotubes: Deformation and damage, super-elasticity and super-hardness, *Nanotechnolgoy* **16**, 1035.
28. Wang, Z., Zhao, Y. and Tait, K. et al. (2004) A quenchable superhard carbon phase synthesized by cold compression of carbon nanotubes, *PNAS* **101**, 13699.
29. Zhang, B. and Guo, W. (2005) Cracking diamond anvil cells by compressed nanographite sheets near the contact edge, *Appl. Phys. Lett.* **87**.

Author Index

Subject Index

Mechanics

SOLID MECHANICS AND ITS APPLICATIONS

Series Editor: G.M.L. Gladwell

Aims and Scope of the Series

The fundamental questions arising in mechanics are: *Why?*, *How?*, and *How much?* The aim of this series is to provide lucid accounts written by authoritative researchers giving vision and insight in answering these questions on the subject of mechanics as it relates to solids. The scope of the series covers the entire spectrum of solid mechanics. Thus it includes the foundation of mechanics; variational formulations; computational mechanics; statics, kinematics and dynamics of rigid and elastic bodies; vibrations of solids and structures; dynamical systems and chaos; the theories of elasticity, plasticity and viscoelasticity; composite materials; rods, beams, shells and membranes; structural control and stability; soils, rocks and geomechanics; fracture; tribology; experimental mechanics; biomechanics and machine design.

1. R.T. Haftka, Z. Gürdal and M.P. Kamat: *Elements of Structural Optimization*. 2nd rev.ed., 1990 ISBN 0-7923-0608-2
2. J.J. Kalker: *Three-Dimensional Elastic Bodies in Rolling Contact*. 1990 ISBN 0-7923-0712-7
3. P. Karasudhi: *Foundations of Solid Mechanics*. 1991 ISBN 0-7923-0772-0
4. *Not published*
5. *Not published.*
6. J.F. Doyle: *Static and Dynamic Analysis of Structures*. With an Emphasis on Mechanics and Computer Matrix Methods. 1991 ISBN 0-7923-1124-8; Pb 0-7923-1208-2
7. O.O. Ochoa and J.N. Reddy: *Finite Element Analysis of Composite Laminates*. ISBN 0-7923-1125-6
8. M.H. Aliabadi and D.P. Rooke: *Numerical Fracture Mechanics*. ISBN 0-7923-1175-2
9. J. Angeles and C.S. López-Cajún: *Optimization of Cam Mechanisms*. 1991 ISBN 0-7923-1355-0
10. D.E. Grierson, A. Franchi and P. Riva (eds.): *Progress in Structural Engineering*. 1991 ISBN 0-7923-1396-8
11. R.T. Haftka and Z. Gürdal: *Elements of Structural Optimization*. 3rd rev. and exp. ed. 1992 ISBN 0-7923-1504-9; Pb 0-7923-1505-7
12. J.R. Barber: *Elasticity*. 1992 ISBN 0-7923-1609-6; Pb 0-7923-1610-X
13. H.S. Tzou and G.L. Anderson (eds.): *Intelligent Structural Systems*. 1992 ISBN 0-7923-1920-6
14. E.E. Gdoutos: *Fracture Mechanics*. An Introduction. 1993 ISBN 0-7923-1932-X
15. J.P. Ward: *Solid Mechanics*. An Introduction. 1992 ISBN 0-7923-1949-4
16. M. Farshad: *Design and Analysis of Shell Structures*. 1992 ISBN 0-7923-1950-8
17. H.S. Tzou and T. Fukuda (eds.): *Precision Sensors, Actuators and Systems*. 1992 ISBN 0-7923-2015-8
18. J.R. Vinson: *The Behavior of Shells Composed of Isotropic and Composite Materials*. 1993 ISBN 0-7923-2113-8
19. H.S. Tzou: *Piezoelectric Shells*. Distributed Sensing and Control of Continua. 1993 ISBN 0-7923-2186-3
20. W. Schiehlen (ed.): *Advanced Multibody System Dynamics*. Simulation and Software Tools. 1993 ISBN 0-7923-2192-8
21. C.-W. Lee: *Vibration Analysis of Rotors*. 1993 ISBN 0-7923-2300-9
22. D.R. Smith: *An Introduction to Continuum Mechanics*. 1993 ISBN 0-7923-2454-4
23. G.M.L. Gladwell: *Inverse Problems in Scattering*. An Introduction. 1993 ISBN 0-7923-2478-1

Mechanics

SOLID MECHANICS AND ITS APPLICATIONS

Series Editor: G.M.L. Gladwell

24. G. Prathap: *The Finite Element Method in Structural Mechanics.* 1993 ISBN 0-7923-2492-7
25. J. Herskovits (ed.): *Advances in Structural Optimization.* 1995 ISBN 0-7923-2510-9
26. M.A. González-Palacios and J. Angeles: *Cam Synthesis.* 1993 ISBN 0-7923-2536-2
27. W.S. Hall: *The Boundary Element Method.* 1993 ISBN 0-7923-2580-X
28. J. Angeles, G. Hommel and P. Kovács (eds.): *Computational Kinematics.* 1993 ISBN 0-7923-2585-0
29. A. Curnier: *Computational Methods in Solid Mechanics.* 1994 ISBN 0-7923-2761-6
30. D.A. Hills and D. Nowell: *Mechanics of Fretting Fatigue.* 1994 ISBN 0-7923-2866-3
31. B. Tabarrok and F.P.J. Rimrott: *Variational Methods and Complementary Formulations in Dynamics.* 1994 ISBN 0-7923-2923-6
32. E.H. Dowell (ed.), E.F. Crawley, H.C. Curtiss Jr., D.A. Peters, R. H. Scanlan and F. Sisto: *A Modern Course in Aeroelasticity.* Third Revised and Enlarged Edition. 1995 ISBN 0-7923-2788-8; Pb: 0-7923-2789-6
33. A. Preumont: *Random Vibration and Spectral Analysis.* 1994 ISBN 0-7923-3036-6
34. J.N. Reddy (ed.): *Mechanics of Composite Materials.* Selected works of Nicholas J. Pagano. 1994 ISBN 0-7923-3041-2
35. A.P.S. Selvadurai (ed.): *Mechanics of Poroelastic Media.* 1996 ISBN 0-7923-3329-2
36. Z. Mróz, D. Weichert, S. Dorosz (eds.): *Inelastic Behaviour of Structures under Variable Loads.* 1995 ISBN 0-7923-3397-7
37. R. Pyrz (ed.): *IUTAM Symposium on Microstructure-Property Interactions in Composite Materials.* Proceedings of the IUTAM Symposium held in Aalborg, Denmark. 1995 ISBN 0-7923-3427-2
38. M.I. Friswell and J.E. Mottershead: *Finite Element Model Updating in Structural Dynamics.* 1995 ISBN 0-7923-3431-0
39. D.F. Parker and A.H. England (eds.): *IUTAM Symposium on Anisotropy, Inhomogeneity and Nonlinearity in Solid Mechanics.* Proceedings of the IUTAM Symposium held in Nottingham, U.K. 1995 ISBN 0-7923-3594-5
40. J.-P. Merlet and B. Ravani (eds.): *Computational Kinematics '95.* 1995 ISBN 0-7923-3673-9
41. L.P. Lebedev, I.I. Vorovich and G.M.L. Gladwell: *Functional Analysis.* Applications in Mechanics and Inverse Problems. 1996 ISBN 0-7923-3849-9
42. J. Menčik: *Mechanics of Components with Treated or Coated Surfaces.* 1996 ISBN 0-7923-3700-X
43. D. Bestle and W. Schiehlen (eds.): *IUTAM Symposium on Optimization of Mechanical Systems.* Proceedings of the IUTAM Symposium held in Stuttgart, Germany. 1996 ISBN 0-7923-3830-8
44. D.A. Hills, P.A. Kelly, D.N. Dai and A.M. Korsunsky: *Solution of Crack Problems.* The Distributed Dislocation Technique. 1996 ISBN 0-7923-3848-0
45. V.A. Squire, R.J. Hosking, A.D. Kerr and P.J. Langhorne: *Moving Loads on Ice Plates.* 1996 ISBN 0-7923-3953-3
46. A. Pineau and A. Zaoui (eds.): *IUTAM Symposium on Micromechanics of Plasticity and Damage of Multiphase Materials.* Proceedings of the IUTAM Symposium held in Sèvres, Paris, France. 1996 ISBN 0-7923-4188-0
47. A. Naess and S. Krenk (eds.): *IUTAM Symposium on Advances in Nonlinear Stochastic Mechanics.* Proceedings of the IUTAM Symposium held in Trondheim, Norway. 1996 ISBN 0-7923-4193-7
48. D. Ieşan and A. Scalia: *Thermoelastic Deformations.* 1996 ISBN 0-7923-4230-5

Mechanics

SOLID MECHANICS AND ITS APPLICATIONS

Series Editor: G.M.L. Gladwell

49. J.R. Willis (ed.): *IUTAM Symposium on Nonlinear Analysis of Fracture*. Proceedings of the IUTAM Symposium held in Cambridge, U.K. 1997 ISBN 0-7923-4378-6
50. A. Preumont: *Vibration Control of Active Structures*. An Introduction. 1997 ISBN 0-7923-4392-1
51. G.P. Cherepanov: *Methods of Fracture Mechanics: Solid Matter Physics*. 1997 ISBN 0-7923-4408-1
52. D.H. van Campen (ed.): *IUTAM Symposium on Interaction between Dynamics and Control in Advanced Mechanical Systems*. Proceedings of the IUTAM Symposium held in Eindhoven, The Netherlands. 1997 ISBN 0-7923-4429-4
53. N.A. Fleck and A.C.F. Cocks (eds.): *IUTAM Symposium on Mechanics of Granular and Porous Materials*. Proceedings of the IUTAM Symposium held in Cambridge, U.K. 1997 ISBN 0-7923-4553-3
54. J. Roorda and N.K. Srivastava (eds.): *Trends in Structural Mechanics*. Theory, Practice, Education. 1997 ISBN 0-7923-4603-3
55. Yu.A. Mitropolskii and N. Van Dao: *Applied Asymptotic Methods in Nonlinear Oscillations*. 1997 ISBN 0-7923-4605-X
56. C. Guedes Soares (ed.): *Probabilistic Methods for Structural Design*. 1997 ISBN 0-7923-4670-X
57. D. François, A. Pineau and A. Zaoui: *Mechanical Behaviour of Materials*. Volume I: Elasticity and Plasticity. 1998 ISBN 0-7923-4894-X
58. D. François, A. Pineau and A. Zaoui: *Mechanical Behaviour of Materials*. Volume II: Viscoplasticity, Damage, Fracture and Contact Mechanics. 1998 ISBN 0-7923-4895-8
59. L.T. Tenek and J. Argyris: *Finite Element Analysis for Composite Structures*. 1998 ISBN 0-7923-4899-0
60. Y.A. Bahei-El-Din and G.J. Dvorak (eds.): *IUTAM Symposium on Transformation Problems in Composite and Active Materials*. Proceedings of the IUTAM Symposium held in Cairo, Egypt. 1998 ISBN 0-7923-5122-3
61. I.G. Goryacheva: *Contact Mechanics in Tribology*. 1998 ISBN 0-7923-5257-2
62. O.T. Bruhns and E. Stein (eds.): *IUTAM Symposium on Micro- and Macrostructural Aspects of Thermoplasticity*. Proceedings of the IUTAM Symposium held in Bochum, Germany. 1999 ISBN 0-7923-5265-3
63. F.C. Moon: *IUTAM Symposium on New Applications of Nonlinear and Chaotic Dynamics in Mechanics*. Proceedings of the IUTAM Symposium held in Ithaca, NY, USA. 1998 ISBN 0-7923-5276-9
64. R. Wang: *IUTAM Symposium on Rheology of Bodies with Defects*. Proceedings of the IUTAM Symposium held in Beijing, China. 1999 ISBN 0-7923-5297-1
65. Yu.I. Dimitrienko: *Thermomechanics of Composites under High Temperatures*. 1999 ISBN 0-7923-4899-0
66. P. Argoul, M. Frémond and Q.S. Nguyen (eds.): *IUTAM Symposium on Variations of Domains and Free-Boundary Problems in Solid Mechanics*. Proceedings of the IUTAM Symposium held in Paris, France. 1999 ISBN 0-7923-5450-8
67. F.J. Fahy and W.G. Price (eds.): *IUTAM Symposium on Statistical Energy Analysis*. Proceedings of the IUTAM Symposium held in Southampton, U.K. 1999 ISBN 0-7923-5457-5
68. H.A. Mang and F.G. Rammerstorfer (eds.): *IUTAM Symposium on Discretization Methods in Structural Mechanics*. Proceedings of the IUTAM Symposium held in Vienna, Austria. 1999 ISBN 0-7923-5591-1

Mechanics

SOLID MECHANICS AND ITS APPLICATIONS

Series Editor: G.M.L. Gladwell

69. P. Pedersen and M.P. Bendsøe (eds.): *IUTAM Symposium on Synthesis in Bio Solid Mechanics*. Proceedings of the IUTAM Symposium held in Copenhagen, Denmark. 1999 ISBN 0-7923-5615-2
70. S.K. Agrawal and B.C. Fabien: *Optimization of Dynamic Systems*. 1999 ISBN 0-7923-5681-0
71. A. Carpinteri: *Nonlinear Crack Models for Nonmetallic Materials*. 1999 ISBN 0-7923-5750-7
72. F. Pfeifer (ed.): *IUTAM Symposium on Unilateral Multibody Contacts*. Proceedings of the IUTAM Symposium held in Munich, Germany. 1999 ISBN 0-7923-6030-3
73. E. Lavendelis and M. Zakrzhevsky (eds.): *IUTAM/IFToMM Symposium on Synthesis of Nonlinear Dynamical Systems*. Proceedings of the IUTAM/IFToMM Symposium held in Riga, Latvia. 2000 ISBN 0-7923-6106-7
74. J.-P. Merlet: *Parallel Robots*. 2000 ISBN 0-7923-6308-6
75. J.T. Pindera: *Techniques of Tomographic Isodyne Stress Analysis*. 2000 ISBN 0-7923-6388-4
76. G.A. Maugin, R. Drouot and F. Sidoroff (eds.): *Continuum Thermomechanics*. The Art and Science of Modelling Material Behaviour. 2000 ISBN 0-7923-6407-4
77. N. Van Dao and E.J. Kreuzer (eds.): *IUTAM Symposium on Recent Developments in Non-linear Oscillations of Mechanical Systems*. 2000 ISBN 0-7923-6470-8
78. S.D. Akbarov and A.N. Guz: *Mechanics of Curved Composites*. 2000 ISBN 0-7923-6477-5
79. M.B. Rubin: *Cosserat Theories: Shells, Rods and Points*. 2000 ISBN 0-7923-6489-9
80. S. Pellegrino and S.D. Guest (eds.): *IUTAM-IASS Symposium on Deployable Structures: Theory and Applications*. Proceedings of the IUTAM-IASS Symposium held in Cambridge, U.K., 6–9 September 1998. 2000 ISBN 0-7923-6516-X
81. A.D. Rosato and D.L. Blackmore (eds.): *IUTAM Symposium on Segregation in Granular Flows*. Proceedings of the IUTAM Symposium held in Cape May, NJ, U.S.A., June 5–10, 1999. 2000 ISBN 0-7923-6547-X
82. A. Lagarde (ed.): *IUTAM Symposium on Advanced Optical Methods and Applications in Solid Mechanics*. Proceedings of the IUTAM Symposium held in Futuroscope, Poitiers, France, August 31–September 4, 1998. 2000 ISBN 0-7923-6604-2
83. D. Weichert and G. Maier (eds.): *Inelastic Analysis of Structures under Variable Loads*. Theory and Engineering Applications. 2000 ISBN 0-7923-6645-X
84. T.-J. Chuang and J.W. Rudnicki (eds.): *Multiscale Deformation and Fracture in Materials and Structures*. The James R. Rice 60th Anniversary Volume. 2001 ISBN 0-7923-6718-9
85. S. Narayanan and R.N. Iyengar (eds.): *IUTAM Symposium on Nonlinearity and Stochastic Structural Dynamics*. Proceedings of the IUTAM Symposium held in Madras, Chennai, India, 4–8 January 1999 ISBN 0-7923-6733-2
86. S. Murakami and N. Ohno (eds.): *IUTAM Symposium on Creep in Structures*. Proceedings of the IUTAM Symposium held in Nagoya, Japan, 3-7 April 2000. 2001 ISBN 0-7923-6737-5
87. W. Ehlers (ed.): *IUTAM Symposium on Theoretical and Numerical Methods in Continuum Mechanics of Porous Materials*. Proceedings of the IUTAM Symposium held at the University of Stuttgart, Germany, September 5-10, 1999. 2001 ISBN 0-7923-6766-9
88. D. Durban, D. Givoli and J.G. Simmonds (eds.): *Advances in the Mechanis of Plates and Shells The Avinoam Libai Anniversary Volume*. 2001 ISBN 0-7923-6785-5
89. U. Gabbert and H.-S. Tzou (eds.): *IUTAM Symposium on Smart Structures and Structonic Systems*. Proceedings of the IUTAM Symposium held in Magdeburg, Germany, 26–29 September 2000. 2001 ISBN 0-7923-6968-8

Mechanics

SOLID MECHANICS AND ITS APPLICATIONS

Series Editor: G.M.L. Gladwell

90. Y. Ivanov, V. Cheshkov and M. Natova: *Polymer Composite Materials – Interface Phenomena & Processes*. 2001 ISBN 0-7923-7008-2
91. R.C. McPhedran, L.C. Botten and N.A. Nicorovici (eds.): *IUTAM Symposium on Mechanical and Electromagnetic Waves in Structured Media*. Proceedings of the IUTAM Symposium held in Sydney, NSW, Australia, 18-22 Januari 1999. 2001 ISBN 0-7923-7038-4
92. D.A. Sotiropoulos (ed.): *IUTAM Symposium on Mechanical Waves for Composite Structures Characterization*. Proceedings of the IUTAM Symposium held in Chania, Crete, Greece, June 14-17, 2000. 2001 ISBN 0-7923-7164-X
93. V.M. Alexandrov and D.A. Pozharskii: *Three-Dimensional Contact Problems*. 2001 ISBN 0-7923-7165-8
94. J.P. Dempsey and H.H. Shen (eds.): *IUTAM Symposium on Scaling Laws in Ice Mechanics and Ice Dynamics*. Proceedings of the IUTAM Symposium held in Fairbanks, Alaska, U.S.A., 13-16 June 2000. 2001 ISBN 1-4020-0171-1
95. U. Kirsch: *Design-Oriented Analysis of Structures*. A Unified Approach. 2002 ISBN 1-4020-0443-5
96. A. Preumont: *Vibration Control of Active Structures*. An Introduction (2nd Edition). 2002 ISBN 1-4020-0496-6
97. B.L. Karihaloo (ed.): *IUTAM Symposium on Analytical and Computational Fracture Mechanics of Non-Homogeneous Materials*. Proceedings of the IUTAM Symposium held in Cardiff, U.K., 18-22 June 2001. 2002 ISBN 1-4020-0510-5
98. S.M. Han and H. Benaroya: *Nonlinear and Stochastic Dynamics of Compliant Offshore Structures*. 2002 ISBN 1-4020-0573-3
99. A.M. Linkov: *Boundary Integral Equations in Elasticity Theory*. 2002 ISBN 1-4020-0574-1
100. L.P. Lebedev, I.I. Vorovich and G.M.L. Gladwell: *Functional Analysis*. Applications in Mechanics and Inverse Problems (2nd Edition). 2002 ISBN 1-4020-0667-5; Pb: 1-4020-0756-6
101. Q.P. Sun (ed.): *IUTAM Symposium on Mechanics of Martensitic Phase Transformation in Solids*. Proceedings of the IUTAM Symposium held in Hong Kong, China, 11-15 June 2001. 2002 ISBN 1-4020-0741-8
102. M.L. Munjal (ed.): *IUTAM Symposium on Designing for Quietness*. Proceedings of the IUTAM Symposium held in Bangkok, India, 12-14 December 2000. 2002 ISBN 1-4020-0765-5
103. J.A.C. Martins and M.D.P. Monteiro Marques (eds.): *Contact Mechanics*. Proceedings of the 3rd Contact Mechanics International Symposium, Praia da Consolação, Peniche, Portugal, 17-21 June 2001. 2002 ISBN 1-4020-0811-2
104. H.R. Drew and S. Pellegrino (eds.): *New Approaches to Structural Mechanics, Shells and Biological Structures*. 2002 ISBN 1-4020-0862-7
105. J.R. Vinson and R.L. Sierakowski: *The Behavior of Structures Composed of Composite Materials*. Second Edition. 2002 ISBN 1-4020-0904-6
106. Not yet published.
107. J.R. Barber: *Elasticity*. Second Edition. 2002 ISBN Hb 1-4020-0964-X; Pb 1-4020-0966-6
108. C. Miehe (ed.): *IUTAM Symposium on Computational Mechanics of Solid Materials at Large Strains*. Proceedings of the IUTAM Symposium held in Stuttgart, Germany, 20-24 August 2001. 2003 ISBN 1-4020-1170-9

Mechanics

SOLID MECHANICS AND ITS APPLICATIONS

Series Editor: G.M.L. Gladwell

109. P. Ståhle and K.G. Sundin (eds.): *IUTAM Symposium on Field Analyses for Determination of Material Parameters – Experimental and Numerical Aspects.* Proceedings of the IUTAM Symposium held in Abisko National Park, Kiruna, Sweden, July 31 – August 4, 2000. 2003 ISBN 1-4020-1283-7
110. N. Sri Namachchivaya and Y.K. Lin (eds.): *IUTAM Symposium on Nonlinear Stochastic Dynamics.* Proceedings of the IUTAM Symposium held in Monticello, IL, USA, 26 – 30 August, 2000. 2003 ISBN 1-4020-1471-6
111. H. Sobieckzky (ed.): *IUTAM Symposium Transsonicum IV.* Proceedings of the IUTAM Symposium held in Göttingen, Germany, 2–6 September 2002, 2003 ISBN 1-4020-1608-5
112. J.-C. Samin and P. Fisette: *Symbolic Modeling of Multibody Systems.* 2003 ISBN 1-4020-1629-8
113. A.B. Movchan (ed.): *IUTAM Symposium on Asymptotics, Singularities and Homogenisation in Problems of Mechanics.* Proceedings of the IUTAM Symposium held in Liverpool, United Kingdom, 8-11 July 2002. 2003 ISBN 1-4020-1780-4
114. S. Ahzi, M. Cherkaoui, M.A. Khaleel, H.M. Zbib, M.A. Zikry and B. LaMatina (eds.): *IUTAM Symposium on Multiscale Modeling and Characterization of Elastic-Inelastic Behavior of Engineering Materials.* Proceedings of the IUTAM Symposium held in Marrakech, Morocco, 20-25 October 2002. 2004 ISBN 1-4020-1861-4
115. H. Kitagawa and Y. Shibutani (eds.): *IUTAM Symposium on Mesoscopic Dynamics of Fracture Process and Materials Strength.* Proceedings of the IUTAM Symposium held in Osaka, Japan, 6-11 July 2003. Volume in celebration of Professor Kitagawa's retirement. 2004 ISBN 1-4020-2037-6
116. E.H. Dowell, R.L. Clark, D. Cox, H.C. Curtiss, Jr., K.C. Hall, D.A. Peters, R.H. Scanlan, E. Simiu, F. Sisto and D. Tang: *A Modern Course in Aeroelasticity.* 4th Edition, 2004 ISBN 1-4020-2039-2
117. T. Burczyński and A. Osyczka (eds.): *IUTAM Symposium on Evolutionary Methods in Mechanics.* Proceedings of the IUTAM Symposium held in Cracow, Poland, 24-27 September 2002. 2004 ISBN 1-4020-2266-2
118. D. Ieşan: *Thermoelastic Models of Continua.* 2004 ISBN 1-4020-2309-X
119. G.M.L. Gladwell: *Inverse Problems in Vibration.* Second Edition. 2004 ISBN 1-4020-2670-6
120. J.R. Vinson: *Plate and Panel Structures of Isotropic, Composite and Piezoelectric Materials, Including Sandwich Construction.* 2005 ISBN 1-4020-3110-6
121. *Forthcoming*
122. G. Rega and F. Vestroni (eds.): *IUTAM Symposium on Chaotic Dynamics and Control of Systems and Processes in Mechanics.* Proceedings of the IUTAM Symposium held in Rome, Italy, 8–13 June 2003. 2005 ISBN 1-4020-3267-6
123. E.E. Gdoutos: *Fracture Mechanics. An Introduction. 2nd edition.* 2005 ISBN 1-4020-3267-6
124. M.D. Gilchrist (ed.): *IUTAM Symposium on Impact Biomechanics from Fundamental Insights to Applications.* 2005 ISBN 1-4020-3795-3
125. J.M. Huyghe, P.A.C. Raats and S. C. Cowin (eds.): *IUTAM Symposium on Physicochemical and Electromechanical Interactions in Porous Media.* 2005 ISBN 1-4020-3864-X
126. H. Ding, W. Chen and L. Zhang: *Elasticity of Transversely Isotropic Materials.* 2005 ISBN 1-4020-4033-4
127. W. Yang (ed): *IUTAM Symposium on Mechanics and Reliability of Actuating Materials.* Proceedings of the IUTAM Symposium held in Beijing, China, 1–3 September 2004. 2005 ISBN 1-4020-4131-6

Mechanics

SOLID MECHANICS AND ITS APPLICATIONS

Series Editor: G.M.L. Gladwell

128. J.-P. Merlet: *Parallel Robots*. 2006 ISBN 1-4020-4132-2
129. G.E.A. Meier and K.R. Sreenivasan (eds.): *IUTAM Symposium on One Hundred Years of Boundary Layer Research*. Proceedings of the IUTAM Symposium held at DLR-Göttingen, Germany, August 12–14, 2004. 2006 ISBN 1-4020-4149-7
130. H. Ulbrich and W. Günthner (eds.): *IUTAM Symposium on Vibration Control of Nonlinear Mechanisms and Structures*. 2006 ISBN 1-4020-4160-8
131. L. Librescu and O. Song: *Thin-Walled Composite Beams*. Theory and Application. 2006 ISBN 1-4020-3457-1
132. G. Ben-Dor, A. Dubinsky and T. Elperin: *Applied High-Speed Plate Penetration Dynamics*. 2006 ISBN 1-4020-3452-0
133. X. Markenscoff and A. Gupta (eds.): *Collected Works of J. D. Eshelby*. Mechanics of Defects and Inhomogeneities. 2006 ISBN 1-4020-4416-X
134. R.W. Snidle and H.P. Evans (eds.): *IUTAM Symposium on Elastohydrodynamics and Microelastohydrodynamics*. Proceedings of the IUTAM Symposium held in Cardiff, UK, 1–3 September, 2004. 2006 ISBN 1-4020-4532-8
135. T. Sadowski (ed.): *IUTAM Symposium on Multiscale Modelling of Damage and Fracture Processes in Composite Materials*. Proceedings of the IUTAM Symposium held in Kazimierz Dolny, Poland, 23–27 May 2005. 2006 ISBN 1-4020-4565-4
136. A. Preumont: *Mechatronics*. Dynamics of Electromechanical and Piezoelectric Systems. 2006 ISBN 1-4020-4695-2
137. M.P. Bendsøe, N. Olhoff and O. Sigmund (eds.): *IUTAM Symposium on Topological Design Optimization of Structures, Machines and Materials*. Status and Perspectives. 2006 ISBN 1-4020-4729-0
138. A. Klarbring: *Models of Mechanics*. 2006 ISBN 1-4020-4834-3
139. H.D. Bui: *Fracture Mechanics*. Inverse Problems and Solutions. 2006 ISBN 1-4020-4836-X
140. M. Pandey, W.-C. Xie and L. Xu (eds.): *Advances in Engineering Structures, Mechanics and Construction*. Proceedings of an International Conference on Advances in Engineering Structures, Mechanics & Construction, held in Waterloo, Ontario, Canada, May 14–17, 2006. 2006 ISBN 1-4020-4890-4
141. G.Q. Zhang, W.D. van Driel and X. J. Fan: *Mechanics of Microelectronics*. 2006 ISBN 1-4020-4934-X
142. Q.P. Sun and P. Tong (eds.): *IUTAM Symposium on Size Effects on Material and Structural Behavior at Micron- and Nano-Scales*. Proceedings of the IUTAM Symposium held in Hong Kong, China, 31 May–4 June, 2004. 2006 ISBN 1-4020-4945-5
143. A.P. Mouritz and A.G. Gibson: *Fire Properties of Polymer Composite Materials*. 2006 ISBN 1-4020-5355-X
144. Y.L. Bai, Q.S. Zheng and Y.G. Wei (eds.): *IUTAM Symposium on Mechanical Behavior and Micro-Mechanics of Nanostructured Materials*. Proceedings of the IUTAM Symposium held in Beijing, China, 27–30 June 2005. 2007 ISBN 1-4020-5623-0